# CONTROL OF THE CARDIOVASCULAR AND RESPIRATORY SYSTEMS IN HEALTH AND DISEASE

# ADVANCES IN EXPERIMENTAL MEDICINE AND BIOLOGY

A Continuation Order Plan is available for this series. A continuation order will bring delivery of each new volume
immediately upon publication. Volumes are billed only upon actual shipment. For further information please contact
the publisher.

# CONTROL OF THE CARDIOVASCULAR AND RESPIRATORY SYSTEMS IN HEALTH AND DISEASE

Edited by

## C. Tissa Kappagoda and Marc P. Kaufman
University of California, Davis
Davis, California

SPRINGER SCIENCE+BUSINESS MEDIA, LLC

Library of Congress Cataloging-in-Publication Data

---

Control of the cardiovascular and respiratory systems in health and
   disease / edited by C. Tissa Kappagoda and Marc P. Kaufman.
        p.    cm. -- (Advances in experimental medicine and biology ;
   v. 381)
      "Proceedings of the Symposium on Control of the Cardiovascular and
   Respiratory Systems in Health and Disease, held April 8-9, 1994, at
   the University of California", Davis, California--T.p. verso.
      Includes bibliographical references and index.
      ISBN 978-1-4613-5773-5    ISBN 978-1-4615-1895-2 (eBook)
      DOI 10.1007/978-1-4615-1895-2
      1. Respiration--Regulation--Congresses.   2. Heart--Contraction-
   -Regulation--Congresses.   3. Cardiopulmonary system--Congresses.
   4. Lungs--Innervation--Congresses.   5. Heart--Innervation-
   -Congresses.   I. Kappagoda, C. T.   II. Kaufman, Marc P.
   III. Symposium on Control of the Cardiovascular and Respiratory
   Systems in Health and Disease (1994 : University of California,
   Davis)   IV. Series.
      [DNLM: 1. Cardiovascular System--physiology--congresses.
   2. Respiratory Systems--phyology--congresses.   3. Neurotransmitters-
   -congresses.   WG 102 C764 1995]
   QP123.C67   1995
   612.1--dc20
   DNLM/DLC
   for Library of Congress                                    95-37629
                                                                   CIP

Proceedings of the Symposium on Control of the Cardiovascular and
Respiratory Systems in Health and Disease,
held April 8–9, 1994, at the University of California, Davis, Davis, California

ISBN 978-1-4613-5773-5

© 1995 Springer Science+Business Media New York
Originally published by Plenum Press, New York in 1995
Softcover reprint of the hardcover 1st edition 1995

10 9 8 7 6 5 4 3 2 1

# PREFACE

On April 8-9, 1994, a symposium entitled *Control of the Cardiovascular and Respiratory Systems in Health and Disease* was held at the University of California Davis Medical Center in Sacramento. The purpose of this symposium was to honor the careers of Professors Hazel M. and John C.G. Coleridge. Participants in this symposium came from throughout the world. Their attendance at the symposium was a symbol of great respect and affection for the honorees. The Professors Coleridge have made many important contributions to the scientific literature concerning neural control of the cardiovascular and respiratory systems. In addition, they have made remarkable contributions to the lives of other scientists working in this field of investigation. Some of us have known them as mentors, counselors, friends, and supervisors; others have known them as co-investigators. Most importantly, all of us have known them as friends. This book, which contains the proceedings of the symposium, is dedicated to Hazel and John Coleridge.

C. T. Kappagoda
M. P. Kaufman

# ACKNOWLEDGMENTS

We wish to acknowledge the financial support of the following agencies for making this symposium a reality:

- Astra Merck Group (Tarek Ackad, M.D., Ph.D.)
- Boehringer Ingelheim Pharmaceuticals, Inc. (Ms. Kathryn B. Lucas and Mr. Allan Holloway)
- Bristol-Myers Squibb (David L. Cram, Jr., Pharm.D.)
- Marion/Merrrell Dow, Inc. (Mr. Brian Scheffield)
- Merck and Company (Mr. Johnathan Sakakibara)
- Pfizer Laboratories (Mr. Thomas Werth)
- Smith Kline Beecham and Company (Ms. Eileen Abbott)
- The Upjohn Company (Mr. Thomas Nannizzi)
- Zeneca Pharmaceutical Group (Mr. Jack W. Britts)

In addition we would also like to acknowledge the superb organizational skills of Ms. Shirley Martin, who was responsible for handling the logistics relating to the symposium, as well as Rhonda McBride and Elizabeth Walker for their clerical assistance. Finally, we would like to thank John C. Longhurst, M.D., Ph.D., Chief of Cardiovascular Medicine, University of California, Davis, for extending the facilities of the division to ensure a successful conclusion to this project.

# CONTENTS

**Short Communications**

# INTERACTIONS OF NEUROTRANSMITTERS AND ENDOTHELIAL CELLS IN DETERMINING VASCULAR TONE

John T. Shepherd

Mayo Clinic and Mayo Foundation
Rochester, Minnesota

## INTRODUCTION

My interest in research on the cardiovascular system began in 1948. I had joined the Department of Physiology at the Queen's University of Belfast with the title of "Demonstrator", indicating that my responsibilities were to supervise the practical classes. This provided a paid opportunity to spend one year preparing for the basic science section of the examinations to qualify as a Fellow of the Royal College of Surgeons of England.

The chairman of the department was Henry Barcroft, later FRS, who had brought the method of venous occlusion plethysmography to the department, for the measurement of blood flow in human limbs. In his first studies, Barcroft demonstrated the presence of sympathetic nerves to the human forearm muscle blood vessels.

After the second World War started in 1939, fainting was noted to occur in certain blood donors. Sir Thomas Lewis had concluded that vagal slowing of the heart was the primary cause of the sudden decrease in arterial blood pressure. Barcroft and Edholm decided to use plethysmography to study the circulation to the leg muscles, using conscientious objectors to the war who were willing volunteers for the studies. Venous blood was withdrawn until the subjects fainted. At the time of the faint, despite the marked decrease in arterial blood pressure, the blood flow to the leg muscles increased, demonstrating that the resistance vessels had dilated. This did not occur if the sympathetic nerves were interrupted; the dilatation was too large to be attributed solely to a decrease in sympathetic vasoconstrictor activity. Barcroft and Edholm concluded, therefore, that vasodilator nerves to skeletal muscles were activated during the faint; since these muscles form about 40 per cent of the body mass, this could explain the decrease in arterial blood pressure. The nature of the vasodilator nerves remains to be elucidated (see Barcroft and Swan 1953).

Gunshot wounds of major arteries were common in the war, necessitating ligation of major vessels. At the same time, surgeons frequently did a surgical sympathectomy on the assumption that this would enhance the collateral circulation. Therefore, Barcroft asked me to investigate whether these vessels had a sympathetic innervation in humans, and offered me a tenure appointment as Lecturer in the Department.

*Control of the Cardiovascular and Respiratory Systems in Health and Disease*
Edited by C. T. Kappagoda and M. P. Kaufman, Plenum Press, New York, 1995

By measuring the changes in blood flow to the calf of the leg in normal subjects during temporary mechanical occlusion of one femoral artery, before and after sympathetic blockade using a newly developed drug, tetraethylammonium bromide, I succeeded in showing that, in the leg with the femoral artery occluded, the collateral flow increased after blockade, both at rest and after exercise of the leg.

I submitted these studies for the Master of Surgery degree of the Queen's University. The external examiner was a vascular surgeon and fortunately, a leading advocate of surgical sympathectomy, so I received a gold medal. This was my only contribution to surgery, since I had succumbed to the lure and challenges of an Investigator's career. Barcroft comforted me by saying that if I stayed around long enough, he felt certain that physiologists would soon receive a living wage!

Henry Barcroft left The Queen's University in 1948 to accept the Chair of Physiology at St. Thomas Hospital Medical School in London, and was succeeded in Queen's by David Greenfield from St. Mary's Hospital Medical School in London. At that time Ian Roddie and Robert Whelan had joined the department; Autar Paintal, then working at the University of Edinburgh, had demonstrated the A and B receptors in the cat atria and found that type B are stretch receptors responding to changes in atrial filling (1953).

These studies were the stimulus for us to test the effect of increasing central venous pressure on the reflex control of the forearm muscle resistance vessels by passively raising the legs of the horizontal subject. This caused an increase in blood flow in the skeletal muscles of the forearm due to a decrease in sympathetic vasoconstrictor activity; I was the best responder because of my long legs. Since the arterial blood pressure did not change, we concluded that this was due to a reflex mediated by "stimulation of receptors in a low-pressure area of the intrathoracic vascular bed" (Roddie and Shepherd 1956; Roddie, Shepherd and Whelan 1957). Some years later, David Greenfield took a sabbatical leave at the Cardiovascular Research Institute at the University of California in San Francisco. Here he developed a lower body box with a seal around the waist and attached it to a vacuum cleaner for the application of negative pressure. With negative pressures of 15-20 mm/Hg, which did not cause changes in arterial blood pressure, the forearm muscle vessels reflexly constricted, consistent with deactivation of mechanoreceptors in the cardiopulmonary area. Since then, the "Greenfield suck-box" has become the standard method for investigation of these receptors in humans (Brown et al 1966).

An important impetus to the further studies of these cardiopulmonary mechanoreceptors in humans was the demonstration by Coleridge et al (1964) in dog's ventricles, that some of these receptors are mechanoreceptors, capable of signalling changes in the inotropic state of the ventricles as well as changes in its volume. Some of these receptors are subserved by myelinated vagal afferents (Paintal 1955; Brown 1965) and others, including the heart and lungs, unmyelinated (C fiber) afferents (Coleridge et al 1964; Baker et al 1979; Thorén et al 1976).

As the papers in this symposium indicate, the search continues to understand the signal transduction processes responsible for activation, not only of the cardiopulmonary, but also the arterial mechanoreceptors and skeletal muscle ergoreceptors, and the transmitters involved in the processing of the information received from these receptors by the centers in the brain, with the resultant translation of the afferent inputs from the peripheral sensors into the appropriate alterations in autonomic outflow.

## THE VASCULAR NEUROEFFECTOR JUNCTION

The final determinant of the response of the cardiovascular system occurs at the effector organs, where the neurotransmitters released from the nerve endings and their actions on specific receptors can be continuously modified by local factors and by various hormones.

Since the discovery in 1946 by von Euler that norepinephrine is the sympathetic neurotransmitter, a wealth of information has accrued in recent years of sympathetic and parasympathetic co-transmitters and modulators, of non-adrenergic-non-cholinergic nerves, the so-called nitroxidergic nerves, and of sensory-motor nerves. Also identified are numerous receptors on the sympathetic varicosities, some of which when activated can decrease, and others enhance the output of the neurotransmitters.

In addition, the ability of the endothelial cells to form and release dilator and constrictor paracoids has been demonstrated. Thus, the final determinant of the vascular responses to alterations in autonomic outflow depends not only on the action of the neurotransmitters on specific receptors on the vascular smooth muscle, the possibility that the neurotransmitters might activate autonomic prejunctional receptors and that endothelium-derived vasoactive products will contribute to the resultant changes in vessel tone.

In this chapter current knowledge of the interactions between neurotransmitters released from the perivascular nerves and vasoactive autocoids and paracoids formed and released from the endothelial cells is reviewed. A key question is whether the resultant action on the blood vessels is the algebraic sum of their individual actions, or whether complex interactions between the perivascular nerves and endothelial-derived substances play an important role. Obviously, a key question is whether neurotransmitters can reach endothelial cells and vice versa. In large subepicardial and resistance-sized (38-140µm) isolated canine coronary arteries, endothelium-dependent vasodilation occurred following extraluminal administration of acetylcholine (Myers et al 1989). This demonstrates that acetylcholine can diffuse through the vascular wall.

The perivascular nerves and potential interactions with endothelial factors that have to be considered are illustrated in Fig. 1. Concerning the nerves, the role of the sympathetic, parasympathetic and nitroxidergic nerves will be discussed, recognizing that the specific innervation of different vascular beds will differ. While the importance of sensory-motor nerves to many blood vessels has been recognized (Burnstock 1990a), their potential interactions with endothelial-derived vasoactive factors awaits investigation. The sensory-motor nerves store and release vasoactive neuropeptides; substance P, calcitonin gene related-peptide (CGRP), neurokinin A and neurokinin B (Burnstock 1993). The release of these neuropeptides could affect the release of neurotransmitters from the sympathetic and parasympathetic nerves and vice-versa. For example, in guinea pig atria, NE, ATP and NPY, via prejunctional mechanisms can inhibit release of transmitters from the sensory-motor nerves. Also ACh and its co-transmitter VIP can activate the sensory-motor nerves (Amerini et al 1991; Rubino et al 1992; Rubino 1993).

## SYMPATHETIC NEUROTRANSMITTERS (FIG. 2)

In addition to norepinephrine (NE), both adenosine triphosphate (ATP) and neuropeptide Y (NPY) (a 36 amino acid peptide) have been identified in post-ganglionic sympathetic nerves (Lundberg and Hökfelt 1983; Lundberg et al 1983; Pernow and Lundberg 1989; Burnstock 1990a, b, c, d; Mione et al 1990; Warner et al 1991). Norepinephrine (NE) is stored in small and large dense-cored vesicles. In the latter it is co-stored with NPY (Forsgren 1989). In the vesicles the NE is secured from deamination by the monoamine oxidase in the mitochondria. NE and ATP may be derived from different vesicle populations. ATP and NPY may be co-released with norepinephrine and serve as co-transmitters (Morris et al 1986). Like norepinephrine, these activate specific receptors on the vascular smooth muscle cells. The kinetics involved in the release of these transmitters is still debated (Stjärne

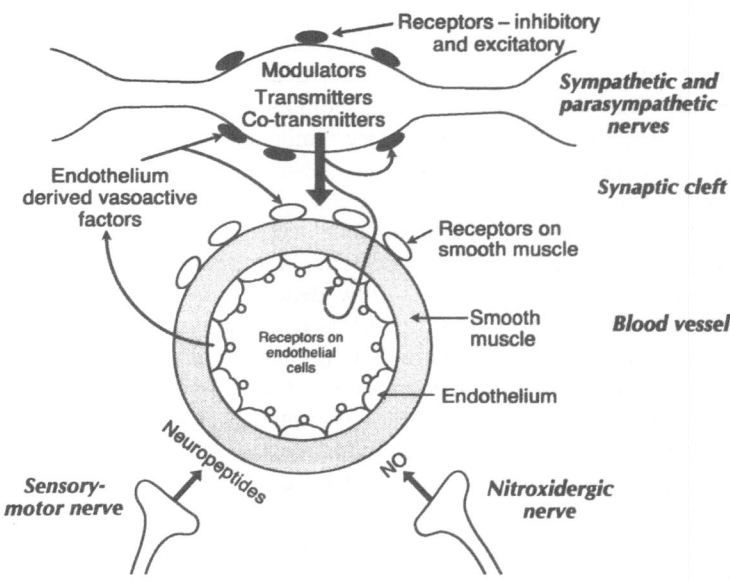

**Figure 1.** The complexity of the vascular neuroeffector junction. The sympathetic and parasympathetic nerves can release various neurotransmitters, co-transmitters and modulators to act on receptors on the vascular smooth muscle. In addition, there are a series of receptors on the nerve terminals, particularly the sympathetic, some of which when activated can increase and others decrease the output of the transmitters from these nerves. Nitroxidergic nerves are present in certain vascular beds and by releasing nitric oxide cause relaxation of the vascular smooth muscle. There are also sensory motor nerves to certain blood vessels that, when activated, release neuropeptides. Numerous receptors are present on the endothelial cells which, if activated, cause a release of vasoactive paracoids. An increase in shear stress also causes a release of nitric oxide from these cells. There is evidence that certain neurotransmitters can reach the endothelial cells to activate specific receptors with a resultant release of nitric oxide. Also, certain endothelium-derived factors can reach the receptors on the sympathetic nerves to alter the output of transmitters. For details, see text.

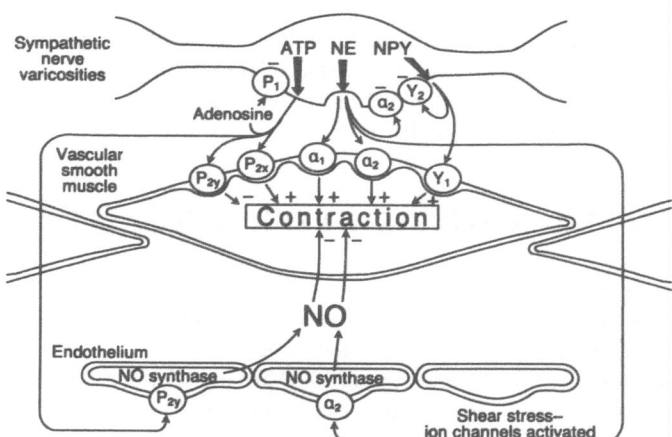

**Figure 2.** Sympathetic neurotransmitters and their actions on receptors on the vascular smooth muscle, the nerves themselves and the endothelial cells. ATP = adenosine triphosphate. $\alpha_1$ and $\alpha_2$ = alpha-adrenoceptors. NE = norepinephrine. $P_1$, $P_2$, $P_{2Y}$ = purinoceptors. NO = nitric oxide. $Y_1$, $Y_2$ = receptors for neuropeptide Y. NPY = neuropeptide Y.

1989; Stjärne et al 1993). The relative release differs in different vascular beds (Burnstock 1990a; Fillenz 1992) and may depend on the pattern and relative frequency of action potentials (Pernow et al 1989; Stjärne et al 1993). Compared to NE, NPY release occurs mainly during strong reflex activation in animals and man.

In the resistance vessels NE and ATP cause synergistic vasoconstriction via alpha$_1$-adrenoceptors and P$_{2X}$ purinoceptors respectively (Burnstock 1990c). By contrast, in the coronary circulation of the isolated rabbit heart, NE and ATP act on beta-adrenoceptors, predominantly of the $\beta_1$ subtype (Macdonald et al 1987) and P$_{2Y}$-purinoceptors respectively to cause vasodilation (Corr and Burnstock 1991). Purinergic receptors and neurotransmitters also participate in endogenous sympathetic vasoconstriction in humans (Taddei et al 1990).

In the coronary arteries and in several other vascular beds the vasoconstrictor action of NPY has been demonstrated (Franco-Cereceda et al 1985; Rudehill et al 1986; Zukowska-Grojec et al 1987). In addition to the vascular constrictor action of NPY acting on its specific receptor, it also augments the vasoconstrictor response to other substances, including NE (Ekblad et al 1984; Pernow 1988; Lundberg et al 1989; Macho et al 1989). This is especially noted when NPY by itself causes minimal or no vasoconstriction. This enhancing action of NPY is said to be endothelium-dependent (Pernow 1988).

## CAN SYMPATHETIC NEUROTRANSMITTERS AFFECT SYMPATHETIC PRESYNAPTIC RECEPTORS?

Numerous receptors have been identified on the sympathetic nerve varicosities, including those for the neurotransmitters themselves; the latter are alpha$_2$-adrenoceptors ($\alpha_2$), purinergic (P$_1$) and NPY (Y$_2$) receptors (Fig. 2). These, if activated by the neurotransmitters, cause a decrease in the output of norepinephrine. If the co-transmitters come from the same source as norepinephrine, their release will also be depressed (Fillenz 1992). It is suggested that stimulation of the presynaptic alpha$_2$-adrenoceptors decreases norepinephrine output by limiting the availability of Ca$^{2+}$ ions for excitation-secretion coupling (Langer 1977). The purinergic receptors could be activated by the rapid conversion in the synaptic cleft of ATP to adenosine by ectoenzymes. This decreases the output of norepinephrine and ATP (Fredholm and Hedqvist 1980; Taddei et al 1990; Ralevic and Burnstock 1991). NPY, acting on prejunctional Y$_2$ receptors on the sympathetic nerves, decreases the output of norepinephrine and ATP (Stjärne 1989). Thus, the potential exists for a negative feedback on transmitter release in vivo.

## CHOLINERGIC NERVES

Acetylcholine is synthesized in the nerve terminals from choline and acetyl coenzyme A by choline acetyltransferase. In certain vascular beds, cotransmitters may also be present in these nerves. For example, in the nerves which innervate salivary glands, vasoactive intestinal polypeptide (VIP) is co-released with acetylcholine (Lundberg 1981; Lundberg and Hökfelt 1983).

## NON-ADRENERGIC NON-CHOLINERGIC NERVES

These nerves have been demonstrated in cerebral and mesenteric arteries, and arteries of the corpora cavernosa. They act by releasing nitric oxide to cause relaxation of the vascular smooth muscle and hence have been termed nitroxidergic nerves (Toda and Okamura 1989; Ignarro et al 1990; Toda and Okamura 1992). Unlike classical neurotransmitters, nitric oxide (NO) is not stored in synaptic vesicles. It is formed from the abundant and recyclable substrate L-arginine by enzyme nitric oxide synthase; it does not require exocytosis for its release from the nerve endings, and can diffuse freely through cell membranes (Madison 1993). The localization of nitric oxide synthase in nerve fibers in the wall of coronary arteries, suggests that the dual origin of NO from these nerves and the endothelium both operate to regulate the tone of the coronary arterial tree (Klimaschewski et al 1992).

# INTERACTIONS BETWEEN AUTONOMIC NERVES

The close proximity of the postganglionic sympathetic and vagal nerves to the heart permits a presynaptic interaction between them. Thus, activation of the sympathetic nerves in the dog releases neuropeptide Y which activates $Y_2$ receptors on the vagal nerves to inhibit the release of acetylcholines (Wahlestedt et al 1986; Potter 1987; Yang and Levy 1992). In the cat this is caused by another amino acid peptide, galanin (Ulman et al 1992; Potter and Ulman 1994). If the vagal nerves are activated, there is the potential for some of the acetylcholine to act on a muscarinic receptor on the sympathetic nerves to reduce the output of norepinephrine (Löffelholz and Muscholl 1970). Acetylcholine also could decrease its own release from cholinergic nerves, since atropine can enhance the release of acetylcholine. This negative feedback modulation of release of the transmitter from vagal nerve terminals is frequency dependent and thus requires a high enough concentration of acetylcholine in the synaptic cleft (Manabe et al 1991).

# ENDOTHELIUM-DERIVED VASOACTIVE FACTORS

The vascular endothelium forms and releases a multitude of biologically active substances, including factors which modulate the tone of the underlying smooth muscle (Rubanyi 1991).

## Endothelium-Derived Relaxing Factors

These are prostacyclin, nitric oxide or a closely related substance, endothelium-derived hyperpolarizing factor and C-type natriuretic factor (Shepherd and Katušić 1991; Suga et al 1992). These can be released by activation of specific receptors on the endothelial cells. In addition, an increase in shear stress on endothelial cells causes a release of nitric oxide. This occurs when blood flow and velocity is increased in the arterial conduit and resistance vessels. The importance of nitric oxide is evident from the fact that in animals, blockade of its formation from L-arginine by the enzyme nitric oxide synthase results in sustained hypertension (Rees et al 1989; Moncada et al 1991). However it appears that the hypertension caused by the nitric oxide synthase inhibitor is associated with an elimination of nitroxidergic neural function rather than an impairment of the basal release of nitric oxide from the endothelium (Toda et al 1993).

## Endothelium-Derived Contracting Factors

These are angiotensin II, endothelin I, thromboxane $A_2$. prostaglandin $H_2$ and superoxide anions. In the normal arterial vessels, relaxing factors predominate while the same stimuli in the venous system cause release of contracting factors (Kifor and Dzau 1987;

Yanagisawa et al 1988; Katušić and Shepherd 1991). Endothelium-derived nitric oxide can inhibit the release of endothelin (Boulanger and Lüscher 1990).

# INTERACTIONS BETWEEN NEUROTRANSMITTERS AND ENDOTHELIAL CELLS

In the coronary arteries of the dog, when vagal fibers are activated, the acetylcholine released activates muscarinic receptors on the endothelial cells to release nitric oxide to cause relaxation of the smooth muscle. Thus the para-sympathetic coronary vasodilation is dependent on endothelium-derived relaxing factor (Broten et al 1992). Likewise, norepinephrine released from the sympathetic nerves, in addition to activating alpha$_1$- and alpha$_2$-adrenoceptors on the vascular smooth muscle to cause its contraction, can reach alpha$_2$-adrenoceptors on the endothelial cells (González et al 1990; Miller 1991). As a consequence, nitric oxide is released and attenuates the vasoconstriction (Fig. 2).

The potential mechanisms by which the sympathetic nervous system has an important role in the vasodilator actions of NO in urethane-anesthetized rats include:

1. sympathetic discharge increases the generation of NO by a direct interaction with adrenergic receptors;

2. neurogenetically derived vasoconstriction enhances the generation of NO;

3. circulating adrenergic receptor agonists also augment generation of NO by direct receptor interaction and production of vasomotor tone;

4. NO may modulate neurogenetically mediated vasoconstriction by presynaptic inhibition of NE release as well as reducing the responsiveness of the smooth muscle to the neurotransmitter;

5. the sympathetic nerves may release NO, as suggested by the finding of NO synthase in perivascular nerves (Bredt et al 1990; Lacolley et al 1991).

By contrast, stimulation of any $\beta_2$ adrenoceptors on the endothelium of the dog coronary artery does not release EDRF (Macdonald et al 1987). It is possible that some of the ATP released from the sympathetic nerves, as well as causing contraction by acting on $P_{2x}$ receptors on the smooth muscle, with a resultant activation of voltage-dependent $Ca^{2+}$ channels, might reach the $P_{2y}$ receptors on the endothelial cells, and as a consequence decrease the vasoconstriction by releasing nitric oxide (or EDRF) (Fig. 2).

# INTERACTIONS BETWEEN ENDOTHELIAL DERIVED PARACOIDS AND THE SYMPATHETIC NERVES

A factor formed in and released from the endothelial cells, which might be NO or CNP (Fig. 3), can reach the sympathetic nerves and reduce the output of norepinephrine (Cohen and Weisbrod 1988). This attenuation results from inhibition of the activation by extracellular calcium of the adrenergic nerves, probably due to an action on voltage-dependent calcium channels (Tesfamarian et al 1989). Endothelin-I inhibits release of norepinephrine by acting on an as yet undefined presynaptic receptor (Wiklund et al 1988; Tabuchi et al 1990; Takagi et al 1992). Other inhibiting presynaptic receptors on sympathetic nerves include those for histamine and 5-hydroxytryptamine (serotonin). These could be activated by histamine released from mast cells and serotonin from platelets.

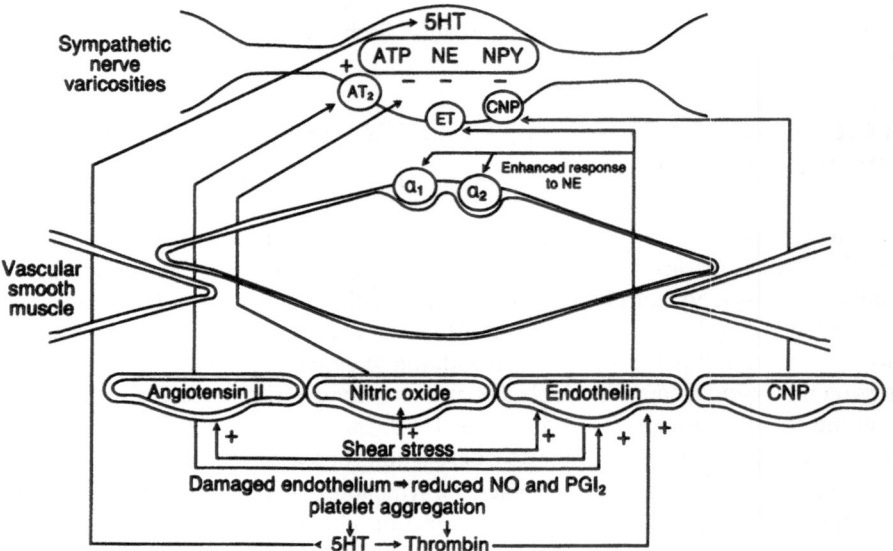

**Figure 3.** Endothelium-sympathetic nerve interactions. Angiotensin II, nitric oxide, endothelin and CNP (C-type neuropeptide) formed in endothelial cells, in addition to acting on the vascular smooth muscle, can interact with the sympathetic nerves. Angiotensin II can act on its receptor on the sympathetic nerves ($AT_2$) to enhance neurotransmitter release; nitric oxide can decrease the neurotransmitter release, and increased shear stress enhances release of nitric oxide from the endothelial cells. Endothelin, acting on its receptor (ET), and CNP, acting on its receptor, also can decrease neurotransmitter release. Endothelin also enhances the response of the alpha receptors on the smooth muscle ($\alpha_1$ and $\alpha_2$) to norepinephrine.

If the endothelium is damaged, less of the vasorelaxants, nitric oxide (NO) and prostacyclin ($PGI_2$) are formed. Since normally these also act normally synergistically to inhibit platelet aggregation, their diminution leads to aggregation and release of 5-hydroxy-tryptamine (5-HT; serotonin). some of this can be taken up by the nerves and released when the nerves are activated. By acting on a specific receptor on the smooth muscle, it enhances the vasoconstriction caused by adenosine triphosphate (ATP), norepinephrine (NE) and neuropeptide Y (NPY).

By contrast, the potential exists for angiotensin II released from endothelial cells to activate angiotensin II receptors on the nerve terminals to increase norepinephrine output (Hughes and Roth 1971; Zimmerman 1977).

## ROLE OF ENDOTHELIAL CELLS IN REGULATION OF BARORECEPTOR ACTIVITY

Arterial baroreceptors play a key role in regulation of arterial blood pressure. Activation of these receptors occur in response to increase in blood pressure, leading to reflex inhibition of sympathetic nerve activity that buffers the rise in pressure. Endothelium-derived vasoactive substances are important modulators of baroreceptor activity. The removal of endothelial cells or inhibition of cyclooxygenase activity with indomethacin, decreases the sensitivity of baroreceptors, whereas addition of prostacyclin to denuded carotid sinus restores function of baroreceptors (Chapleau et al 1991). In contrast, endothelin I suppresses baroreceptors activity (Chapleau et al 1992).

These findings indicate that the balance of excitatory and inhibitory factors released from endothelial cells modulates baroreceptor sensitivity and that during development of vascular diseases (e.g., hypertension and atherosclerosis) dysfunction of endothelial cells causes decrease in baroreceptor sensitivity with subsequent increase in arterial blood pressure.

## POST-JUNCTIONAL RECEPTORS ON THE VASCULAR SMOOTH MUSCLE

Both neuropeptide-Y and endothelin I can enhance the contractions to norepinephrine (Stjärne et al 1986; Pernow et al 1986; Tabuchi et al 1990). Concerning chronic muscular exercise, the increase in coronary blood flow in dogs increases the nitric oxide synthase gene expression in the endothelial cells, presumably as a consequence of the increase in shear stress on these cells with a consequential greater release of L-arginine derived nitrite (Sessa et al 1994). The same events will occur in the conduit arteries and resistance vessels in the active muscles. It is also possible that the increased flow would decrease the formation of endothelin-I, the most potent vasoconstrictor known (Sharefkin et al 1991). This provides a scientific basis for the beneficial effects of exercise on the vascular system.

What influence the increased formation of nitric oxide might have on the action of the sympathetic nerves on the resistance vessels in the active muscles has not been determined. Certainly, preservation of the activity of these nerves is important in preventing muscle edema during exercise by adjusting the pre- to post-capillary vascular resistance (Mellander and Bjornberg 1992).

## VASCULAR DISEASES

In the cardiovascular diseases, including primary hypertension and heart failure where the endothelium-dependent relaxations of arterial vessels is impaired, the result will be less ability to have a feedback to reduce the constrictor action of the neurotransmitters. In vitro studies have shown that contractions to norepinephrine are increased in the absence of the endothelium (Carrier and White 1985). Thus a dysfunction in the synthesis or release of NO may be an important factor in genetically and experimentally induced models of hypertension. This could represent a basic abnormality between the sympathetic outflow and the endothelium-derived autocoids and paracoids (Lacolley et al 1991).

In vivo studies in dogs demonstrated that damage of the endothelium of a coronary artery by a balloon-tipped catheter leads to platelet adhesion and aggregation and accumulation of 5-hydroxytryptamine (5-HT) in the sympathetic nerves. The similar structure of norepinephrine and 5-HT means that both are substrates for the amine-uptake pump on the membrane of the sympathetic nerve terminals. On activation of the nerves, the 5-HT is released along with the other neurotransmitters, and can stimulate pre- and postjunctional 5-HT receptors. The former action would inhibit release of NE and other neurotransmitters, the latter would directly contract the vascular smooth muscle (Cohen 1988).

## REFERENCES

Aarnio P, CGA McGregor and V Miller. Autonomic Modulation of Contractions to Endothelin-I in Canine Coronary Arteries. *Hypertension* 1993, 2: 680-686.

Amerini S, A Rubino, S Filippi, F Ledda and L Mantelli. Modulation by Adrenergic Transmitters of the Efferent Function of Capsaicin-Sensitive Nerves in Cardiac Tissue. *Neuropeptides* 1991, 20: 225-232.

Baker DG, HM Coleridge and JCG Coleridge. Vagal Afferent C Fibres from the Ventricle. In: Hainsworth R, C Kidd and RJ Linden (eds). *Cardiac Receptors*. Cambridge, University Press, 1979. pp. 117-137.

Barcroft H and HJC Swan. Sympathetic Control of Human Blood Vessels. *Monographs of the Physiological Society* Number 1, 1953. 165 p.

Boulanger C and TF Lüscher. Release of Endothelin from the Porcine Aorta — Inhibition by Endothelium-Derived Nitric Oxide. *J Clin Investig* 1990, 85: 587-590.

Bredt DS, PM Hwang and SH Snyder. Localization of Nitric Oxide Synthase Indicating A Neural Role for Nitric Oxide. *Nature* 1990, 347: 768-770.

Broten TP, JK Miyashiro, S Moncada and EO Feigl. Role of Endothelium-Derived Relaxing Factor in Parasympathetic Coronary Vasodilatation. *Amer J Physiol (Heart Circ Physiol 31)* 1992, 262: H1579-H1584.

Brown AM. Mechanoreceptors In or Near the Coronary Arteries. *J Physiol (Lond)* 1965, 177: 203-214.

Brown E, J S Goei, ADM Greenfield and GC Plassaras. Circulatory Responses to Simulated Gravitational Shifts of Blood in Man Induced by Exposure of the Body Below the Ileal Crests to Sub-Atmospheric Pressure. *J Physiol (London)* 1966, 183: 6007-627.

Burnstock G. Local Mechanisms of Blood Flow Control by Perivascular Nerves and Endothelium. *J Hypertension* 1990a, 8 (Suppl 7): S95-S106.

Burnstock G. Overview of Purinergic Mechanisms. *Ann NY Acad Sci* 1990b, 603: 1-14.

Burnstock G. Noradrenaline and ATP as Cotransmitters in Sympathetic Nerves. *Neurochem Int* 1990c, 17: 357-368.

Burnstock G. Cotransmission. The Fifth Heymans Lecture, Ghent, February 17, 1990-. *Arch Int Pharmacodyn Ther* 1990d, 304: 7-33.

Burnstock G. Introduction: Changing Face of Autonomic and Sensory Nerves in the Circulation. In: Edvinsson L and R Uddman (eds). *Vascular Innervation and Receptor Mechanisms. New Perspectives*. USA, Academic Press, Inc., 1993. pp. 1-22.

Carrier GO and RE White. Enhancement of Alpha$_1$- and Alpha$_2$-Adrenergic Agonist-Induced Vasoconstriction by Removal of Endothelium in Rat Aorta. *J Pharmacol Exp Ther* 1985, 232: 682-687.

Chapleau MW, G Hajduczok amd FM Abboud. Suppression of Baroreceptor Discharge by Endothelin at High Carotid Sinus Pressure. *Am J Physiol* 1992, 263: R103-R108.

Chapleau MW, G Hajduczok amd FM Abboud. Paracrine Modulation of Baroreceptor Activity by Vascular Endothelium. *News In Physiological Sciences* 1991, 6: 210-214.

Cohen RA. Platelet 5 Hydroxytrptamine and Vascular Adrenergic Nerves. *News In Physiological Sciences* 1988, 3: 185-189.

Cohen RA, KM Zitnay and RM Weisbrod. Accumulation of 5-Hydroxytryptamine Leads to Dysfunction of Adrenergic Nerves in Canine Coronary Artery Following Intimal Damage in Vivo. *Circ Res* 1987, 61: 829-833.

Cohen RA and RM Weisbrod. Endothelium Inhibits Norepinephrine Release from Adrenergic Nerves of Rabbit Carotid Artery. Am J Physiol (Heart Circ Physiol 23) 1988, 254: H871-H878.

Coleridge HM, JCG Coleridge and C Kidd. Cardiac Receptors in the Dog, with Particular Reference to Two Types of Afferent Endings in the Ventricular Wall. *J Physiol (London)* 1964, 174: 323-339.

Corr L and G Burnstock. Vasodilator Response of Coronary Smooth Muscle to the Sympathetic Co-Transmitters Noradrenaline and Adenosine 5'-Triphosphate. *Brit J Pharmacol* 1991, 104: 337-342.

Daly RN and JP Hieble. Neuropeptide Y Modulates Adrenergic Neurotransmission by an Endothelium-Dependent Mechanism. *Eur J Pharmacol* 1987, 138: 445-446.

Edvinsson C, EL Ekblad, R Häkanson and R Wahlestedt. Neuropeptide Y Potentiates the Effect of Various Vasoconstrictor Agents in Rabbit Blood Vessels. *Brit J Pharmacol* 1984, 83: 519-525.

Ekblad EL, C Edvinsson, R Wahlestedt, R Uddman, R Häkanson and F Sundler. Neuropeptide Y Co-Exists and Co-Operates With Noradrenaline in Perivascular Nerve Fibres. *Regul Pept* 1984, 8: 225-235.

Fillenz M. Transmission: Noradrenaline. In Burnstock G and CHV Hoyle (eds). *Autonomic Neuroeffector Mechanisms*. Chur, Harwood Academic Publishers, 1992. pp. 323-365.

Forsgren S. Neuropeptide Y-like Immunoreactivity in Relation to the Distribution of Sympathetic Nerve Fibers in the Heart Conduction System. *J Mol Cell Cardiol* 1989, 21: 279-290.

Franco-Cereceda A, JM Lundberg and C Dahlöf. Neuropeptide Y and Sympathetic Control of Heart Contractility and Coronary Vascular Tone. *Acta Physiol Scand* 1985, 124: 361-369.

Fredholm BB and P Hedqvist. Modulation of Neurotransmission by Purine Nucleotides and Nucleosides. *Biochem Pharmacol* 1980, 29: 1635-1643.

González C, C Martin, E Hamel, E Galea, B Gómez, S Lluch and C Estrada. Endothelial Cells Inhibit the Vascular Response to Adrenergic Nerve Stimulation by a Receptor-Mediated Mechanism. *Can J Physiol* 1990, 68: 104-109.

Hughes J and RH Roth. Evidence that Angiotensin Enhances Transmitter Release During Sympathetic Nerve Stimulation. *Brit J Pharmacol* 1971, 41: 239-255.

Ignarro LJ, GM Bush, KS Wood, JM Fukuto and J Rajfer. Nitric Oxide and Cyclic GMP Formation Upon Electric Field Stimulation Cause Relaxation of Corpus Cavernosum Smooth Muscle. *Biochem Biophys Res Commun* 1990, 170: 843-850.

Inoue A, M Yanagisawa, S Kimuva, Y Kasuya, T Miyauchi, K Goto and T Masaki. The Human Endothelin Family. Three Structural and Pharmacologically Distinct Isopeptides Predicted by Three Separate Genes. *Proc Natl Acad Sci USA* 1989, 86: 2863-2867.

Joyner MJ, RL Lennon, DJ Wedel, SH Rose and JT Shepherd. Blood Flow to Contracting Human Muscles. Influence of Increased Sympathetic Activity. *J Applied Physiol* 1990, 68: 1453-1457.

Katušić ZS and JT Shepherd. Endothelium-Derived Vasoactive Factors I. Endothelium-Dependent Contraction. *Hypertension* 1991, 18 (Suppl III): III**86**-III**92**.

Kifor I and VJ Dzau. Endothelial Renin — Angiotensin Pathway: Evidence for Intracellular Synthesis and Secretion of Angiotensin. *Circ Res* 1987, 60: 422-428.

Klimaschewski L, W Kummer, B Mayer, JY Cournand, U Preissler, B Philippin and C Heym. Nitric Oxide Synthase in Cardiac Nerve Fibers and Neurons of Rat and Guinea Pig Heart. *Circ Res* 1992, 71: 1533-1537.

Lacolley PJ, SJ Lewis and MJ Brody. Role of Sympathetic Nerve Activitiy in the Generation of Vascular Nitric Oxide in Urethane-Anesthetized Rats. *Hypertension* 1991, 17: 881-887.

Langer SZ. Presynaptic Receptors and Their Roles in the Regulation of Transmitter Release. *Brit J Pharmacol* 1977, 60: 481-491.

Löffelholz K and E Muscholl. Inhibition by Parasympathetic Nerve Stimulation of the Release of the Adrenergic Neurotransmitter. *Naunyn Schmiedebergs Archives* 1970, 267: 181-184.

Lundberg JM. Evidence for Coexistence of Vasoactive Intestinal Polypeptide (VIP) and Acetylcholine in Neurons of Cat Exocrine Glands. Morphological, Biochemical and Functional Studies. *Acta Physiol Scand* 1981, 112 (Suppl 496): S1-S57.

Lundberg JM and T Hökfelt. Coexistence of Peptides and Classical Neurotransmitters. *Trends Neurosci* 1983, 62: 325-333.

Lundberg JM, J Pernow, A Franco-Cerecedo and A Rudehill. Neuropeptide Y (NPY) in Relation to Sympathetic Vascular Control and Effects of Antihypertensive Drugs. In: Nobin A, C Owman and B Arneklo-Nobin (eds). *Neuronal Messengers in Vascular Function*. Amsterdam, Elsevier Science Publishers (Biomedical Division), 1987. pp. 415-434.

Lundberg JM, J Pernow and JS Lacroix. Neuropeptide Y: Sympathetic Cotransmitter and Modulator. *News In Physiological Sciences* 1989, 4: 13-17.

Lundberg JM, L Terenius, T Hökfelt and M Goldstein. High Levels of Neuropeptide Y in Peripheral Noradrenergic Nerves in Various Mammals Including Man. *Neurosci Lett* 1983, 42: 167-172.

Macdonald PS, PN Dubbin and GJ Dusting. β-Adrenoceptors on Endothelial Cells Do Not Influence Release of Relaxing Factors in Dog Coronary Arteries. *Clinical and Exp Pharmacol and Physiol* 1987, 14: 525-534.

Macho P, R Pérez, JP Huidobro-Toro and RJ Domenech. Neuropeptide Y (NPY): A Coronary Vasoconstrictor and Potentiator of Catecholamine-Induced Coronary Constriction. *European J Pharmacol* 1989, 167: 67-74.

Madison DV. Pass the Nitric Oxide (Commentary). *Proc Natl Acad Sci USA* 1993, 90: 4329-4331.

Manabe N, FF Foldes, A Töröcsik, H Nagashima, PL Goldiner and ES Vizi. Presynaptic Interaction Between Vagal and Sympathetic Innervation in the Heart: Modulation by Acetylcholine and Noradrenaline Release. *J Auton Nerv Syst* 1991, 32: 233-242.

Mellander S and J Björnberg. Regulation of Smooth Muscle Tone and Capillary Pressure. *News In Physiological Sciences* 1992, 7: 113-119.

Miller VM. Interactions Between Neural and Endothelial Mechanisms in Control of Vascular Tone. *News In Physiological Sciences* 1991, 6: 60-63.

Mione MC, V Ratevic and G Burnstock. Peptides and Vasomotor Mechanisms. *Pharmacol Therap* 1990, 46: 429-468.

Moncada S, RMJ Palmer and EA Higgs. Nitric Oxide: Physiology, Pathophysiology, and Pharmacology. *Pharmacol Rev* 1991, 43: 109-142.

Morris MJ, AE Russell, V Kapoor, MD Caine, JM Elliott, MJ West, LMH Wing and JP Chalmers. Increases in Neuropeptide Y Concentrations During Sympathetic Activation in Man. *J Auton Nerv Sys* 1986, 17: 143-149.

Myers PR, PF Banitt, R Guerra Jr and DG Harrison. Characteristics of Canine Coronary Resistance Arteries: Importance of Endothelium. *Am J Physiol* 257 (Heart Circ Physiol 26) 1989, H603-H610.

Paintal AS. A Study of Right and Left Atrial Receptors. *J. Physiol (London)* 1953, 120: 596-610.

Paintal AS. A Study of Ventricular Pressure Receptors and Their Role in the Bezold-Jarisch Reflex. *Quart J Exp Physiol* 1955, 40: 348-363.

Pernow J. Co-release and Functional Interactions of Neuropeptide Y and Noradrenaline in Peripheral Sympathetic Vascular Control. *Acta Physiol Scand* 1988, 133 (Suppl 568): 1-56.

Pernow J and JM Lundberg. Modulation of Noradrenaline and Nuropeptide Y (NPY) Release in the Pig Kidney In Vivo: Involvement of $\alpha_2$, NPY and Angiotensin Receptors. *Nauryn Schmiedeberg's Archives of Pharm* 1989, 340: 379-385.

Pernow J, J Schwieler, T Kahan, P Hjemdahl, J Oberle, BG Wallin and JM Lundberg. Influence of Sympathetic Discharge Pattern on Norepinephrine and Neuropeptide Y Release. *Amer J Physiol (Heart Circ Physiol 26)* 1989, 257: H866-H872.

Potter EK. Presynaptic Inhibition of Cardiac Vagal Postganglionic Nerves by Neuropeptide Y. *Neurosci Lett* 1987, 83: 101-106.

Potter EK and LG Ulman. Neuropeptides in Sympathetic Nerves Affect Vagal Regulation of the Heart. *News In Physiological Sciences* 1994 (in press).

Ralevic V and G Burnstock. Roles of $P_2$-Purinoceptors in the Cardiovascular System. *Circulation* 1991, 84: 1-14.

Rees DD, RMJ Palmer and S Moncada. Role of Endothelium-Derived Nitric Oxide in the Regulation of Blood Pressure. *Proc Natl Acad Sci USA* 1989, 86: 3375-3378.

Roddie IC and JT Shepherd. The Reflex Nervous Control of Human Skeletal Muscle Blood Vessels. *Clin Sci* 1956, 15: 433-440.

Roddie IC, JT Shepherd and RF Whelan. The Vasomotor Nerve Supply to the Skin and Muscles of the Human Forearm. *Clin Sci* 1957, 16: 67-74.

Rubanyi GM. Cardiovascular Significance of Endothelium-Derived Vasoactive Factors. Mount Kisco (NY), Futura Publishing Co., Inc., 1991. 357 p.

Rubino A. Non-Adrenergic Non-Cholinergic (NANC) Neural Control of the Atrial Myocardium. *Gen Pharmacol* 1993, 24: 539-545.

Rubino A, S Amerini, F Ledda and L Mantelli. ATP Modulates the Efferent Function of Capsaicin-Sensitive Neurones in Isolated Guinea-Pig Atria. *Brit J Pharmacol* 1992, 105: 516-520.

Rudehill A, A Solleri, A Franco-Cereceda and JM Lundberg. Neuropeptide Y (NPY) and the Pig Heart: Release and Coronary Vasoconstrictor Effects. *Peptides* 1986, 7: 821-826.

Sessa WC, K Pritchard, N Seyedi, J Wang and TH Hintze. Chronic Exercise in Dogs Increases Coronary Artery Vascular Nitric Oxide Production and Endothelial Cell Nitric Oxide Synthase Gene Expression. *Circ Res* 1994, 74: 349-353.

Shepherd JT and ZS Katušić Endothelium-Derived Vasoactive Factors I. Endothelium-Dependent Relaxation. *Hypertension* 1991, 18 (Suppl III): III**76**-III**85**.

Stjärne L. Basic Mechanisms and Local Modulation of Nerve Impulse-Induced Secretion of Neurotransmitters from Individual Sympathetic Nerve Varicosities. *Rev Physiol Biochem Pharmacol* 1989, 112: 1-138.

Stjärne L, J-X Bao, FG Gonon, M Msghina and A Stjärne. A Nonstochastic String Model of Sympathetic Neuromuscular Transmission. *News In Physiological Sciences* 1993, 8: 253-260.

Suga S-I, K Nakao, H Hoh, Y Komatsu, Y Ogawa, N Hama and H Imura. Endothelial Production of C-type Natriuretic Peptide and Its Marked Augmentation by Transforming Growth Factor-$\beta$. *J Clin Investig* 1992, 90:1145-1149.

Tabuchi Y, M Nakamuru, H Rakugi, M Nagano, K Higashimori, H Mikami and T Ogihara. Effects of Endothelin on Neuroeffector Junction in Mesenteric Arteries of Hypertensive Rats. *Hypertension* 1990. 15: 739-743.

Taddei S, R Pedrinelli and A Salvetti. Sympathetic Nervous System-Dependent Vasoconstriction in Humans. Evidence for Mechanistic Role of Endogenous Purine Compounds. *Circulation* 1990, 82: 2061-2067.

Takayi et al 1992

Tesfamarian BR, RM Weisbrod and RA Cohen. The Endothelium Inhibits Activation by Calcium of Vascular Neurotransmission. *Am J Physiol 257 (Heart Circ Physiol 26)* 1989, H1871-H1877.

Thorén P, DE Donald and JT Shepherd. Role of Heart and Lung Receptors with Non-Medulated Afferents in Circulatory Control. *Circ Res* 1976, 38 (Suppl 2): 2-9.

Toda N, Y Kitamura and T Okamura. Neural Mechanism of Hypertension by Nitric Oxide Synthase Inhibition in Dogs. *Hypertension* 1993, 21: 3-8.

Toda N and T Okamura. Modification by L-N$^G$- Monomethyl Arginine (L-NMMA) of the Response to Nerve Stimulation in Isolated Dog Mesenteric and Cerebral Arteries. *J Pharmacol (Jpn)* 1989, 52: 170-173.

Toda N and T Okamura. Regulation by Nitroxidergic Nerves of Arterial Tone. *News In Physiological Sciences* 1992, 7: 148-152.

Ulman LG, EK Potter and DI McCloskey. Effects of Sympathetic Activity and Galanin on Cardiac Vagal Activity in Anesthetized Cats. *J Physiol (London)* 1992, 448: 225-235.

Vanhoutte PM and MN Levy. Prejunctional Modulation of Adrenergic Neurotransmission in the Cardiovascular System. Am J Physiol (Heart Circ Physiol) 198, 238: H275-H281.

von Euler US. A Specific Sympathomimetic Ergone in Adrenergic Nerve Fibres (Sympathin) and Its Relation to Adrenaline and Noradrenaline. *Acta Physiol Scand* 1946, 12: 73-97.

Wahlestedt C, H Yanaihara and R Häkanson. Evidence for Different Pre- and Post-junctional Receptors for Neuropeptide Y and Related Peptides. *Regul Peptides* 1986, 13: 307-318.

Wali FA and E Greenidge. Evidence that ATP and Noradrenaline Are Released During Electrical Field Stimulation of the Rat Isolated Seminal Vesicle. *Pharmacological Research* 1989, 21: 397-404.

Warner MR, P de Senanayake, CM Ferrano and MN Levy. Sympathetic Stimulation-Evoked Overflow of Norepinephrine and Neuropeptide Y from the Heart. *Circ Res* 1991, 69: 455-465.

Wiklund NP, A Ohlen and B Cederquist. Inhibition of Adrenergic Neuroeffector Transmission by Endothelin in the Guinea-Pig Femoral Artery. *Acta Physiol Scand* 1988, 134: 311-312.

Yang T and Levy MN. Sequence of Excitation as a Factor in Sympathetic-Parasympathetic Interactions in the Heart. *Circ Res* 1992, 71: 898-905.

Yanagisawa M, H Kurihara, S Kimuri, Y Tomobe, M Kobayashi, Y Mitsui, Y Yatrasak, K Gobo and T Masaki. A Novel Potent Vasoconstrictor Peptide Produce by Vascular Endothelial Cells. *Nature* 1988, 332: 411-441.

Zimmerman BG. Actions of Angiotensin on Adrenergic Nerve Endings. *Federation Proceedings* 1978, 37: 199-202.

Zukowska-Grojec Z, ES Marks and M Haass. Neuropeptide Y is a Potent Vasoconstrictor and a Cardiodepressant in Rat. *Amer J Physiol (Heart Circ Physiol)* 1987, 253: H1234-H1239.

# 2

# SOME RECENT ADVANCES IN STUDIES ON J RECEPTORS

A. S. Paintal

The DST Centre for Visceral Mechanisms
Vallabhbhai Patel Chest Institute
Delhi University
Delhi-110007, India

## EARLIER ADVANCES IN J RECEPTOR PHYSIOLOGY

Twenty years ago, John and Hazel Coleridge made a major advance in the field of patho-physiology of the lungs when they presented, at the Krogh Centenary symposium held in Srinager in 1974, evidence showing that the J receptors, which they called pulmonary C-fibres, were clearly stimulated by rise in left atrial pressure and that the degree of this stimulation was related to the degree of rise in pressure (Coleridge & Coleridge, 1977). This quantitative relation between the degree of resulting pulmonary congestion and stimulation of the receptors was a considerable advance over what had been observed qualitatively, earlier (Paintal, 1955;1969). A few years later, another important advance was made when it was shown that increasing cardiac output by about two times stimulated the J receptors effectively enough to produce marked reflex effects, notably the J reflex, thereby strengthening the view that the J receptors must be clearly stimulated during moderate exercise in man (Anand & Paintal, 1980). Six years later another advance was made when the Coleridges' and their collaborators demonstrated the relation between the degree of congestion of the lungs as seen from histological evidence to the activity of the J receptors (Roberts, Bhattachharaya, Shultz, Coleridge & Coleridge, 1986). In the same year evidence was presented to show that these receptors produced sensations of irritation in the throat of man, leading to the production of dry cough when they were stimulated by lobeline or high altitude pulmonary edema (Paintal, 1986a). A short while later Anand, Raj, Singh & Paintal (1989) provided preliminary evidence to show that increase in permeability of the pulmonary capillaries produced by phosgene gas in cats caused much greater stimulation of the J receptors by phenyl diguanide (PDG). They stated "The amount of increase in permeability could be roughly estimated from the phenyl diguanide dose - J receptors response curves obtained under control conditions." They extended this conclusion to its application in man (Anand et al., 1989).

# RECENT ADVANCES IN J RECEPTOR PHYSIOLOGY

However, the above preliminary conclusion on estimating increase in permeability were not readily accepted for various reasons mainly arising from the fact that the concentration of PDG in the capillary blood was not known and it was also not known to what degree PDG sensitized the J receptors to their natural stimulus namely pulmonary congestion/interstitial edema which was produced by phosgene gas. Fortunately, however, a major advance was made recently by the discovery of the principle of relative dilution of multiple solutes in flowing fluids (Paintal & Anand, 1991). This principle states"If *n* number of substances each weighing the same and *with identical physical properties* contained in the same injectate are injected into a blood vessel, then the concentration at any particular time of each one of the substances at a point downstream from the injection site will be identical". It follows that if the weight of one of two substances is twice that of the second, then the concentration of the first substance will be twice that of the second one. Thus it is easy to calibrate the concentrations of the different substances if the quantity of each one of substances in the solution is known. Since thermal particles can be regarded as a solute, it follows that one can use the fall (or rise) in the temperature of the solution of known temperature in the blood as an indicator substance for which a measuring device is available. Thus as shown in Fig.1 a new method was developed for measuring in vivo the concentration of PDG in the blood of the pulmonary artery (Paintal & Anand, 1992).

Using the above method for estimating the concentration of injected PDG the important new observation that emerged was that PDG (and probably other solutes) move out of the pulmonary capillaries by a process of diffusion and not through filtration forces. This came as a surprise because it had been generally assumed that the greater the pressure of blood in the pulmonary capillaries, the greater would be the movement of solutes out of the capillaries. In fact it was impressive to see the opposite i.e. an increase in stimulation of the J receptors (using the same dose of PDG) on reducing the pulmonary capillary pressure by partially occluding the inferior vena cava leading to a rise in the mean concentration of PDG in the blood owing to the reduction of blood flow during the period of partial occlusion of the inferior vena cava (Paintal & Anand, 1992).

It was also observed that increasing left atrial pressure did not obviously enhance the excitation of the J receptors of PDG. These results showing that the excitation of the receptors

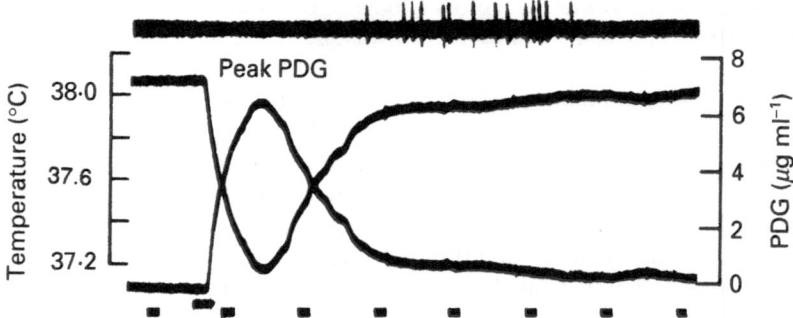

**Figure 1.** Method for recording the rise in concentration of PDG in the blood of the pulmonary artery. The downward curve shows the fall in temperature on injection of 0.5 ml of 100 µg m$^{-1}$ PDG solution into the right atrium at signal (bottom). This curve was inverted to show the rise in concentration of PDG (calibration on the right). The topmost trace shows the train of impulses from a J receptor. The lowest trace is of time marks and signal (From Paintal & Anand, 1992).

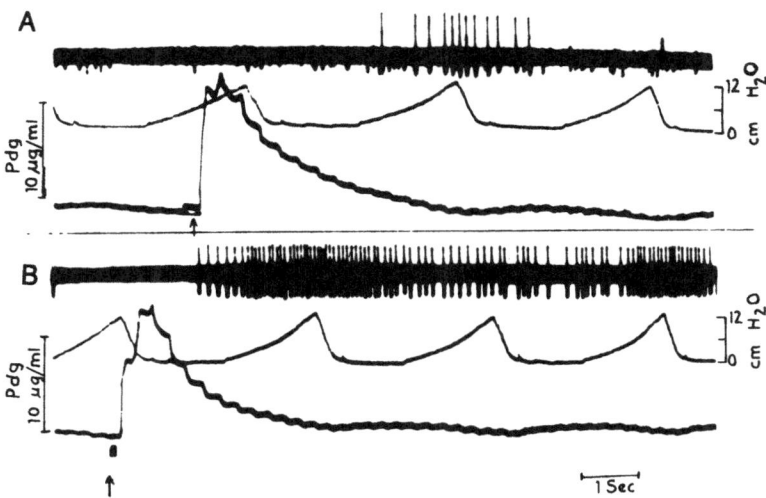

**Figure 2.** Responses of a J receptor to injections of 12 µg kg⁻¹ PDG before (A) and after (B) increase in capillary permeability produced by phosgene. PDG was injected at arrows leading to rise in the concentration of PDG (calibration on the left). The mean concentration of PDG over 3 s was 5 µg ml⁻¹ in both A and B. The middle trace is of intratracheal pressure (calibration on the right). (From Anand et al,1993).

by PDG depended on the concentration of PDG in capillary blood and not on the pulmonary capillary pressure led to the conclusion that the movement of PDG (or other excitants) out of the capillaries into the interstitum does not depend on filtration forces but on the forces of diffusion (Paintal & Anand, 1992).

## EFFECT OF INCREASE IN PERMEABILITY OF PULMONARY CAPILLARIES

The second advance made recently is the demonstration that the excitation of the J receptors by chemical excitants such as PDG, nicotine and capsaicin is greatly enhanced after the permeability of the pulmonary capillaries increases (Anand et al, 1993). This fact is easily appreciated by comparing the responses of the receptors to the same dose of the excitant leading to the same increase in the mean concentration of the excitant (PDG) in the blood of the pulmonary capillaries before (Fig.2A) and after (Fig.2B) increase in permeability of the capillaries. Fig. 3 shows the progressive increase in the responses of the J receptor to PDG after phosgene inspite of insignificant changes in the concentration of PDG in the blood. Fig.3 suggests that the enhanced responses to PDG might be due to the gradual stimulation of the receptor by phosgene itself. However, a closer look at Fig. 3 shows that the enhanced responses to PDG started occurring before any increase in the natural activity of the receptors due to the interstitial oedema produced by phosgene. Such increases in the responses of the receptors to PDG before any increase in natural activity produced by phosgene was seen in most receptors, in fact, in some the increase occurred 10 to 19 min before the increase in natural activity set in. Moreover in several receptors the enhanced responses to PDG were seen during the silent periods of the intermittent natural activity of the receptors which is a characteristic feature of the activity during pulmonary oedema (Paintal, 1969).

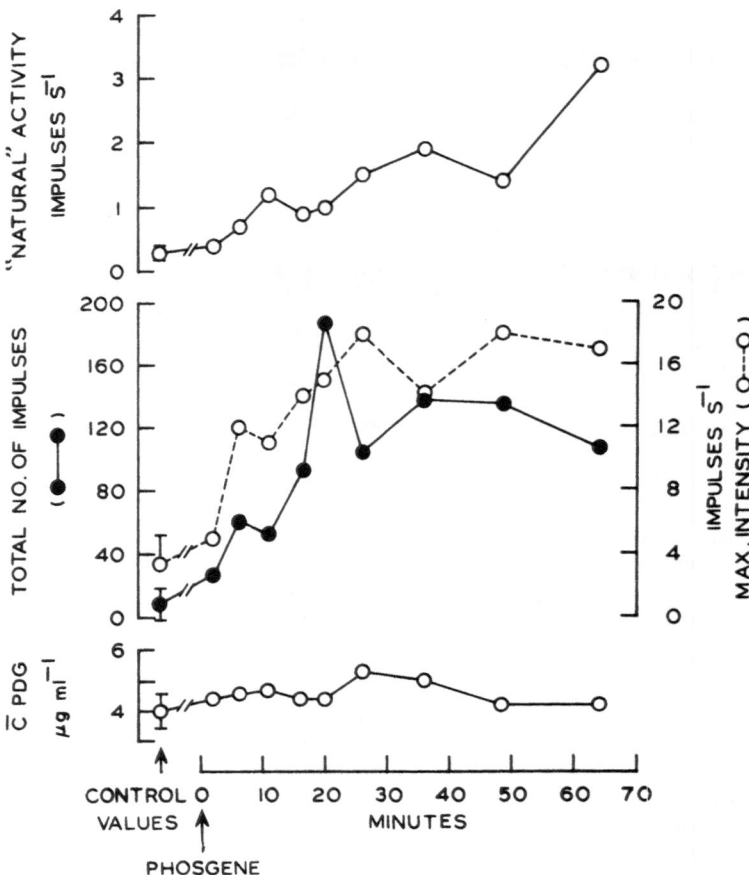

**Figure 3.** Responses of a J receptor to control injections of PDG before (Bars, ±1 S.D.) and at various times after giving phosgene at arrow (0 time). The edema induced activity (topmost trace) was measured over 10 s before each of nine injections of PDG. Note insignificant variations in the concentration of PDG (lowest curve) and the marked increase in the total number of impulses and the maximum intensity of discharge produced by PDG. (From Anand et al, 1993).

## DISTINCTION OF PERMEABILITY EDEMA FROM HAEMODYNAMIC EDEMA

The above results show that the responses of the J receptors to J receptor excitants such as PDG which do not sensitize the receptors to their natural stimulus is greatly increased during permeability edema. This is consistent with the fact that when the natural activity of the receptors is increased by raising pulmonary capillary pressure, there is no enhancement of the responses of the receptors to such excitants such as PDG (Paintal & Anand, 1992). This difference in the responses could provide a method for distinguishing between permeability edema and haemodynamic edema which occurs not uncommonly in patients with left ventricular failure. However, it is not possible to record impulses from J receptors in man so far. One would therefore have to use an indirect index of J receptor activity such as intensity of respiratory reflex effects or the intensity of sensations produced by stimulation of J receptors (see below). One would also have to ensure that the excitant used does not sensitize the receptors to the natural stimulus. Thus in the case of haemodynamic edema one should expect that since the increased level of background natural activity due to the interstitial edema would not enhance the responses of the receptors to a non-sensitizing excitant, that responses of the receptors to the excitant would not increase thereby in the presence of the same concentration of the excitant in the blood of the pulmonary capillaries. This would lead to the respiratory reflex effects or the sensations generated remaining unchanged. On the other hand in permeability edema the reflex effects and the sensations would be enhanced. This important possible outcome needs to be pursued urgently.

## SENSATIONS PRODUCED BY J RECEPTORS

The third significant advance made most recently relates to the sensations produced by impulses from the J receptors in man. Although evidence regarding this had been obtained earlier by comparing the sensations produced by injecting lobeline intravenously with those produced by natural stimulation of J receptors during pulmonary edema of high altitude in man (Paintal, 1986a) it is only very recently that it has been established that the sensations produced by lobeline in normal human subjects is due entirely to stimulation of J receptors. This was achieved by correlating the responses of the three groups of pulmonary receptors, the slowly adapting stretch receptors (SAR), rapidly adapting receptors (RARs) and the J receptors of cats to injections of lobeline with the sequence of events relating to the sensations produced by lobeline in man. Lobeline clearly sensitized the RARs of cats. Assuming that lobeline sensitized the RARs of man greatly in the same way as it *sensitized* the RARs of cats it was found that such assumed sensitisation that would occur during forced expiration to FRC did not facilitate the sensations produced by lobeline. Similarly, facilitation was also absent during deep inspiration. Moreover, it was found that injection of lobeline while the breath was held i.e. at a time when the activity of the RARs would be reduced and manifestation of sensitization would be low, as seen in cats, (Raj et al, 1994) did not reduce the sensations produced by lobeline injected during breath-holding. Thus it appears that neither the RARs nor the SARs are involved in the production of the sensations produced by lobeline.

On the other hand there was considerable evidence to show that the J receptors were responsible for the sensations produced by lobeline. The most convincing evidence was that the sensations and reflex respiratory effects appeared almost simultaneously. Actually the mean difference between the reflex effects and the sensations was 0.3s (n=26) and this was attributable to the reaction time of the subjects in signaling the sensations. In fact it was remarkable that in ten subjects there was no difference in the latencies for the reflex effects and the sensations. These results therefore strongly suggest that since it is known that the respiratory reflex effects of injecting lobeline into the pulmonary artery is due to impulses from J receptors (Jain, Subramanium, Julka & Guz., 1972) it follows that the sensations in the throat and chest that occur at the same time as the reflex effects must also be due to the J receptors.

From the above it is obvious that the sensations must be perceived in the brain as a result of a direct flow of impulses from the J receptors to the brain. The sensations cannot be the result of some secondary sensory input from the diaphragm - a view that had been entertained in the case of certain respiratory sensations for several years by some investigators (See Campbell & Guz, 1981). It follows that the use of vagal block as a tool for investigating patients with dyspnea (Guz et al, 19970) should prove of great value once again as demonstrated recently by Davies et al (1987). Incidentally the data of Davies et al (1987) reveals the large excitatory effect of J receptor inputs on ventilation inspite of markedly reduced end tidal $PCO_2$ to about 23mm Hg. This has also been observed in cats (Anand et al, 1982) (See also Trzebski, 1983).

It should be noted that it has already been pointed out (Paintal, 1986a) that receptors downstream from the lungs cannot be involved at all in the production of the sensations because as observed by Stern, Bruderman & Braun, 1966) no cough (i.e. sensations) is produced by injecting lobeline into the left ventricle. They also observed that no cough was produced when lobeline was injected into a branch of the pulmonary artery. However, cough appeared when lobeline was injected into the main trunk of the pulmonary artery. These observations led them to conclude that the receptors involved were located in the main pulmonary artery or its bifurcation i.e. in the same region where Coleridge and Kidd (1961) had located their pulmonary artery baroreceptors (but see Paintal, 1972). Thus their observations suggest that the receptors involved in the production of the sensations are not located in the lungs. However, this conclusion is not justified because as shown by Pórszász Such & Pórszász-Gibiszer (1957) drugs injected into a *branch* of the pulmonary artery of dogs bypass the receptors in the lungs completely and as argued earlier (Paintal, 1986b) this is probably the reason why injections of lobeline into a branch of the pulmonary artery did not cause any cough in their subjects i.e. it bypassed the J receptors. Thus the observations of Stern et al (1966) can be reconciled with the conclusion that lobeline produced cough by stimulating the J receptors.

Here it would be appropriate to reconcile the observations of Tatar, Webber & Widdicombe (1988) with the present ones. On the face of it they are contradictory because their observations suggest that the J receptors of cats not only do not cause cough but that they actually inhibit cough. Tatar et al (1988) produced cough in cats by irritating the receptors in the trachea mechanically. The strength of the cough was estimated from the electromyograms of the muscles producing the cough. They found that when the J receptors were stimulated by injecting phenyl biguanide the contractions of the muscles were greatly reduced. However, the reason why the cough generated by mechanical irritation of the tracheal mucosa is inhibited by stimulation of the J receptors is because it is known that the J reflex (Paintal, 1970) inhibits all somatic muscles including the muscles responsible for producing cough as well. On the other hand it turns out that the observations of Tatar et al (19988) may be clinically rather important because their observations suggest that when the J receptors of man are markedly stimulated as may occur during severe pulmonary congestion, cough will be inhibited, thus reducing the patient's ability to expectorate the secretions accumulating in his trachea.

**Table 1.** Comparison of effects produced by lobeline, in twenty six subjects (Raj et al, 1994), and capsaicin in three subjects (Winning et al, 1986)

| Effects of drug | Lobeline doses | | Capsaicin doses | |
|---|---|---|---|---|
| | 12 µg kg$^{-1}$ (threshold) | 24 µg kg$^{-1}$ | 0.5 µg kg$^{-1}$ (threshold) | 2-4 µg kg$^{-1}$ |
| 1. Non-painful sensations | Present in throat and chest | Present in throat and chest | Absent | Absent |
| 2. Burning/pain sensations | Absent | Present in 12% in throat and chest | Present in chest | Present in chest |
| 3. Cough | Absent | Present in all subjects | Absent | Present in one subject only |
| 4. Sensation after local anaesthesia | Present | Present | Abolished | Abolished |
| 5. Respiratory reflex | Present | Present | Absent | Absent |

# DIFFERENCES BETWEEN EFFECTS OF LOBELINE AND CAPSAICIN

Table I summarises the differences between the effects of injecting lobeline and capsaicin in man. As shown in this Table capsaicin did not produce any sensations of irritation in the throat even when it was injected at the lowest doses (see Winning et al, 1986). Whenever it did produce any sensation it was a raw burning sensation in the chest. In contrast lobeline at threshold doses never produced painful sensations, but sensations of irritation in the throat and upper chest. In much higher doses it produced pain in only 12% of the subjects. Another difference between the two substances is that whereas lobeline produced reflex respiratory effects typical of stimulation of J receptors (see Jain et al, 1972), capsaicin did not produce any respiratory reflex effects even in high doses. Finally the most important difference between the two substances is that the sensations produced by capsaicin (burning) in the chest were abolished after prior local anaesthesia of the upper airways. In contrast local anaesthesia of the airways did not abolish the sensations in the throat and chest and dry coiugh produced by injecting lobeline. Thus it is clear that different sets of sensory receptors in different locations are involved in producing the respective sensations (Raj et al, 1994). Lobeline produces the J receptor induced sensations whereas capsaicin produces the burning sensations probably through stimulation of the mucosal receptors (hitherto unidentified) in the trachea. There are two possible ways by which the latter receptors could be simulated following intravenous injections of capsaicin, either directly through the blood stream or they could be stimulated after capsaicin diffuses out of the alveoli and acts on the mucosal receptors in a manner similar to that of a capsaicin aerosol.

# NATURE OF SENSATIONS

At threshold doses, lobeline produced a variety of sensations in the throat and upper chest. These sensations were drescribed mainly as pressure and choking. Other descriptions

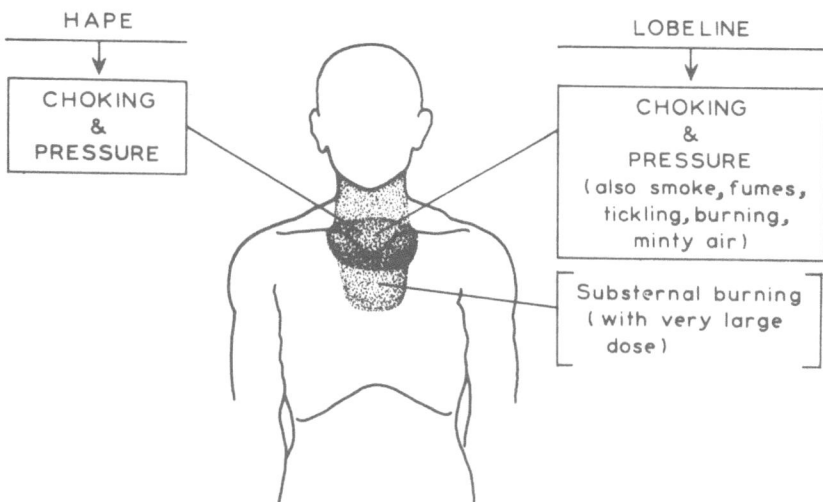

**Figure 4.** The nature of J receptor induced sensations and the regions where they are felt after injections of lobeline or during high altitude pulmonary edema (HAPE), darker shading indicates greater intensity of sensations.

**Table 2.** Frequency of occurrence of dry cough, breathlessness and chest "pain" in three groups of subjects with high altitude pulmonary edema

|  | Menon (1965) | Lal (1967) | Paintal (1986) |
|---|---|---|---|
| Total no. of subjects studied | 101 | 213 | 36 |
| Symptoms | Frequency (%) | | |
| Dry cough | 51 | 80 | 75 |
| Breathlessness | 84 | 79 | 86 |
| Chest "pain" i.e. chest discomfort = pressure or choking | 66 | 63 | 72 |

were minty air, smoke, tickling and burning (Raj et al, 1994). It is noteworthy that only 12% of their normal subjects experienced a burning type of sensation in the throat and upper chest when lobeline upto 24 ug kg$^{-1}$ was injected. In contrast about 90% of (6 out of 7) felt substernal burning in the series reported by Eckenhoff & Comroe (1951). However, Echenhoff & Comroe injected 5 to 7.5 mg lobeline in their subjects. This would be roughly four times the dose injected by Raj et al (1994). This sensation of substernal burning is the only kind that can could be regarded as painful.

On the other hand the so-called pain reported by Lal (1967) in (Table 2) cannot be regarded as true pain because Lal described the sensation of 'pain' as being essentially chest discomfort consisting of heaviness (i.e. pressure) and choking. Since choking sensations are usually felt in the throat it follows that the sensations in HAPE are also localised in the throat as is the case following injections of lobeline as shown in Fig.4 which depicts the location-intensity of the sensations.

The same conclusion relating to the so-called pain sensation in HAPE may be drawn from the observations of Menon (1965) who stated that "all patients complained of chest discomfort but only 66 had actual pain in the chest. It was invariably felt in the substernal region and was never very severe. It was more in the nature of an unpleasant ache that was constantly present". From the above it appears that none of the HAPE subjects of Menon (1965) or Lal (1967) experienced actual pain but felt chest discomfort consisting of pressure or choking.

Paintal (1986 a) gave no details in his study but since his subjects were drawn from a similar population of healthy soldiers ten years later it may be assumed that the 'pain' recorded during the clinical examination in his Table I must have been sensations of pressure and choking.

The J receptors are stimulated by both pulmonary edema (Paintal, 1969) and by lobeline. Therefore it is not surprising that the J receptor induced sensations experienced during HAPE are similar to those produced after injections of lobeline. These sensations are essentially sensations of pressure and choking in the throat and upper chest as shown in Fig.4. It would appear that these sensations probably constitute the underlying sensations felt during breathlessness which the majority (83%) of the subjects experienced during HAPE in the three studies listed in Table 2. It is interesting that in an earlier study (Raj, Singh & Paintal, 1986) six out of twenty normal subjects when questioned stated that the sensations they felt after lobeline injections resembled those generated during moderate exercise which by increasing pulmonary blood flow stimulates the J receptors (Anand & Paintal,1980) and must therefore produce the J reflex which gives rise to the feeling of muscle weakness (Paintal, 1986 c). It should be noted that the J reflex manifests itself clearly during pulmonary congestion (Pickar, Hill & Kaufman, 1993) (see also Kalia 1977). The physiological importance of this reflex has been recently elaborated by Coleridge & Coleridge, 1994).

## BREATHLESSNESS

It is generally agreed that breathlessness is the awareness of increased breathing during exercise (see Howell & Campbell, 1966). What is the evidence that breathlessness and the accompanying sensations of choking and pressure in the throat and upper chest in certain normal people during appropriate exertion are J receptor induced? This question can be answered by consideration of the following lines of reasoning.

There can be little doubt that the breathlessness experienced by young subjects (soldiers) at high altitudes with established HAPE are J receptor induced (see Paintal, 1986a and above data). As can be calculated from the data in Table 2 about 80% of the subjects with breathlessness had the accompanying sensations of choking and pressure in the throat and upper chest characteristics of J receptor stimulation by lobeline. One can assume that in doubtful cases of HAPE with minimal evidence of alveolar infiltration in the radiographs, the breathlessness and sensations in the throat and chest are also J receptor induced. Currently cases of such soldiers at high altitudes with only minimal evidence of lung infiltration in the radiographs are not admitted as indoor patients for the usual treatment of HAPE. These are the subclinical cases of HAPE with dry cough and breathlessness who form part of the normal population of soldiers at high altitudes. It follows that in all these "normal" people the breathlessness and the accompanying sensations in the throat and chest which appear after exertion should also be J receptor induced. From this it is reasonable to extrapolate that similar breathlessness and accompanying sensations in certain people at lower altitudes down to sea level after greater exertion would also be J receptor induced. So one can further assume that when certain normal individuals get breathlessness plus a feeling of choking and pressure in the throat and upper chest after running for a bus (or after climbing flights of stairs quickly that it is because of the inflow of impulses from the J receptors that these sensations are produced.

What about the role of slowly and rapidly adapting pulmonary stretch receptors in producing the sensations of breathlessness? It is certain that the slowly adapting ones must do the opposite i.e. produce relief of the sensations of breathlessness as shown by the observations of Fowler (1954) and recently confirmed by Flume et al, (1994). The observations of the latter are particularly noteworthy because the early relief of the sensations, produced by breath holding after rapid *expiration* when the rapidly adapting receptors would be stimulated suggest that the latter might also relieve the sensations of breathlessness i.e. they might be functionally similar to the slowly adapting receptors in this respect. (See Paintal, 1983).

In conclusion it is satisfying to think that as a result of this symposium in honour of John and Hazel Coleridge, at long last, we are coming close to understanding the J receptor induced sensations of breathlessness in health and disease - a fitting tribute to their many contributions in this field of physiology (See Coleridge & Coleridge, 1984).

## SUMMARY

While describing recent advances in studies on J receptors it was shown that the discovery of the principle of the relative dilution of multiple solutes in flowing fluids paved the way for developing a new method for measuring *in vivo* the concentration of injected drugs in the blood of the pulmonary artery. This led to the finding that excitatory solutes move out of the capillaries through a process of diffusion not through filtration.

Increase in the permeability of the capillaries causes a marked increase in the responses of the J receptors to excitants by causing greater movement of the excitants to the receptors. This information is likely to yield a method for distinguishing permeability edema from hamodynamic edema in man.

The most recent advance relates to the evidence showing conclusively that the sensations and dry cough produced by injecting lobeline intravenouly in man is due to the stimulation of the J receptors. The slowly and rapidly adapting receptors play little or no role in this. The nature of the sensations felt is somewhat variable, most commonly it is choking and pressure localised in the throat and upper chest. Similar sensations are felt by subjects with high altitude pulmonary edema (HAPE). From this data it is extrapolated that the same kinds of sensations that accompany breathlessness after moderate or severe exercise at sea level are also J receptor induced.

## REFERENCES

Anand. A. & Paintal, A.S. (1980). Reflex effects following selective stimulation of J receptors in the cat. *Journal of Physiology*. 299, 553-572.

Anand, A., Loeschcke, H.H., Marek, W. & Paintal, A.S. (1982). Significance of the respiratory drive by impulses from J receptors. *Journal of Physiology* 325, 14p.

Anand, A. , Paintal, A.S., Raj, H. & Singh, V.K. A (1989). A method for estimating changes in pulmonary capillary permeability in animals and man from the responses of J receptors to drugs. *Journal of Physiology*, 412, 36P.

Anand, A., Paintal, A.S. & Whitteridge, D. (1993). Mechanisms underlying enhanced responses of J rceptors of cats to excitants in pulmonary oedema. *Journal of Physiology*, 471, 535-547.

Campbell, E.J.M. & Guz.A. (1981). Breathlessness. In *Regulation of Breathing*, Part II, ed. Hornbein, T.F., pp 1181 - 1195. Marcel Dekker Inc., New York.

Coleridge, H.M. & Coleridge, J.C.G. (1977). Afferent vagal C-fibers in the dog lung: their discharge during spontaneous breathing, and their stimulation by alloxan and pulmonary congestion. In *Respiratory Adaptations, Capillary Exchange and Reflex Mechanisms*, ed. Paintal, A.S. & Gill-Kumar, P., pp 396-405, Vallabhbhai Patel Chest Institute, Delhi.

Coleridge, H.M. & Coleridge, J.C.G. (1994). Pulmonary reflexes: Neural mechanisms of pulmonary defense, *Annual Review of Physiology*, 56, 69-91.

Coleridge, J.C.G. & Coleridge, H.M. (1984) Afferent vagal C fibre innervation of the lungs and airways and its functional significance. *Reviews of Physiology, Biochemistry and Pharmacology*, 99, 1-110.

Coleridge, J.C.G. and Kidd, C. (1961). Relationship between pulmonary arterial pressure and impluse activity in pulmonary arterial baroreceptor fibres. *Journal of Physiology* 158, 197-205.

Davies, S.F., Mcquaid, K.R., Iber, C., Mcarthur,, C.D. Path, M.J., Beebe, D.S. & Helseth, H.K.. (1987) Extreme dyspnea from unilateral pulmonary venous obstruction. Demonstration of a vagal mechanism and relief by right vagotomy. *American Review of Respiratory Diseases*, 136, 184-188.

Eckenhoff, J.E. & Comroe, J.H. Jr. (1951). Blocking action of tetraethylammonium on lobelin-induced thoracic pain. *Proceedings of the Society for experimental Biology and Medicine* 76, 725-726.

Flume, P.A., Eldridge, F.L., Edwards, L.J., & Houser, L.M. (1994). The Fowler breathholding study revisited: continuous rating of respiratory sensation, *Respiration Physiology*, 95, 53-66.

Fowler, W.S. (1954). Breaking point of breath-holding. *Journal of Applied Physiology* 6, 539-545.

Guz, A., Noble, M.I.M., Eisele, J.H. & Trenchard, D. (1970). Experimental results of vagal block in cardiopulmonary diesease. In *Breathing*: Hering-Breuer Centenary Sumposium, ed. Porter, R.pp. 315-329, Churchill, London.

Howell, J.B.L. & Campbell, E.J.M. (1966). *Breathlessness*, pp. xiii-xiv, Blackwell Scientiic Publications, Oxford.

Jain, S.K., Subramanian, S. Julka, D.B. & Guz, A. (1972). Search for evidence of lung chemoreflexes in man: study of respiratory and circulatory effects of phenyldiguanide and lobeline. *Clinical Science* 42, 163-177.

Kalia, M. (1973). Effects of certain cerebral lesions on the J reflex. *Pflugers Archives gesamte Physiology* 343, 297-308.

Kalia, M. (1977). The J reflex produced by pulmonary congestion in the cat. *Proceedings of the International Union of Physiological Sciences* 13, 364.

Lal, M. (1967). Clinical aspects of high altitude pulmonary oedema. *Indian Journal of Chest Diseases*, 9, 82-89.

Menon, N.D. (1965). High-altitude pulmonary edema. *New England Journal of Medicine* 273, 66-73.

Paintal, A.S. (1955). Impulses in vagal afferent fibres from specific pulmonary deflation receptors. The response of these receptors to phenyl diguanide, potato starch, 5hydroxytryptamine and nicotine, and their role in respiratory and cardiovascular reflexes. *Quarterly Journal of experimental Physiology* 40, 89-111.

Paintal, A.S. (1969). Mechanism of stimulation of type J pulmonary receptors. *Journal of Physiology* 203, 511-532

Paintal, A.S. (1970). The mechanism of excitation of type J receptors, and the J reflex. In *Breathing: HeringBreuer Centenary Symposium*, ed. Porter, R. pp. 59-71, Churchill, London.

Paintal, A.S. (1972). Cardiovascular receptors. In *Handbook of Sensory Physiology*, Vol. III/I ed. Neil, E. pp 1-45, Springer-Verleg, Berlin.

Paintal, A.S. (1977). Thoracic receptors connected with sensation. *British Medical Bulletin*, 33, 169-174.

Paintal, A.S. (1983). Lung and airway receptors. In *Control of Respiration*, ed. Pallot, D.J. pp. 78-107, Croom Helm, London.

Paintal, A.S. (1986a). The visceral sensations - some basic mechanisms. *Progress in Brain Research* 67, 3-19.

Paintal, A.S. (1986b). Pocity vyvolane drazdenim receptorov J. *Bratislavske lekarske listy*. 85, 415-423.

Paintal, A.S. (1986c). The significance of dry cough, breathlesness and muscle wakness. *The Indian Journal of Tuberculosis*, 33, 51-55.

Paintal, A.S. & Anand, A., (1991). Estimating in vivo the blood concentration of chemical substances injected intravenously in anaesthetized cats: use in studies on sensory receptors. *Journal of Physiology* 438, 247 p

Paintal, A.S. & Anand, A. (1992). Factors affecting movement of excitatory substances from pulmonary capillaries to J receptors of anaesthetized cats. *Journal of Physiology*, 449, 155-168.

Pickar, J.G., Hill, J.M. & Kaufman, M.P. (1993). Stimulation of vagal afferents inhibits locomotion in mesencephalic cats, *Journal of Applied Physiology*, 74, 103110.

Pórszász, J. Such, G. & Porszasz-Gibiszer, K. (1957). Circulatory and respiratory chemoreflexes. I. Analysis of the site of action and receptor types of capsaicin. *Acta physiologica Hungarica* 12, 189-205.

Raj, H., Singh, V.K., Anand, A. & Paintal, A.S. (1994). Sensory origin of lobeline induced sensations: A correlative study in man and cat. *Journal of Physiology*, In press.

Raj, H., Singh, V.K. & Paintal, A.S. (1986). Sensations produced by J receptors. *Proceedings of the International Union of Physiological Sciences*, 16, 308.

Roberts, A.M., Bhattacharya, J., Schultz, H.D., Coleridge, H.M. & Coleridge, J.C. G( 1986). Stimulation of pulmonary vagal afferent C-fibers by lung edema in dogs. *Circulation Research*, 58, 512-522.

Stern, S., Bruderman, I. & Braun, K. (1966). Localization of lofeline-sensitive receptors in the pulmonary circulation in man. *American Heart Journal*, 71, 651-655.

Tatar, M., Webber, S.E. & Widdicombe, J.G. (1988). Lung Cfibre receptor activation and defensive reflexes in aesthetized cats. *Journal of Physiology*, 402, 411-420.

Trzebski, A. (1983). Respiratory reflexes. In *Control of Respiration*, ed. Pallot, D.J. pp 108-156. Croom Helm, London.

Winning, A.J., Hamilton, R.D., Shea, S.A., & Guz, A. (1986). Respiratory and cardiovascular effects of central and peripheral intravenous injections of capsaicin in man: evidence for pulmonary chemosensitivity. *Clinical Science*, 71, 519-526.

# 3

# PATHOPHYSIOLOGY OF BRONCHIAL ASTHMA

John Widdicombe

Department of Physiology
St George's Hospital Medical School
Cranmer Terrace, London, SW17 0RE

The primary pathology of asthma is inflammation of the lower airway mucosa. This may lead to secondary contraction of bronchial smooth muscle, both processes narrowing the airways and causing obstruction. The relative importance of mucosal thickness and of smooth muscle contraction is subject to much discussion, and the balance probably varies between patients and during the course of the disease.

John and Hazel Coleridge have over the years extensively studied reflexes set up from the airway mucosa, especially those due to C-fibre receptors. Many of the reflex responses feed back to the airways, and probably contribute to the obstruction in conditions like asthma (Coleridge and Coleridge, 1986; Widdicombe, 1989). The receptors responsible for these reflexes have an additional role in asthma if they set up local neurogenic inflammation (Fig 1).

## NEUROGENIC INFLAMMATION

Neurogenic inflammation can occur in many tissues, including skin and various viscera. It is due to an axon reflex in sensory nerve fibres, probably mainly those of C-fibre receptors, which ramify in the tissues (McDonald, 1987, 1992). Activation of some of the receptor terminals will cause a spread of impulses along the entire receptor complex, with release of sensory neuropeptides at the nerve terminals and probably along the course of the nerves. The neuropeptides in turn will provoke motor responses characteristic of inflammation. In the case of the airways these responses may include smooth muscle contraction, mucus secretion, vasodilation, and possibly changes in epithelial function. Extensive studies show that for the airways of rodents, such as rat and guinea-pig, activation of mucosal receptors by irritants such as cigarette smoke, sulphur dioxide and capsaicin (the pungent extract of chili peppers) causes the local responses of neurogenic inflammation (Lundberg et al., 1984; Lundberg, 1990). Release of sensory neuropeptides has been demonstrated, and these have been shown to set up the various appropriate motor responses (Lundberg et al., 1987; Saria et al., 1988). If the C-fibre complexes are destroyed by excessive doses of

*Control of the Cardiovascular and Respiratory Systems in Health and Disease*
Edited by C. T. Kappagoda and M. P. Kaufman, Plenum Press, New York, 1995

27

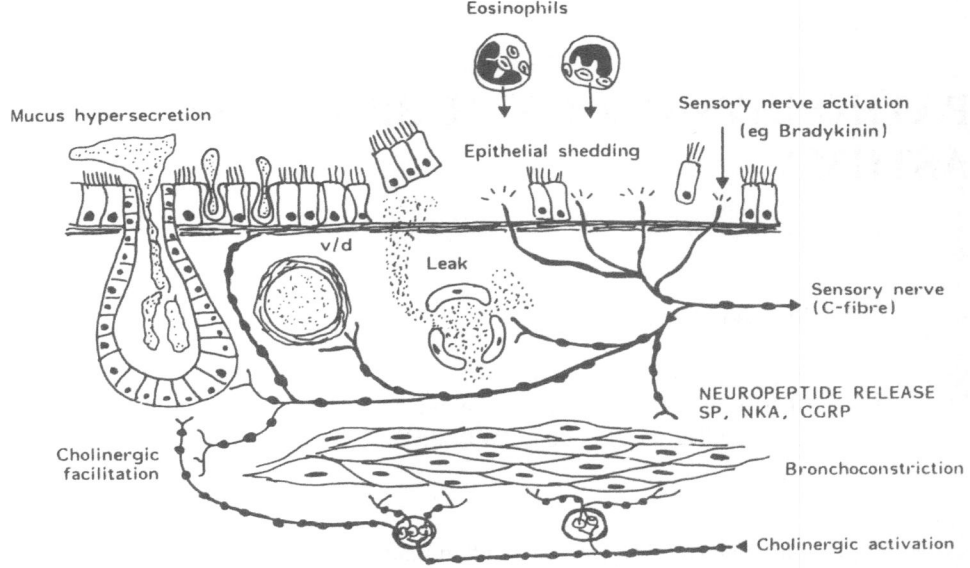

**Figure 1.** Possible axon reflex mechanisms (neurogenic inflammation) in asthma. From Barnes (1992b).

capsaicin or inhibited by local anaesthesia, the local responses of neurogenic inflammation are abolished.

Although neurogenic inflammation is well established in rodents, a number of fundamental problems remain to be resolved. These include:

1. The process does not seem to have been established in man, although there may be methodological difficulties in displaying it. Therefore its relevance to asthma is speculative.

2. Although sensory nerves in the epithelium are well established in a number of species (Das et al., 1978), their connections with deeper effector tissues have not been mapped out; this structural analysis is essential before we can understand the underlying anatomical basis of neurogenic inflammation. Two types of afferent endorgan are involved in the pathophysiology of mucosal inflammation and allergic conditions such as asthma. These are the C-fibre receptors and the rapidly adepting receptors (RAR's) (Coleridge and Coleridge, 1986; Widdicombe, 1989). RARs have branches that spread both to the epithelium and to the airway smooth muscle (Mortola et al., 1975; Sant' Ambrogio et al., 1978), but these studies need to be extended to include the C-fibre receptor complexes, and given an anatomical basis. It is usually assumed that the C-fibre receptors underlie neurogenic inflammation, but the RARs may also be involved.

3. The same receptors that are thought to cause neurogenic inflammation also set up central nervous reflexes that feed back to airway effector tissues. A fundamental question, not yet resolved, is whether the dispersion of nerve impulses in the sensory complex that evokes neurogenic inflammation also inevitably sends impulses up to the central nervous system to induce reflex changes in the same effector tissues. Work with other tissues, such as the skin, suggests that axon reflex responses can be provoked without central nervous reflexes (Perl, 1984). We do not know if this applies to the airways.

4. Following from this, we do not know the balance of neurogenic inflammatory changes in the airways compared with central nervous responses for the same effector tissues. Although methods could be developed to address this problem, since different mediators are involved in axon reflexes and central nervous reflexes and therefore different antagonists should be effective, such studies have been little applied to experimental animals and not directly to man.

5. Although some of the reflex changes induced by stimulation of both types of receptors are the same, in particular those on airway smooth muscle and secretory tissues, others are different such as those on respiration (Coleridge and Coleridge, 1986; Karlsson et al., 1988). As indicated below, probably both groups of receptors are involved in pathophysiology of asthma, but little is known about the relative balance between the responses to the two receptor groups.

## TYPES OF RECEPTOR

After Paintal (1955) had extensively studied the pulmonary C-fibre receptors, originally called J-receptors, the Coleridges and their colleagues made a major advance in our understanding of these receptors by showing that a second group of C-fibre endings occurred in the walls of the bronchi, now called bronchial C-fibre receptors (Coleridge and Coleridge, 1984). They showed that the receptor properties of the two groups had much in common (Table 1), but that there were important differences in the sensitivity to various inflammatory

**Table 1.** Stimuli to C-fibre receptors and RARs

|  | C-fibre receptors | | RARs |
|  | Pulmonary | Bronchial |  |
|---|---|---|---|
| Mechanical | Inflation<br>Foreign bodies | Foreign bodies | Inflation<br>Deflation<br>Dust<br>Mucus<br>Foreign bodies |
| Chemical | Irritant gases<br>Cigarette smoke<br>Capsaicin<br>Volatile anaesthetics | Irritant gases | Irritant gases<br>Cigarette smoke<br>Capsaicin<br>Volatile anaesthetics |
| Mediators | Acetylcholine<br>Histamine<br>Serotonin<br>Prostaglandins<br>Bradykinin | Histamine<br>Serotonin<br>Prostaglandins<br>Bradykinin | Acetylcholine<br>Histamine<br>Serotonin<br>Prostaglandins |
| Diseases | Microembolism<br>Pulmonary oedema<br>Pulmonary congestion<br>Pneumonia | Pulmonary congestion | Anaphylaxis<br>Microembolism<br>Atelectasis<br>Broncho-constriction<br>Pulmonary oedema |

In general the three groups of receptor respond to the same stimuli. However sensitivities vary greatly. The main differences in response relate to mechanical stimuli. The lists of chemical and mediator stimuli are incomplete, and not all agents have been tested on all groups of receptors.

**Table 2.** Reflex responses to receptor stimulation

| C-fibre receptors | | |
| --- | --- | --- |
| Pulmonary | Bronchial | RARs |
| Apnoea | Apnoea | Cough |
| Tachypnoea | Tachypnoea | Tachypnoea |
| Cough inhibition | | Augmented breaths |
| | | |
| Bronchoconstriction | Bronchoconstriction | Bronchoconstriction |
| Mucus secretion | Mucus secretion | Mucus secretion |
| Laryngoconstriction | | Laryngoconstriction |
| Vasodilation | | Vasodilation |
| Somatic inhibition | | |

The main differences between the reflex responses are respiratory. The lists may be incomplete because not all reflexes have been studied for the three groups of receptor.

mediators. The central nervous reflex responses from the two groups of receptors also had much in common (Table 2), although some differences have been identified. The distinctions between the two groups of C-fibre receptor are important because it is presumably the bronchial endings that are mainly involved in asthma.

RAR's in the airways have been known since 1929, when Keller and Loeser (1929) first described them. They were studied in detail by Knowlton and Larrabee (1946), and by many others since (Coleridge and Coleridge, 1986). As first suggested by Keller and Loeser they are now thought to cause coughing. The role of C-fibre receptors in coughing will be mentioned later. Many of the other central nervous reflex changes produced by activation of RARs are similar to those from C-fibres. Tables 1 and 2 list some of the stimuli that affect the three groups of receptor, and some of the reflex responses to their activation.

## Epithelial Receptors

It is assumed that the first neurological target in causing asthma lies in the epithelium, since inhaled allergens and irritants that cause or trigger an asthmatic attack will first hit this tissue. There are many nerves in the epithelium (Karlsson et al., 1988) but we do not know to what extent they are part of C-fibre receptor complexes or of the RARs. Indirect evidence based upon the relative density of nerve fibres, and histological evidence that the epithelial nerves connect to vagal myelinated fibres, suggest that they are part of the RAR complex which causes coughing(Das et al., 1978, 1979). However this does not rule out the possibility that some of them are C-fibre receptors.

Epithelial nerves have been beautifully delineated by Baluk et al. (1992) (Fig 2). They mapped them in the epithelium of the rat airways, and showed that most lie close to the basal layer of the epithelium, but they had processes extending towards the lumen which often ended in a splayed terminals. They could be displayed by immunological methods for sensory neuropeptides such as substance P (SP). On stripping the epithelium these nerves disappeared (Fig 2b). This is an observation of great importance in relation to neurogenic inflammation. If the nerves mediate this process, and the removal of epithelium destroys virtually all of the nerves, then the phenomenon should disappear. Pictures such as Fig 2 give no reason to believe that the nerves have neuropeptide-containing branches that penetrate to deeper tissues. This is relevant to asthma because in this condition the epithelium can be greatly damaged or even destroyed (Laitinen et al., 1985; Laitinen and Laitinen, 1991), yet we know that reflexes such as the cough reflex are enhanced and that prominent inflammatory conditions exist in the airways. Either the sensory nerve processes deep to the

**Figure 2.** A region of the tracheal mucosa directly above a cartilaginous ring of a rat. Nerves are exhibited by immunofluorescence for substance P. the density and orientation of the intraepithelial plexus of SP-IR axons are similar to those found between the cartilaginous rings. The intraepithelial nerve plexus is absent in a region of the epithelium (*), which was accidentally removed during processing. Bar, 25 μm. From Baluk et al. (1992).

epithelium do not contain SP but may still mediate neurogenic inflammation, possibly by other neuropeptides, or although very few in number they can still exert important neurogenic effects.

## CENTRAL NERVOUS REFLEXES

Excitation of airway C-fibre receptors (of both types, pulmonary and bronchial) and of RARs sets up an array of central nervous reflexes which may contribute to the pathophysiology of bronchial asthma (Table 2).

1. Both cause contraction of airway smooth muscle which will augment airways obstruction (Coleridge et al., 1982; Roberts et al., 1981, 1988; Coleridge and Coleridge, 1986) (Fig. 3). This reflex seems to be largely mediated by the parasympathetic transmitter acetylcholine. The same motor nerves may also release vasoactive intestinal polypeptide (VIP) and nitric oxide (NO) both of which are bronchodilators (Barnes, 1992a, b; Belvesi et al., 1992. 1993). The reflex smooth muscle contraction has been demonstrated both by use of the posterior tracheal muscle isolated in situ (Roberts et al., 1981) and by changes in total lung resistance reflecting mainly bronchial smooth muscle contraction (Tomori and Widdicombe, 1969).

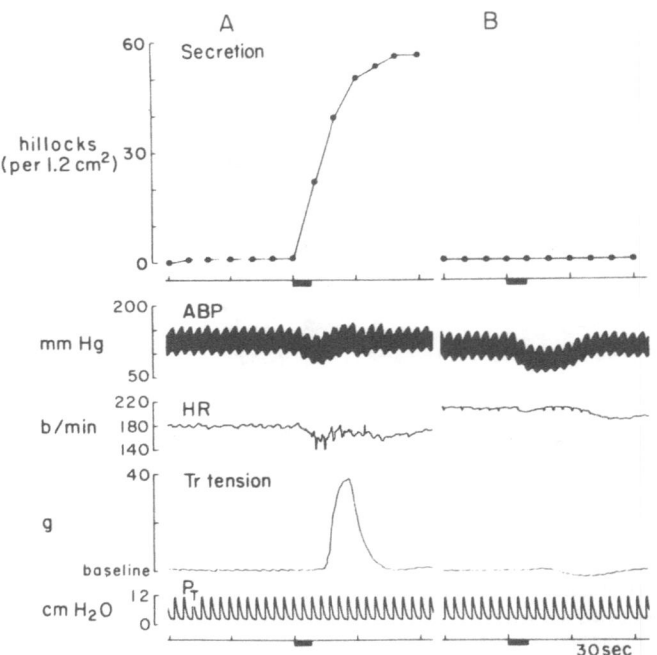

**Figure 3.** Effects on tracheal submucosal gland secretion, tracheal tension, arterial blood pressure and heart rate evoked by injecting 3 μg capsaicin into the bronchial artery of a dog (at signal shown in (a) and (c)) to stimulate bronchial C-fibre receptors and abolition of the effects by vagal cooling. Temperature of vagus nerves was A, 37°C, and B, 0°C. Recurrent and pararecurrent nerves were already cut so that cooling of midcervical vagus nerves had no effect on the segment's afferent nerve supply in the superior laryngeal nerves. From above down: mucus secretion in hillocks per 1.2 mm² of mucosa; arterial blood pressure (ABP) in mm Hg; heart rate (HR) in beats.min⁻¹; tracheal tension in g above baseline set at 75 g; tracheal pressure in cm $H_2O$. From Coleridge and Coleridge (1986).

2. Both groups of the C-fibre receptors and the RARs reflexly activate lower airway submucosal gland secretion of mucus (Coleridge and Coleridge, 1986) (Fig. 3). This secretion presumably contributes to airways obstruction. The mechanism is predominantly vagal parasympathetic, mediated by acetylcholine, since it is blocked by atropine. The extent to which other parasympathetic motor transmitters are involved has not been determined, although VIP can stimulate secretion by tracheal submucosal glands (Shimura and Takishima, 1994).

3. Both groups of sensory receptor probably cause tracheo- bronchial vasodilation (Sahin et al., 1987; Coleridge et al., 1992) (Fig. 4). Motor pathways for these responses are not clear, although at least for the upper trachea parasympathetic dilator fibres seem to be activated (Laitinen et al., 1987). Since the lower airway vasculature has a powerful sympathetic vasoconstrictor control (Matran, 1991), a decrease in sympathetic tone is probably part of the motor response. The airway vasodilation will thicken the mucosa and thus contribute to airways obstruction (Corfield et al., 1991). If it is associated with increased interstitial liquid volume, then the resulting oedema will have a similar effect. Pulmonary C-fibre receptors and probably RARs also cause a nasal vasodilation with a tendency to airway blockage (Lung and Widdicombe, 1987). In asthma this could be harmful, since airflow will be diverted to the mouth with greater inhalation of potentially

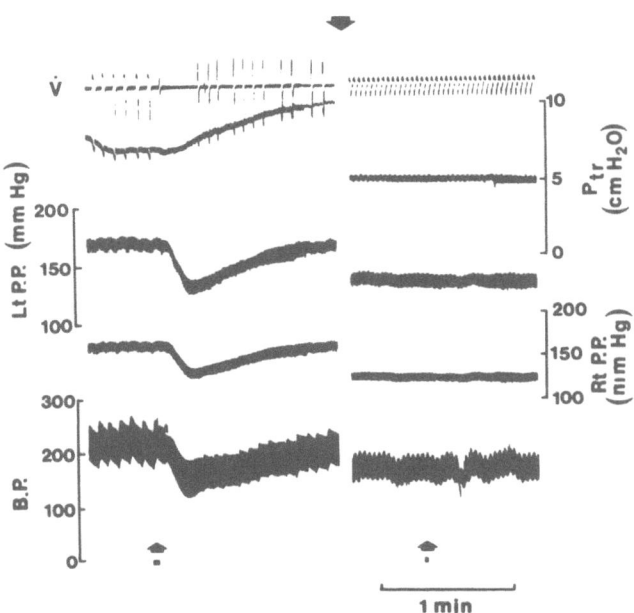

**Figure 4.** Responses to intravenous injection of capsaicin (50 mg kg$^{-1}$) in an anaesthetized dog. Traces from above down: tidal airflow (V), tracheal pressure $P_{tr}$), left tracheal perfusion pressure (Lt P.P.), right perfusion pressure (Rt P.P.), and systemic arterial blood pressure (B.P.). On the left: during spontaneous breathing, capsaicin causes apnoea, tracheal smooth muscle contraction, decrease in left and right perfusion pressures, and a fall in blood pressure, by stimulation of lung C-fibre receptors. On the right: during paralysis and artificial ventilation and after bilateral cervical vagosympathectomy, all responses were abolished. Drug injections are at the arrows. From Sahin et al. (1987).

injurious agents and a weakening of the air conditioning system normally provided by the nose.

4. Pulmonary C-fibre receptors constrict the larynx (Stransky et al., 1973), and increased upper airway resistance can contribute to human airways obstruction. A similar phenomenon probably occurs when RARs are activated, although there is little evidence that bronchial C-fibre receptors cause the same effect.

5. Pulmonary C-fibre receptors cause hypotension and brady- cardia, and inhibit spinal reflexes (Coleridge and Coleridge, 1986). These changes are probably part of a general nociceptive response. Similar responses do not seem to have been described for the bronchial C-fibre receptors or the RARs.

6. RAR's cause coughing (Coleridge and Coleridge, 1986), which is enhanced in asthma. C-fibre receptors cause apnoea and rapid shallow breathing and the pulmonary receptors inhibit cough (Karlsson et al., 1988; Tatar et al., 1988).

7. There has been considerable discussion as to whether the pulmonary C-fibre receptors cause unpleasant sensation (Paintal, 1986). Undoubtly this can arise from the lungs and the most likely candidates are these receptors. However experiments have to be done in man to test this concept and this is not yet possible in a definitive way. Inflating collapsed lung in man causes an unpleasant tearing sensation (Burger and Macklem, 1968) and, since this stimulus acts on RARs but has little effect on C-fibre receptors, it seems likely that the former are responsible.

Thus both groups of receptors may cause unpleasant sensation. This is probably similar to many other tissues, such as the skin and skeletal muscle and various viscera, which have both C-fibre receptors and receptors with small myelinated fibres in the afferent nerves; both groups of receptors can cause an unpleasant sensation and the quality of sensation depends upon the interaction of the two groups (Perl, 1984).

## CONCLUSIONS

It was predominantly the studies of John and Hazel Coleridge that analysed the reflex actions of the pulmonary and bronchial C-fibre receptors. Their studies were primarily physiological, but the pathophysiological implications are considerable. This is both because the reflexes are likely to play an important role in airways obstruction in asthma, and because activation of the receptors will lead also to neurogenic inflammation, the two processes interacting in their contribution to the obstruction.

Perhaps without knowing it, the Coleridges have been illustrating the classical Roman signs of inflammation. These are:

1. *Rigor.* This is stiffness associated with inflammation, and in the case of the airways induced by the reflex broncho- constriction studied by the Coleridges.
2. *Rubor.* This is the redness of inflamed tissue; the Coleridges indirectly studied it by demonstrating a reflex vasodilation associated with activation of airways receptors.
3. *Tumor.* This is thickening of inflamed tissue, usually due to oedema. Although the Coleridges have not measured this, it is an implicit result of the mucosal vasodilation they described.
4. *Calor.* The warming of inflamed tissues may not apply to the airways, which at least distal to the larynx are probably at body temperature, certainly when breathing is through the nose. However in some conditions increased mucosal blood flow could lead to warming of the mucosa, and would result from the reflex vasodilation described by the Coleridges.
5. *Dolor.* This is the dolorous or unpleasant sensation from the lungs, felt in an asthmatic attack. It cannot be easily measured in experimental animals but on indirect evidence is likely to result from stimulation of the airway receptors studied by the Coleridges.

It would be nice to suggest that the Coleridges have added a sixth signs to those of inflammation as identified by Roman doctors. This is:

6. *Mucor.* This would be the reflex secretion of mucus that occurs in airways inflammation and as analysed by the Coleridges. However the word mucor has been appropriated by the botanists, who use it to describe the genus *Mucor* of fungi, which includes the mould-plants. It is stretching imagination too much to say that in asthma the airways are mouldy. However they are certainly messy and mucky and, as the name mucosa implies, are mucoid or slimy.

## REFERENCES

Baluk P, Nadel, J.A. & McDonald D.M. (1992) Substance P-immunoreactive sensory axons in the rat respiratory tract: A quantitive study of their distribution and role in neurogenic inflammation. *J. Comp Neurol* **319**, 586-598.
Barnes, P.J. (1992a) Modulation of neurotransmission in airways. *Physiol Rev* **72**, 699-729.

Barnes, P.J. (1992b) Neural mechanisms in asthma. *Br Med Bull* **48**, 149-168.

Belvisi, M.G., Miura, M., Stretton, D. & Barnes, P.J. (1993) Vasoactive intestinal peptide and nitric oxide modulate cholinergic neurotransmission in guinea pig trachea. *Eur J Pharmacol* **231**, 97-102.

Belvisi, M.G., Stretton, C.D. & Barnes, P.J. (1992) Nitric oxide is the endogenous neurotransmitter of bronchodilator nerves in humans. *Eur J Pharmacol* **210**, 221-222.

Burger, E.J. & Macklem, P.T. (1968) Airway closure demonstration by breathing 100% $O_2$ at low lung volumes and by $N_2$ washout. *J Appl Physiol* **25**, 139-144.

Coleridge, H.M. & Coleridge, J.C.G. (1986) Reflexes evoked from the tracheobroncial tree and lungs. In: *Handbook of Physiology, 3. The respiratory system, vol II, control of breathing*, edited by Cherniack, N.S. & Widdicombe, J.G. American Physiological Society, Bethesda, pp. 395-429.

Coleridge, H.M., Coleridge, J.C.G., Green, J.F. & Parsons, G.H. (1992) Pulmonary C-fibre stimulation by capsaicin evokes reflex cholinergic bronchial vasodilation in sheep. *J Appl Physiol* **72**, 770-778.

Coleridge, J.C.G. & Coleridge, H.M. (1984) Afferent vagal C-fiber innervation of the lungs and airways and its functional significance. *Rev Physiol Biochem Pharmacol* **99**, 1-110.

Coleridge, J.C.G., Coleridge, H.M., Roberts, A.M., Kaufman, M.P. & Baker, G.G. (1982) Tracheal contraction and relaxation initiated by lung and somatic afferents in dogs. *J Appl Physiol*, **52**, 984-990.

Corfield, D.R., Hanafi, Z., Webber, S.E. & Widdicombe, J.G. (1991) Changes in mucosal thickness and blood flow in sheep. *J Appl Physiol* **71**, 1282-1288.

Das, R.M., Jeffery, P.K. & Widdicombe, J.G. (1978) The epithelial innervation of the lower respiratory tract of the cat. *J Anat* **126**, 123-131.

Das, R.M., Jeffery, P.K. & Widdicombe, J.G. (1979) Experimental degeneration of intraepithelial nerve fibres in cat airways. *J Anat* **128**, 259-263.

Karlsson, J-A., Sant'Ambrogio, G. & Widdicombe, J.G. (1988) Afferent neural pathways in cough and reflex bronchoconstriction. *J Appl Physiol* **65**, 1007-1023.

Keller, C.J. & Loeser A. (1929) Der zentripetale Lungen-vagus. *Z Biol* **89**, 373-395.

Knowlton, G.C. & Larrabee, M.G. (1946) A unitary analysis of pulmonary volume receptors. *Am J Physiol* **147**, 100-114.

Laitinen, L.A. & Laitinen, A. (1991) Overview of the pathology of asthma. In: *Pharmacology of asthma, handbook of Experimental Pharmacology*, edited by Page, C.P. & Barnes, P.J. Springer-Verlag, New York, p. 1-25.

Laitinen, L.A., Heino, M., Laitinen, A., Kava, T. & Haahtela, T. (1985) Damage of the airway epithelium and bronchial reactivity in patients with asthma. *Am Rev Respir Dis* **131**, 599-606.

Laitinen, L.A., Laitinen, M.V.A. & Widdicombe, J.G. (1987) Parasympathetic nervous control of tracheal vascular resistance in the dog. *J Physiol* **385**, 135-146.

Lundberg, J.M. (1990) Peptide and classical transmitter mechanisms in the autonomic nervous system. *Arch Int Pharmacodyn* **303**, 9-19.

Lundberg, J.M., Brodin, E., Hua, X-Y. & Saria, A. (1984) Vascular permeability changes and smooth muscle contraction in relation to capsaicin sensitive substance P afferents in the guinea-pig. *Acta Physiol Scand* **120**, 217-227.

Lundberg, J.M., Saria, A. Lundblad, L., Angard, A., Martling, C-R., Theodorsson-Norheim, E., Stjarne, P. & Hokfelt, T. (1987) Bioactive peptides in capsaicin sensitive C-fibre afferents of the airways: functional and pathophysiological implications. In: *The airways: neural control in health and disease*, edited by Kaliner, M.A. Barnes, P.J. Marcel Dekker, New York, p. 417-445.

Lung, M.A. & Widdicombe, J.G. (1987) Lung reflexes and nasal vascular resistance in the anaesthetized dog. *J Physiol* **386**, 465-474.

Matran, R. (1991) Neural control of lower airway vasculature. *Acta Physiol Scand* **142**, 1-54.

McDonald, D.M. (1987) Neurogenic inflammation in the respiratory tract: Actions of sensory nerve mediators on blood vessels and epithelium of the airway mucosa. *Am Rev Respir Dis* **136**, S65-S71.

McDonald, D.M. (1992) Effects of infection on neurogenic micro- vascular permeability in the airway. *Am Rev Respir Dis* **146**, S40-S44.

Mortola, J.P., Sant'Ambrogio, G. & Clement, M.G. (1975) Localization of irritant receptors in the airways of the dog. *Respir. Physiol.* **24**, 107-114.

Paintal, A.S. (1955) Impulses in vagal afferent fibres from specific pulmonary deflation receptors. The response of these receptors to phenyl diguanide, potato starch, 5-hydroxytryptamine and nicotine, and their role in respiratory and cardiovascular reflexes. *Q J Exp Physiol* **40**, 89-111.

Paintal, A.S. (1986) The visceral sensations - some basic mechanisms. In: *Visceral sensation*, edited by Cervero, F. & Morrisson, J.F.B. Elsevier, Amsterdam, p. 3-19.

Perl, E.R. (1984) Pain and nociception. In: *Handbook of Physiology, Section 1, The Nervous System, Vol. 3, Sensory Processes, Part 2,* Ed Davian-Smith, I. American Physiological Society, Bethesda, pp.915-975.

Roberts, A.M., Kaufman, M.P., Baker, D.G., Brown, J.K., Coleridge, H.M. & Coleridge, J.C.G. (1981) Reflex tracheal contraction induced by stimulation of bronchial C-fibres in dogs. *J Appl Physiol* **51**, 485-493.

Roberts, A.M., Coleridge, H.M., & Coleridge, J.C.G. (1988) Reciprocal action of pulmonary vagal afferents on tracheal smooth muscle tension in dogs. *Respir Physiol* **72**, 35-46.

Sahin, G., Webber, S.E. & Widdicombe, J.G. (1987) Lung and cardiac reflex actions on the tracheal vasculature in anaesthetized dogs. *J Physiol,* **387**, 47-57.

Sant'Ambrogio, G., Remmers, J.E., De Groot, W.A., Callas, G. & Mortola, J.P. (1978) Localization of rapidly adapting receptors in the trachea and main stem bronchus of the dogs. *Respir Physiol* **33**, 359-366.

Saria, A., Martling, C.R., Yan, Z., Theodorsson-Norheim, E., Gamse, R. & Lundberg, J.M. (1988) Release of multiple tachykinins from capsaicin-sensitive nerves in the lung by bradykinin, histamine, dimethyl-phenylpiperazinium, and vagal nerve stimulation. *Am Rev Respir Dis* **137**, 1330-1335.

Shimura S. & Takishima T. (1994) Airway Submucosal Gland Secretion. In: Airway Secretion: Physiological bases for the control of mucous hypersecretion. Ed Takishima, T. and Shimura, S. Marcel Dekker, New York. 325-398.

Stransky, A., Szereda-Przestaszewska, M. & Widdicombe, J.G. (1973) The effects of lung reflexes on laryngeal resistance and motoneurone discharge. *J Physiol* **231**, 417-438.

Tatar, M., Webber, S.E., & Widdicombe, J.G. (1988) Lung C-fibre receptor activation and defensive reflexes in anaesthetized cats. *J Physiol* **402**, 411-420.

Tomori, Z. & Widdicombe, J.G. (1969) Muscular, bronchomotor and cardiovascular reflexes elicited by mechanical stimulation of the respiratory tract. *J Physiol* **200**, 25-49.

Widdicombe, J.G. (1989) Nervous receptors in the tracheobronchial tree airway smooth muscle reflexes. In: *Airway smooth muscle in health and disease,* edited by Coburn, R.F. Plenum Press, New York, p. 35-53.

# 4

# UPPER AIRWAY INFLUENCES ON BREATHING

Giuseppe Sant'Ambrogio

Department of Physiology and Biophysics
University of Texas Medical Branch
Galveston, Texas 77555-0641

The upper airway may be defined as the extrathoracic portion of the airway and comprises the nose, the nasopharynx, the oropharynx, the larynx and the cranial half of the trachea. Its function is not limited to respiration but includes olfaction, air conditioning, chewing, swallowing and vocalization. The non respiratory activities are often in conflict with breathing and their coexistence demands a well developed control mechanism. The abundant nerve supply to the structures of the upper airway seems to be in line with this requirement.

The larynx is the portion of upper airway that has attracted most of the studies on the sensory supply and reflex responses. Only recently the nose has been the object of long overdue attention with significant benefit to our knowledge. These two portions of the upper airway will be the object of this short review.

The superior laryngeal nerve is the main source of laryngeal afferent activity. A clear respiratory modulation can be clearly noted when recording from the peripheral cut end of this nerve in several mammals. This respiratory related activity is present even when the larynx is functionally by-passed, as during tracheostomy breathing, but is markedly accentuated when the experimental animals breathe through the upper airway and especially when they perform inspiratory efforts against the occluded upper airway (Fig.1). This indicates that the presence of airflow and especially changes in pressure constitute additional stimuli that elicit afferent activity (Mathew et al., 1984).

Recording from single fibers of the SLN and analyzing their pattern of activation during upper airway breathing, tracheal breathing, upper airway occlusion and tracheal occlusion, respiratory-modulated receptors have been identified as 'flow', pressure and 'drive' receptors (Sant'Ambrogio et al., 1983). The experimental set-up used in anesthetized dogs is illustrated schematically in Fig.2 which also depicts the behavior of the three types of respiratory-modulated receptors.

Flow receptors are stimulated, in most circumstances, during upper airway breathing and remain silent during tracheal breathing as well as during occlusion of the upper airway or the trachea (Fig.2). The proper stimulus for these endings is a decrease in temperature of the laryngeal mucosa (Fig.3). They remain silent at body temperature and start firing at lower temperatures. The receptive field of these endings is localized in a discrete region of the

*Control of the Cardiovascular and Respiratory Systems in Health and Disease*
Edited by C. T. Kappagoda and M. P. Kaufman, Plenum Press, New York, 1995

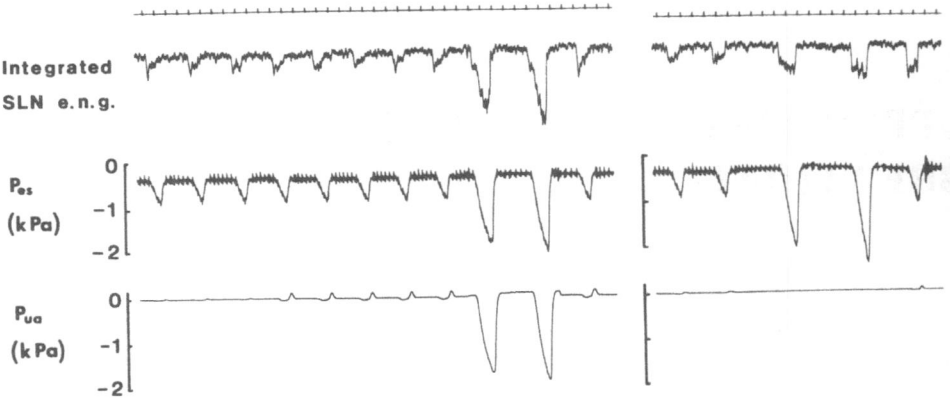

**Figure 1.** Effect of diversion of breathing and airway occlusion on the afferent activity of the superior laryngeal nerve (SLN) in an anesthetized dog. From top to bottom: Time marker in seconds; integrated SLN electroneurogram (e.n.g.); esophageal pressure (Pes) and upper airway pressure (Pua) in kiloPascals (kPa). Breathing was diverted from a tracheostomy to the upper airway after 3 breaths (left), as indicated by the pressure signal in the upper airways. Two upper airway obstructions (left) and two tracheal occlusion (right) are shown. There is a much greater increase in SLN activity with upper airway occlusion than with tracheal occlusion.

vocal folds, at the level of the vocal process of the arytenoids (Sant'Ambrogio et al.,1985). A conspicuous presence of cold receptors has been found within the nasal cavity of cats (Glebovsky and Bayev, 1984; Wallois et al., 1991), rats (Tsubone, 1989) and guinea pigs (Tsubone and Sekizawa, 1993).

Laryngeal pressure receptors are markedly stimulated during upper airway occlusion at end inspiration or expiration depending on whether the proper stimulus is positive

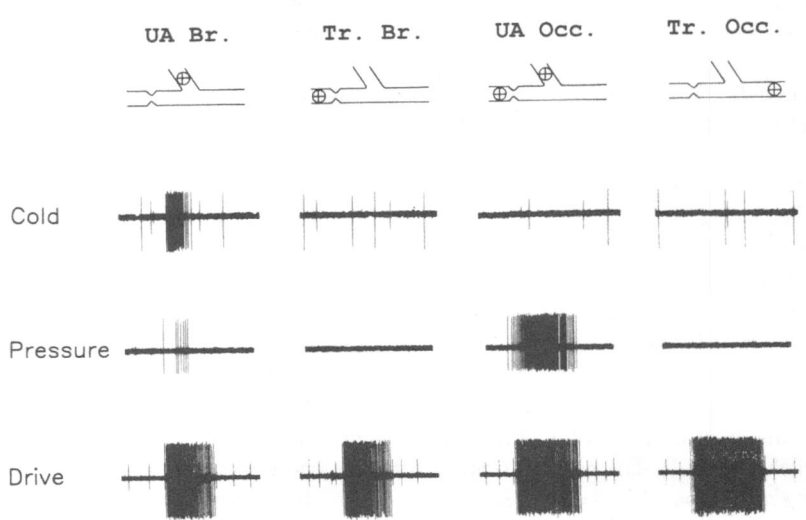

**Figure 2.** Behavior of laryngeal cold, pressure and "drive" receptors during four experimental conditions. UABr = Upper airway breathing; Tr.Br. = Tracheostomy breathing; UA.Occ. = Upper airway occlusion; Tr.Occ. = Tracheal occlusion.

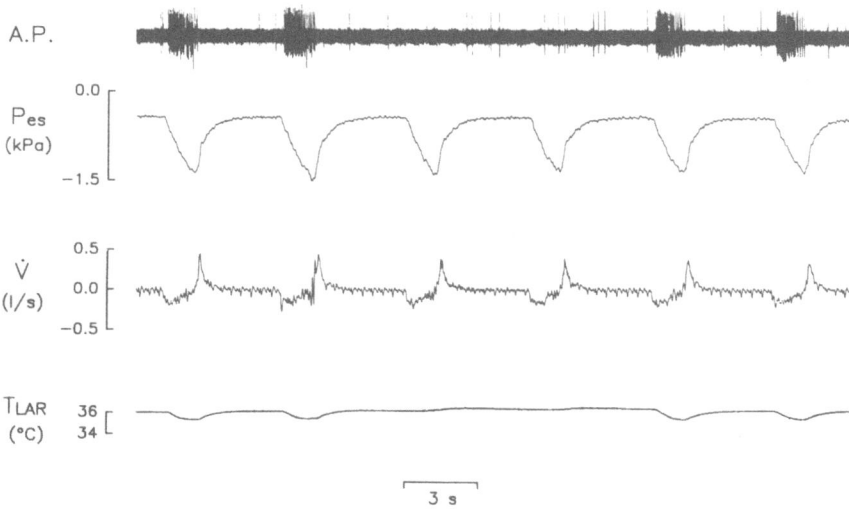

**Figure 3.** Behavior of a laryngeal cold receptor during inhalation of warm air. The dog is spontaneously breathing through the upper airway. A.P.= action potentials from a thin filament of the SLN; Pes = esophageal pressure; V (l/sec) = airflow in liters/sec; Lar. Temp = laryngeal temperature. Note that the inspiratory modulation of this receptor disappears when laryngeal temperature during inspiration is raised by inhalation of warm air (From Sant'Ambrogio et al., 1985).

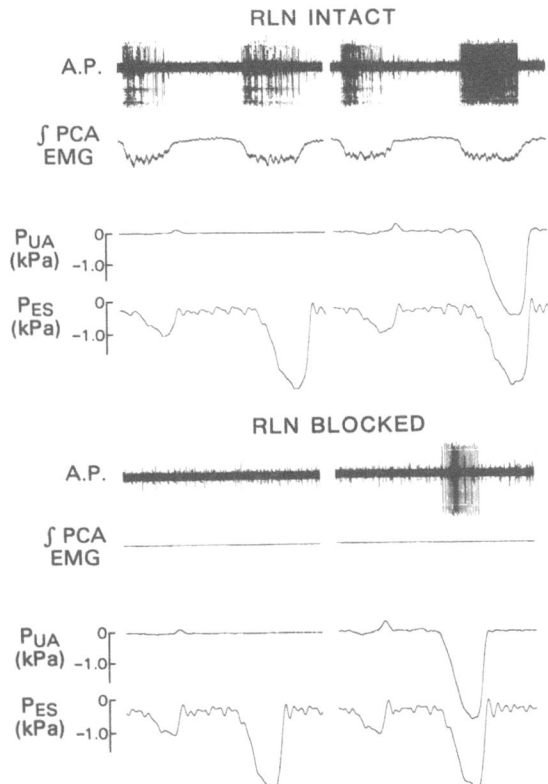

**Figure 4.** Behavior of a "drive" receptor during tracheal breathing and tracheal occlusion (left panels), upper airway breathing and upper airway occlusion (right panels). Paralysis of the posterior cricoarytenoid muscle, induced by cold block of the recurrent laryngeal nerve, eliminates, or markedly reduced the receptor discharge (lower panels). The receptor is now activated by negative pressure in the upper airway when the upper airway is occluded (lower right panel). Traces from above: A.P. = action potentials; PCA EMG = electromyogram of the posterior cricoarytenoid muscle; Pua = pressure in the upper airway; Pes = pressure in the esophagus (Widdicombe et al., 1988).

(distending) or negative (collapsing) pressure (Fig.2). Usually these endings are active during upper airway breathing and most of them, at least in the dog, are responsive to negative pressure. In addition to their primary responsiveness to transmural pressure, several of these receptors show a residual activity even when the larynx is out of the circuit, i.e. during tracheal breathing and occlusion, i.e., in the absence of any pressure stimulus. The presence of receptors responsive to changes in upper airway pressure has also been demontrated within the nasal cavity of rats (Tsubone, 1990) and cats (Wallois et al.,1991).

Laryngeal "drive" receptors are stimulated at each breath, even in the absence of airflow and pressure. Reversible laryngeal paralysis, by cold blocking either the recurrent nerve or the external branch of the SLN, can effectively be used to demonstrate that it is the contraction of intrinsic laryngeal muscles that frequently stimulates these endings (Fig. 4). Tracheal tug, motion due to the action of chest wall muscles transmitted to the larynx through the trachea, can also be identified as an excitatory influence for some of these endings. The term "drive" introduced to identify these endings (Sant'Ambrogio et al., 1983) describes the activating stimulus exerted by the respiratory center that *drives* both laryngeal and chest wall pump muscles.

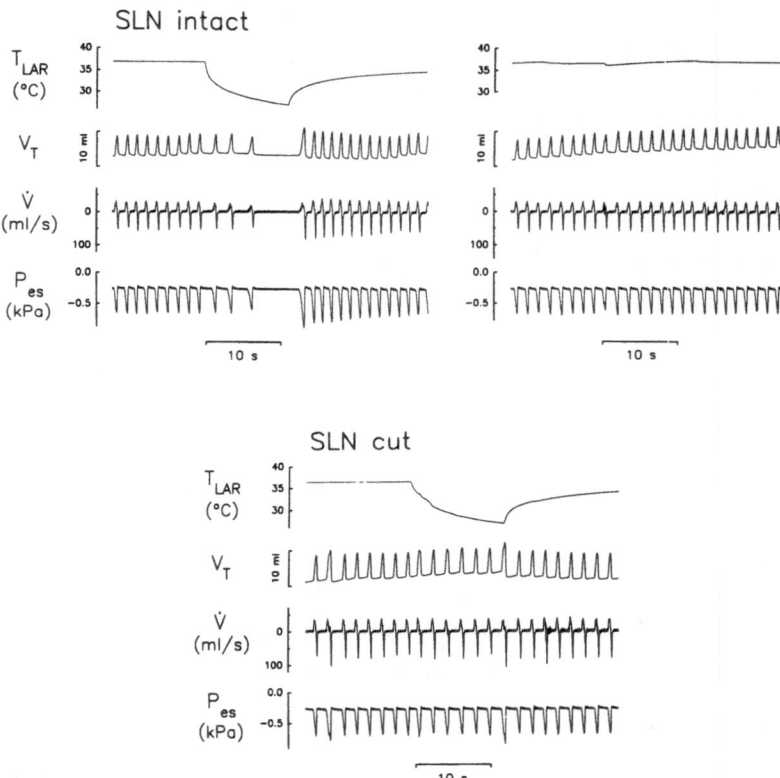

**Figure 5.** Effect of cold and warm airflows passed in a expiratory direction through the isolated in situ upper airway of a 12 day-old anesthetized puppy breathing spontaneously through a tracheostomy. In each panel traces are, from top to bottom: laryngeal temperature, tidal volume, airflow and esophageal pressure. In the top left panel the upper airway is cooled with cold air, as shown by the decrease in laryngeal temperature, and there is a marked ventilatory depression. The passage of warm air at the same flow does not cause any visible effect on ventilation (top right panel). When the cold air challenge is repeated after bilateral section of the SLNs (bottom panel) there is no effect on ventilation.

Recordings from trigeminal afferents of the PNN, ION and EN have revealed the existence of nasal "drive" endings which reflect the respiratory activity of nasal muscles, especially the alae nasi (Tsubone, 1987; Wallois et al., 1991)

Experiments have been recently carried out to evaluate the effects of laryngeal cold, pressure and drive receptors on the breathing pattern and the maintenance of upper airway patency (Sant'Ambrogio, F.B. et al., 1985). Laryngeal cooling was found to have a depressive influence on the activity of the posterior cricoarytenoid muscle in anesthetized dogs (Mathew et al., 1986) and a bronchoconstrictive effect in cats (Jammes et al., 1983). Moreover, laryngeal cooling induced a marked ventilatory depression in newborn puppies (Fig. 5) and adult guinea pigs (Mathew et al., 1990; Orani et al., 1991).

Laryngeal cooling, besides having a stimulatory influence on cold receptors inhibits both pressure and drive endings (Sant'Ambrogio et al., 1986), which could therefore contribute, or even be responsible, for the response. However, the role of an independent cold receptor stimulation could be inferred from experiments on guinea pigs and newborn puppies in which l-menthol, a specific stimulant on cold receptors (Sant'Ambrogio, F.B.,

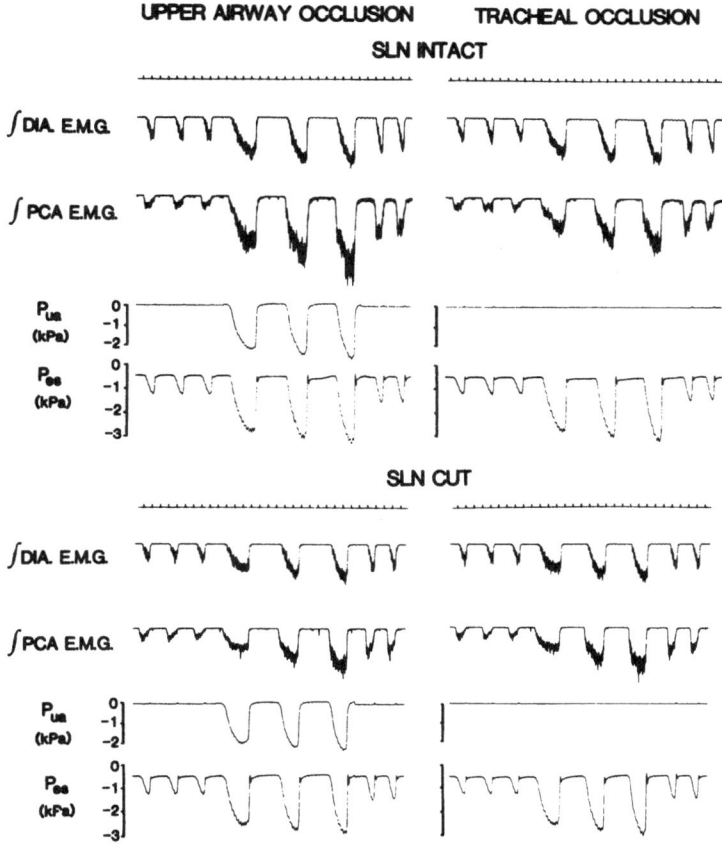

**Figure 6.** Effects of upper airway occlusions and tracheal occlusions on diaphragmatic (DIA) and posterior cricoarytenoid (PCA) e.m.g. activities. In each panel from top to bottom: time in seconds, integrated e.m.g. activities, upper airway (Pua) pressure and esophageal (Pes) pressure. Note that the activation of the PCA is much higher during upper airway occlusion than during tracheal occlusion (upper panels) and that the difference is virtually abolished by SLN section (bottom panels). Modified from Sant'Ambrogio, F.B. et al., 1985).

1991) was also seen to depress ventilation in the absence of any change in temperature (Orani et al., 1991; Sant'Ambrogio, F.B., 1992). Whereas in the dog, adults and newborns. the prevailing, or unique, afferent pathway for these reflex responses was recognized to be in the SLN, in guinea pigs a substantial contribution was attributed to trigeminal nasal afferents (Orani et al., 1991).

The reflex function of "drive" receptors could be inferred by comparing the response to tracheal occlusion before and after section or block of the SLN (Sant'Ambrogio, F.B., 1985), i.e., in a condition in which the larynx is subjected to the activity of its own muscles and the tracheal tug, in the absence of pressure and cold stimuli. No changes in breathing pattern or in upper airway muscle activity were found between the two conditions and thus no reflex effects could be assigned to these receptors. However, since many "drive" receptors also have some pressure sensitivity, at least part of the response to pressure (upper airway occlusion) may be attributed to them. In any event it must be recognized that, due to their association with laryngeal muscles, they could play an important role in the fine coordination of these muscles, as required in vocalization.

The role of laryngeal pressure receptors in the regulation of breathing pattern and upper airway muscle activity has been evaluated either by comparing the responses to upper airway and tracheal occlusions (Sant'Ambrogio, F.B. et al., 1985), or analyzing the responses obtained with the application of pressure to the isolated upper airway in situ (Mathew et al., 1982; van Lunteren et al., 1984). In either case the results indicated the marked effects that collapsing pressure in the upper airway has in activating various upper airway dilating muscles and in prolonging inspiratory duration without changing peak inspiratory activity (Fig. 6). Both responses should be considered beneficial to the maintenance of upper airway patency. These responses are abolished or markedly reduced by sectioning the SLN or by anesthetizing the laryngeal mucosa. However, in the experiments of van Lunteren et al., (1984) and in those of Horner et al., (1991) conducted in healthy human subjects, a significant contribution of nasal afferents was also found.

The importance of laryngeal afferents for the preservation of airway patency is also suggested by the results obtained in human subjects performing forced vital capacities,

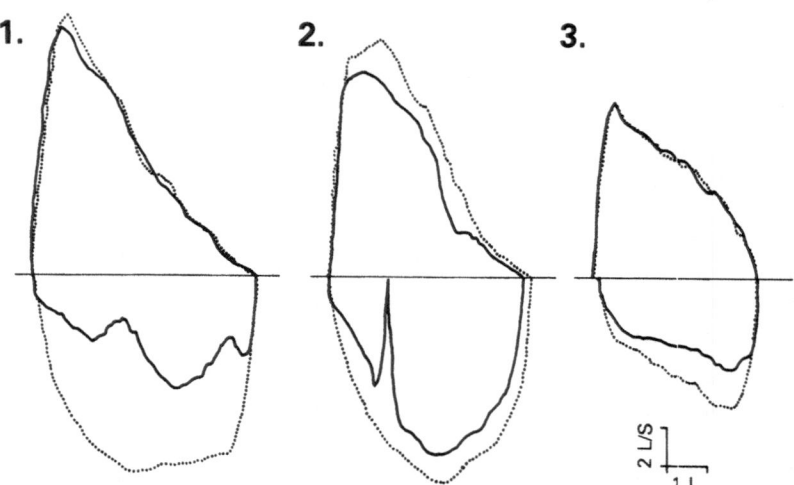

**Figure 7.** Effect of laryngeal airway anesthesia on a maximal forced inspiratory and expiratory maneuver in 3 subjects. For each subject, the flow volume loop just prior to anesthesia (dotted line) is superimposed on the first loop performed after airway anesthesia (solid line). The loops are aligned on the volume axis at TLC. From Kuna et al., 1988.

inspiratory and expiratory, in control conditions and after total anesthesia of the larynx (Kuna et al., 1988). When the two conditions are compared it becomes apparent that the subjects, after topical anesthetization, had become less capable to sustain the maximal inspiratory flows produced in control conditions (Fig. 7). These results suggest that the afferent activity of the larynx plays a significant role in overcoming, or preventing, oropharyngeal obstruction, through a reflex recruitment of upper airway muscles in conditions such as obstructive sleep apnea.

In addition to receptors discharging in close association with the breathing cycle there are receptors that are either silent or randomly active in control condition (Fig. 8). These endings can, however, be promptly recruited when the laryngeal mucosa is exposed to mechanical and/or chemical irritation (Anderson et al., 1990a). In fact, these receptors respond to several tussigenic stimuli and should be recognized as providing the triggering mechanism for the cough reflex from the larynx (Karlsson et al., 1988).

Receptors with similar characteristic have been described in cats (Wallois et al., 1991) and rats (Tsubone and Kawata, 1991) within the nasal cavity.

A significant proportion of pressure and "drive" receptors can be stimulated by solution of low osmolality (Fig. 9) and inhibited by carbon dioxide (Anderson et al.,1990a;

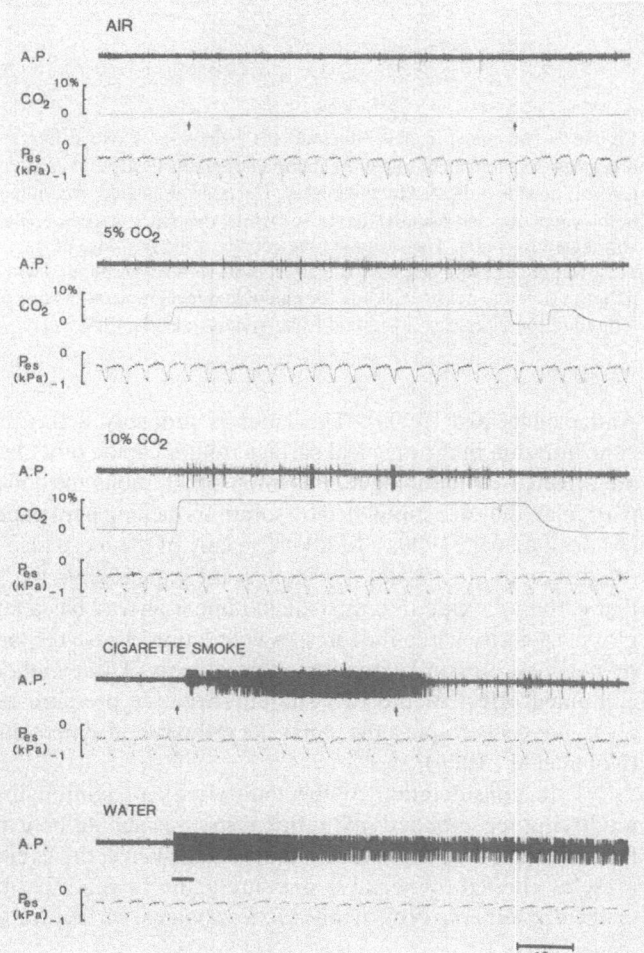

**Figure 8.** Effect of warm air, 5 and 10% carbon dioxide, cigarette smoke, and water on a laryngeal irritant receptors. In each panel traces area; action potentials recorded from the SLN, carbon dioxide concentration (first 3 panels) and esophageal pressure. Whereas air does not affect receptor discharge, carbon dioxide causes a clear increase in activity, especially at the concentration of 10%. Cigarette smoke (between arrows) and water (time of delivery indicated by the thick line) are very potent stimuli for this receptor.

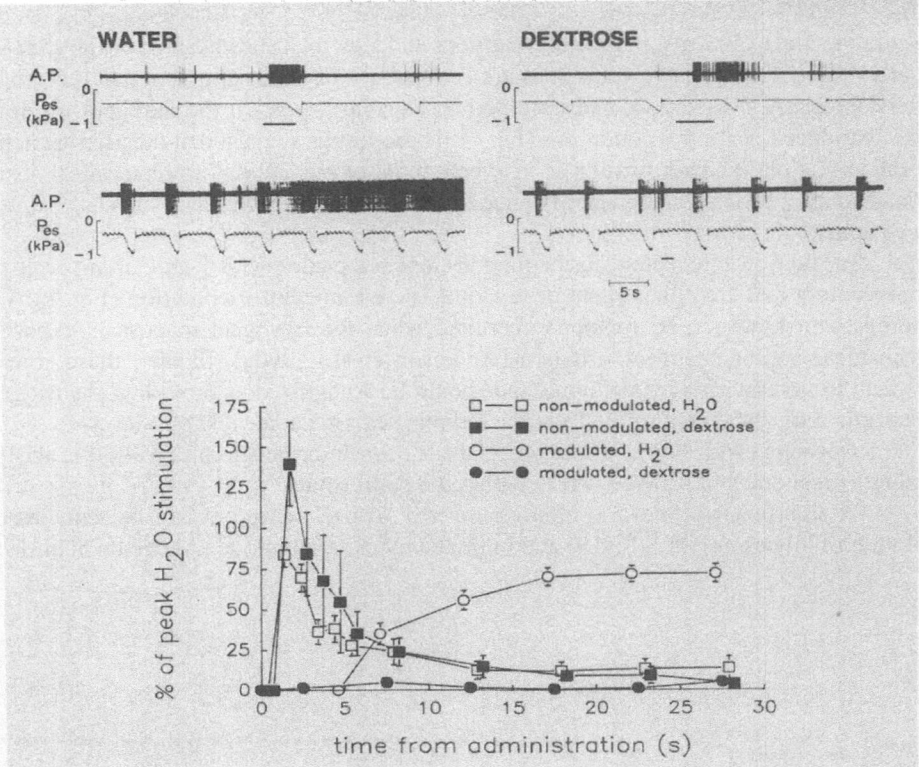

**Figure 9.** The upper panels illustrate the behavior of two different types of water-responsive laryngeal receptors. In each panel traces are: action potentials (A.P.) and esophageal pressure (Pes). Application of the test solutions is indicated by thick lines. The non-modulated irritant receptor (upper traces) was stimulated by both water and iso-osmolal dextrose, while the respiratory modulated receptor (lower traces) was only stimulated by water. The average time course of the response of the two types of receptors is represented in the lower panel. Note that the irritant receptors (n = 12) responded with a short delay, short duration discharge to both water and dextrose, while the modulated receptor (n = 21) responded only to water with a longer delay, long duration discharge. Modified from Anderson et al., 1990.

Anderson et al., 1990b). The latter is probably acting by increasing the hydrogen ion concentration in the mucosal surface liquid. On the other hand, laryngeal 'irritant' receptors are affected by the ionic composition of solutions introduced into the laryngeal lumen. In particular, they are stimulated by solutions lacking permeant anions (Fig. 9), such as chloride (Anderson et al., 1990a). Moreover, many of them can also be stimulated by carbon dioxide (Anderson et al., 1990b). Of interest the experiments by Nolan et al. (1990) who showed that carbon dioxide into the isolated upper airway has a strong excitatory effect on genioglossus activity while it depresses ventilation. These responses are similar to those induced by negative pressure across the upper airway. These authors suggested the possibility of a combined effect of the two stimuli (negative pressure and carbon dioxide) in cases of obstructive sleep apnea that "may act reflexly and synergistically to reverse the obstruction" (Nolan et al., 1990).

    In consideration of the multiplicity of stimuli that are capable of activating or modifying receptor activity, reflex responses should be expected to depend on a variety of factors rather than a single stimulus. For instance, the excitatory influence on upper airway muscles elicited by negative pressure could be modified by changes in osmolality and/or concentration of hydrogen ions of the laryngeal surface liquid. Similarly, the reflex reactions to irritants in the larynx could be accentuated by a decrease in chloride concentration or an increase in hydrogen ions concentration of the surface liquid.

# REFERENCES

Anderson, J. W., F. B. Sant'Ambrogio, O. P. Mathew, and G. Sant'Ambrogio. Water-responsive laryngeal receptors in the dog are not specialized endings. *Respir. Physiol.* 79: 33-44, 1990a.

Anderson, J. W., F. B. Sant'Ambrogio, G. P. Orani, G. Sant'Ambrogio, and O. P. Mathew. Carbon dioxide-responsive laryngeal receptors in the dog. *Respir. Physiol.* 82: 217-226, 1990b.

Glebovsky, V. D. and A. V. Bayev. Stimulation of nasal cavity mucose trigeminal receptors with respiratory airflows. *Sechenov Physiol. J. USSR* 70: 1534-1541, 1984.

Horner, R.L., Innes, J.A., Holden, J.B., and Guz, A. Afferent pathway(s) for pharyngeal dilator reflex to negative airway pressure in man; a study using upper airway anaesthesia. *J.Physiol. (London)* 436:31-44, 1991.

Jammes, Y., P. Barthelemy, and S. Delpierre. Respiratory effects of cold air breathing in anesthetized cats. *Respir. Physiol.* 54: 41-54, 1983.

Karlsson, J.- A., G. Sant'Ambrogio, and J. Widdicombe. Afferent neural pathways in cough and reflex bronchoconstriction. *J.Appl.Physiol.* 65: 1007-1023, 1988.

Kuna, S. T., G. E. Woodson, and G. Sant'Ambrogio. Effect of laryngeal anesthesia on pulmonary function testing in normal subjects. *Am. Rev. Respir. Dis.* 137: 656-661, 1988.

Mathew, O. P., Y. K. Abu-Osba, and B. T. Thach. Genioglossus muscle responses to upper airway pressure changes: afferent pathways. *J.Appl.Physiol.* 52: 445-450, 1982.

Mathew, O. P., J. W. Anderson, G. P. Orani, F. B. Sant'Ambrogio, and G. Sant'Ambrogio. Cooling mediates the ventilatory depression associated with airflow through the larynx. *Respir. Physiol.* 82: 359-368, 1990.

Mathew, O. P., F. B. Sant'Ambrogio, and G. Sant'Ambrogio. Effects of cooling on laryngeal reflexes in the dog. *Respir. Physiol.* 66: 61-70, 1986.

Mathew, O. P., G. Sant'Ambrogio, J. T. Fisher, and F. B. Sant'Ambrogio. Respiratory afferent activity in the superior laryngeal nerves. *Respir. Physiol.* 58: 41-50, 1984.

Nolan, P., A. Bradford, R. G. O'Regan, and D. McKeogh. The effects of changes in laryngeal airway $CO_2$ concentration on genioglossus muscle activity in the anesthetized cat. *Exptl. Physiol.* 75: 271-274, 1990.

Orani, G. P., J. W. Anderson, G. Sant'Ambrogio, and F. B. Sant'Ambrogio. Upper airway cooling and *l*-menthol reduce ventilation in the guinea pig. *J.Appl.Physiol.* 70: 2080-2086, 1991.

Sant'Ambrogio, F. B., J. W. Anderson, and G. Sant'Ambrogio. Effect of *l*-menthol on laryngeal receptors. *J.Appl.Physiol.* 70: 788-793, 1991.

Sant'Ambrogio, F. B., J. W. Anderson, and G. Sant'Ambrogio. Menthol in the upper airway depresses ventilation in newborn dogs. *Respir. Physiol.* 89: 299-307, 1992.

Sant'Ambrogio, F. B., O. P. Mathew, W. D. Clark, and G. Sant'Ambrogio. Laryngeal influences on breathing pattern and posterior cricoarytenoid muscle activity. *J.Appl.Physiol.* 58: 1298-1304, 1985.

Sant'Ambrogio, G., O. P. Mathew, J. T. Fisher, and F. B. Sant'Ambrogio. Laryngeal receptors responding to transmural pressure, airflow and local muscle activity. *Respir. Physiol.* 54: 317-330, 1983.

Sant'Ambrogio, G., O. P. Mathew, Sant'Ambrogio, F. B. and J.T. Fisher. Laryngeal cold receptors. *Respir. Physiol.* 59:35-44, 1985.

Sant'Ambrogio, G., F. B. Sant'Ambrogio and O. P. Mathew. Effects of cold air on laryngeal machanoreceptors in the dog. *Respir. Physiol.* 64:45-56, 1986.

Tsubone, H. Different sensory receptors in the nasal mucosa, their response to cold, chemicals, touch and pressure stimuli and to airway occlusions. *J. Clin. Exp. Med.* 142:897-898, 1987.

Tsubone, H. Nasal "flow" receptors of the rat. *Respir. Physiol.* 75: 51-64, 1989.

Tsubone, H. Nasal 'pressure' receptors. *Jpn. J. Vet. Sci.* 52: 225-232, 1990.

Tsubone, H., and Kawata, M. Stimulation to the trigeminal afferent nerve of the nose by formaldehyde, acrolein and acetaldehyde gases. *Inhal. Toxicol.* 3:211-222, 1991.

Tsubone, H., and S. Sekizawa. Nasal ethmoidal cold receptors and their reflex effects. *Proceedings of the 32nd IUPS Congress,* Glasgow 1993, 282.4/P.

Van Lunteren, E., W. B. Van de Graaff, D. M. Parker, J. Mitra, M. A. Haxhiu, K. P. Strohl, and N. S. Cherniack. Nasal and laryngeal reflex responses to negative upper airway pressure. *J.Appl.Physiol.* 56: 746-752, 1984.

Wallois, F., J. M. Macron, V. Jounieaux, and B. Duron. Trigeminal nasal receptors related to respiration and to various stimuli in cats. *Respir. Physiol.* 85: 111-125, 1991.

# 5

# ACTIVATION OF PROTEIN KINASE C MEDIATES INSULIN REGULATION OF THE Na-K PUMP IN CULTURED SKELETAL MUSCLE

S. R. Sampson[*]

Health Sciences Research Center and
The Otto Meyerhoff Center for the Study of Drug-Receptor Interactions
Department of Life Sciences
Bar-Ilan University
Ramat-Gan 52900, Israel

At first glance, the inclusion of this lecture in this symposium might seem to be a bit superfluous - what possible connection is there between regulation of $Na^+$-$K^+$ pump activity in skeletal muscle and control of the respiratory and cardiovascular systems? A brief reflection will, of course, provide the answer. Skeletal muscle represents approximately 40% of the body weight and is a major source of $K^+$. Indeed, during muscle exercise muscle activity can lead to a doubling of $[K^+]o$, and a responsive and active $Na^+$-$K^+$ pump is absolutely essential for continued muscle and cardiovascular function. In addition, the pump plays an important role in control of cardiac and smooth muscle function. Hence, knowledge of the regulatory and signaling mechanisms involved in $Na^+$-$K^+$ pump activity may also provide information regarding important regulatory mechanisms of respiratory and cardiovascular activity.

The $Na^+$-$K^+$ pump consists of a complex of 2-$\alpha$ and 2-$\beta$ subunits, each subunit being encoded by specific genes and subject to independent regulation. There are 3 separate isoforms of each of the $\alpha$-and $\beta$-subunits, and the isoforms are tissue-specific. Skeletal muscle *in vivo* expresses both the $\alpha$-1 and $\alpha$-2 as well as the $\beta$-1 and $\beta$-2 isoforms. The $\alpha$-subunits are the sites of Na-K pumping, ouabain binding and ATP hydrolysis, and the $\beta$-subunits, while apparently not involved in the metabolic activity of the $Na^+$-$K^+$ pump, nevertheless are important for normal activity and in regulation of pump abundance.

The general sequence of the pump cycle is as follows: $Na^+$ ions are bound to their site in the internal side of the $\alpha$-subunit, following which the $\alpha$-subunit is phosphorylated. This phosphorylation induces a transformational change which results in the transfer and

---

[*] Incumbent of the Louis Fisher Chair in Cellular Pathology

*Control of the Cardiovascular and Respiratory Systems in Health and Disease*
Edited by C. T. Kappagoda and M. P. Kaufman, Plenum Press, New York, 1995

release of Na$^+$ ions to the outside of the membrane. K$^+$ ions are then bound to the external side of the α-subunit, the subunit is dephosphorylated and the protein returned to its original conformation. This transfers K$^+$ ions to the cytoplasm. The actual mechanisms of phosphorylation and the its importance as a regulatory site in pump activity is as yet unclear.

The Na$^+$-K$^+$ pump is subject to both long- and short-term regulation by alterations in intracellular and extracellular ion concentrations, hormones and growth factors. Work in our laboratory during the last decade has focused on the physiological expression of Na$^+$-K$^+$ pump activity in skeletal muscle in culture under various conditions. Recently, we have turned our attention to the study of mechanisms involved in the moment-to-moment regulation of pump activity. As a model of a short-term regulatory factor, we have used insulin. Skeletal muscle is one of the most important target organs for this hormone, which plays a major role in the control of plasma K$^+$ concentrations by stimulating Na$^+$-K$^+$ transport within minutes of application to the tissue (7,8). Whereas insulin effects on this transport systems in skeletal muscle have been known for many years, the initial signals and the mechanisms underlying these effects on the Na$^+$-K$^+$-pump in this tissue are as yet unclear. Evidence has been presented both in support of and against the possibility that increased pump activity may be secondary to elevated Na$^+$ influx due to stimulation of amiloride-sensitive Na/H transport (7,26,30,31). And, while some have suggested that insulin increases the number of pump sites in the membrane (10,24,25), others have concluded that insulin only increases the rate of ouabain binding and not the total number of Na$^+$-K$^+$ pump units (8,31). Recently, evidence has been presented from studies on intact muscle that insulin translocates α-2 isoforms of the catalytic subunit from the cytosol to the membrane, where α-1 isoforms normally predominate (16). Mature skeletal muscle cells in culture derived from rat newborn express only the α-1 and β-1 isoforms of the respective subunits (Fig. 1).

One other possible mechanism of insulin action could involve the activation of protein kinase C (PKC), a phenomenon shown to be one of the initial events in certain other insulin-stimulated processes (11,22,28). In addition, Na$^+$-K$^+$ ATPase has been shown to be a substrate for PKC (2,6,20). PKC seems to have a role in phosphorylation of membrane ion channels (see 19), but its function in muscle growth and differentiation, while recognized as important, is not entirely clear. Translocation of PKC occurs on differentiation (21), and activation of PKC is reported to be a major initial signal in insulin-stimulated glucose transport in a variety of cells (11,23), including skeletal muscle (33). There are now recognized several isoforms of PKC in a variety of tissues (see 1). In addition, phorbol esters have been shown to stimulate Na$^+$/K$^+$ pump activity in some cells (14). In this study, we have examined the possibility that activation of PKC may be one of the early signal transduction mechanisms in stimulation of the Na/K$^+$ pump in cultured rat skeletal muscle by insulin.

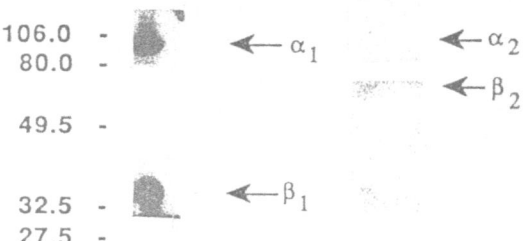

**Figure 1.** Western blot of Na$^+$-K$^+$ pump isoforms in cultured rat skeletal muscle SDS-Page was performed on cultures at age 6 days in vitro (Yosef, Klip and Sampson, unpublished observations).

# METHODS

## Rat Muscle Cell Cultures

Skeletal muscle cultures were prepared from thigh muscles obtained from 1 to 2-day neonatal rats as described (3,5). The limbs were removed, muscles isolated, and washed in phosphate-buffered saline (PBS) and then transferred to a Ca-free, 0.25% trypsin solution containing EDTA (1 mM) for incubation with continuous stirring at 37°C. Cells were collected after serial trypsinization (successive 10 min periods) until all tissue was dispersed, and then centrifuged for 5 min at 500 g. Pellets were resuspended in growth medium and myoblasts were diluted with growth medium (DMEM-high glucose, 83%: horse serum, 15%: chick embryo extract 2%) to a concentration of $0.8x10^6$ cells/mL for plating in 35 mm collagen-coated plastic tissue culture dishes (1.5 mL/dish). Cultures were grown in water-saturated atmosphere of 95% air-5% $CO_2$ at 37°C. Unless otherwise indicated, experiments were done on cultures of 5-7 days *in vitro* (DIV).

## Electrophysiological Recording

The techniques for intracellular recording have been described (3-5). Briefly, growth medium was replaced with a phosphate-buffered recording medium (pH 7.35) of the following composition (mM): NaCl, 135: KCl, 2.7: $Na_2PO_4$, 8: $KH_2PO_4$, 1.4: $MgCl_2$, 0.5: $CaCl_2$, 0.7: glucose, 22. The dishes containing the cultures were placed in a copper heating sleeve for temperature control from a circulating water bath, and this assembly was placed on the stage of an inverted phase-contrast microscope (Nikon). Recordings were made with glass microelectrodes containing a fiberglass fiber and filled with 2.8 M KCl. The mean value and standard error of resting $E_m$ were determined from a minimum of nine myotubes from each of two dishes under any given experimental condition. Recordings were routinely begun 10 min after any changes in medium. Data were analyzed statistically by Student's T-test for unpaired samples or ANOVA (GraphPad, San Diego, CA).

## [86]Rb Uptake

[86]Rb uptake was measured according to the method already described (3). Rate of uptake was measured in 1 ml of $K^+$-free PBS containing 2 mM RbCl containing 2.5 µCi of [86]Rb. Mean values were determined from measurements on 3 dishes under each experimental condition in each experiment. [86]Rb-uptake specifically related to $Na^+$-$K^+$ pump activity was determined by subtraction of the amount taken up in the presence of 1 mM ouabain from that in its absence.

## Measurement of [³H]-ouabain Binding

Specific binding of [³H]-ouabain was measured as described (4; see also 15). Cultures were washed three times with 2 ml K-free phosphate buffer (pH 7.4) and then incubated in 1.0 ml of buffer containing 75 nM [³H]-ouabain at 37°C. Non-specific binding was determined in the presence of 1 mM unlabeled ouabain. After incubation for 30 min, cells were washed and then solubilized in 1% Triton x100, and radioactivity was assayed by standard procedures. Control studies established that ouabain binding by non-muscle elements was less that 5-10% of the total.

## Binding of [³H]-PDBu

Cultures were washed with PBS and then incubated for 30 min in PBS containing 1% bovine serum albumin and various concentrations of [³H]-PDBu. After incubation, cells were washed 4 times in cold PBS and disrupted in 0.1 M NaOH. Radioactivity was counted as above.

## Phorbol Ester Treatment of Cells

Phorbol ester treatment of cells was accomplished by addition of the appropriate compound directly to the cultures in growth medium to the desired concentration. Except where otherwise stated, cells were treated for 30 min, following which the medium was changed to K-free phosphate-buffered saline (PBS) for measurements of Rb-uptake or ouabain-binding, or to regular PBS for $E_m$ measurements.

## Drugs

Tetrodotoxin (TTX), amiloride and the phorbol esters phorbol 12-myristate 13-acetate (PMA), phorbol 20-oxo-20-deoxy12,13-dibutyrate (PDBu) and 4-α-phorbol 12,13-didecanoate (PDD) were all obtained from Sigma Chemical Co. (St. Louis, MO, USA). ⁸⁶Rb, [³H]-PDBu and [³H]-ouabain were purchased from Amersham Ltd. (UK). Human Recombinant Insulin (Insulin Leo Neutral, Nordisk Gentofte A/S, Denmark) was obtained by prescription from the local pharmacy.

## RESULTS

Studies in our laboratory have established that activity of the Na-K pump can be determined by measurements of the ouabain-sensitive component of resting membrane potential ($E_m$), ouabain-sensitive uptake of ⁸⁶Rb and the specific binding of [³H]-ouabain to whole cell preparations of cultured skeletal muscle (3-5). With these techniques, we have characterized developmental properties of the pump as well as long-term effects of hormones and growth factors, and effects of certain drugs on physiological expression of pump activity. Accordingly, we have used these parameters to examine short-term regulation of $Na^+$-$K^+$ pump activity by assessing effects of insulin and phorbol esters.

## Effects of Insulin and Phorbol Esters on $Na^+$/$K^+$ Pump Activity

In a previous study we characterized the dose-effect relation and time course of peak effects of insulin on $Na^+$-$K^+$ pump activity (27). Dose-effect studies showed that a noticeable increase is obtained with 10 mU/ml and maximum effects with 80-100 mU/ml. With regard to the time course, both insulin and phorbol esters increase ⁸⁶Rb uptake as early as 10-15 min after addition to reach peak effects of nearly a 50-100% increase in pump activity after 30 - 60 min. Activity remains elevated for up to 3 hr (the longest time studied in our experiments) (Fig. 2). We also examined effects of insulin and phorbol esters on [³H] ouabain binding, which was also increased as early as 15 min after addition of insulin and remained elevated for up to 3 hr.

Alterations in activity of the $Na^+$-$K^+$ pump are reflected in changes in $E_m$ of cultured myotubes (5,17), and increases in pump activity induced by insulin and phorbol esters also are associated with hyperpolarization of $E_m$. In order to determine if, indeed, the insulin- and phorbol ester induced hyperpolarization of cultured myotubes is due to stimulation of $Na^+$-$K^+$

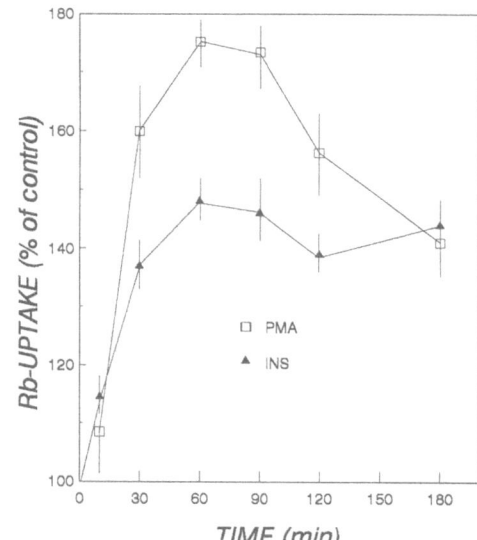

**Figure 2.** Time course of effects of insulin and PMA on Na$^+$-K$^+$ pump activity in cultured skeletal muscle. Values represent the mean ± S.E. of triplicate values obtained in each of 3-5 experiments.

pump activity, we examined effects of ouabain given both before and after the stimulants. As shown in Fig. 3, the increases in $E_m$ induced by both phorbol esters and insulin are both reversed and prevented by ouabain.

To attempt to elucidate the mechanism underlying this stimulation of pump activity, we studied the possibility that it is mediated via increased Na-entry (3,32). Myotubes were pre-treated for 15-30 min with TTX or amiloride prior to administration of insulin or phorbol esters. Amiloride caused a slight but not significant decrease in resting [86]Rb uptake, whereas TTX caused a significant decrease in resting [86]Rb uptake. In each case, insulin and PMA caused [86]Rb uptake to increase significantly. Thus, in control cells insulin caused activity to

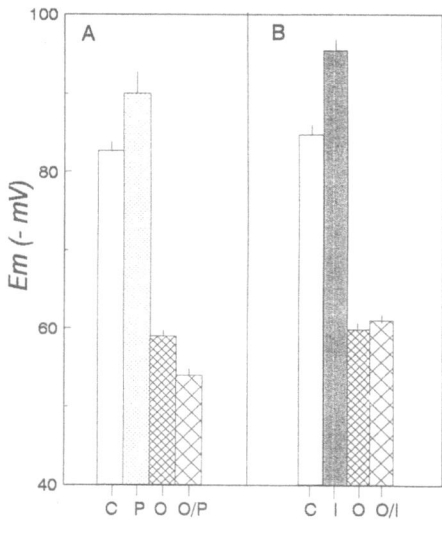

**Figure 3.** Effects of PMA (A) and insulin (B) on resting membrane potential ($E_m$) of cultured skeletal myotubes. Both PMA (P) and insulin (I) caused $E_m$ to increase. The increase was reversed by treatment with ouabain (O) and prevented by pretreatment with ouabain (O/P; O/I). Values represent the mean $E_m$ ± s.e. measured from 9 cells in 2 dishes in each of 3 experiments.

increase by about 50% above the resting level. In amiloride-treated cells, insulin increased pump activity by nearly 50%, and in TTX-treated cells the increase induced by insulin was nearly 100% (p < 0.001), in both cases the level of pump activity in the presence of insulin being nearly as high as that produced by insulin in control myotubes. Similarly, blockade of Na-entry via TTX- and amiloride-sensitive pathways were without effect on PMA-induced increases in NaK pump activity (27)

As PKC has been reported to be involved in the activation of other insulin-stimulated processes (9,11,12,13,28,33), we examined the possibility that the insulin-induced effects on the $Na^+/K^+$ pump might involve PKC activation. This was done in 3 ways. In one, we studied effects of PMA and insulin, given separately and together, on Rb-uptake. In this study, insulin alone caused an increase of nearly 100% in Rb-uptake, while PMA alone caused an increase of nearly 50%. When given together, Rb-uptake was only slightly greater than that obtained with PMA alone. In the second, we studied effects of PKC depletion by 20-hr treatment with PMA (150 nM) on responses to insulin and PMA. In control cells, insulin and PMA caused increases of 60 and 80%, respectively. In PMA-treated cells, however, insulin had no effect on Rb-uptake, and the increase by PMA was reduced to 35%. Finally, we examined effects of the PKC inhibitor staurosporine on the increase in $^{86}Rb$-uptake induced by both insulin and PMA. Staurosporine (100 nM for 4 hr) had no effect on resting Rb uptake, but reduced or abolished effects of both PMA and insulin.

As the evidence presented thus far suggests that PKC activation stimulates $Na^+$-$K^+$ pump activity, we sought to obtain additional evidence regarding this point as well as the activation of PKC by insulin. In addition to phorbol esters, diacylglycerol (DAG), which is liberated by other signal transduction cascades, acts directly on PKC. In addition, certain PKC isoforms are $Ca^{2+}$-dependent and, as such, can be activated directly by endogenously-released $Ca^{2+}$. Accordingly, we examined effects of DAG and the Ca-ionophore A23187 on pump activity. As seen in Fig 4, both substances caused pump activity to increase significantly.

Activation of PKC involves its mobilization from intracellular sites to the membrane. Mobilization to the membrane can be assessed by measurement of the binding of [$^3H$]PDBu, which itself activates PKC. We found (Fig. 5) that treatment of myotubes with insulin

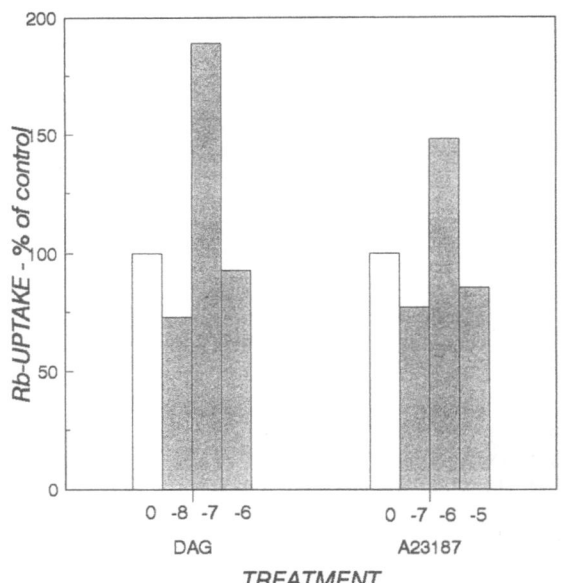

**Figure 4.** Effects of diacyl glycerol and A23187 on $Na^+$-$K^+$ pump activity in cultured skeletal muscle. Values represent the mean ± S.E. of triplicate values obtained in each of 3 experiments.

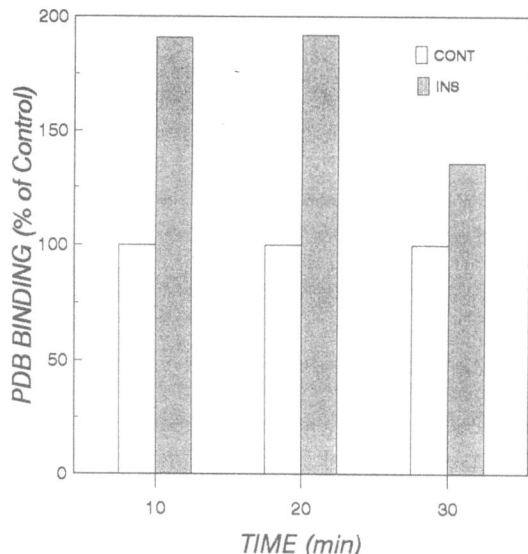

**Figure 5.** Time-course of insulin effects on the binding of [$^3$H]PDBu by skeletal muscle cells in culture. Each point is the mean ± S.E. of values obtained from triplicate samples in 3 experiments.

increases membrane binding of [$^3$H]PDBu as early as 5-10 min after administration, indicating that an early event in the response to insulin is translocation of PKC to the membrane.

## DISCUSSION

In the present study on primary cultures of rat skeletal muscle, we have found that both insulin and phorbol esters increase both the ouabain-suppressible component of $^{86}$Rb-uptake and the amount of [$^3$H]-ouabain bound by whole cell preparations. In addition, our findings on the effect of ouabain to block hyperpolarization of cultured myotubes by both substances demonstrate that the increase in $E_m$ produced by them is due to Na$^+$/K$^+$-pump stimulation.

It has been concluded on the basis of studies on intact rat skeletal muscle *in vivo* and *in vitro*, that insulin increases the rate of [$^3$H]-ouabain binding but not the number of active pump units. We found that both the number of ouabain-binding sites and the rate of Rb-uptake were increased for up to 3 hr. Our findings are thus difficult to reconcile with a mechanism attributable to only an increase in rate of ouabain binding but are instead consistent with the suggestions of several other groups (10,16,24,25) that insulin may also activate, mobilize or otherwise translocate Na$^+$/K$^+$-pump units from intracellular to membrane locations.

The results of studies on phorbol ester effects on Na$^+$/K$^+$-pump activity indicate that PKC activation increases pump activity. These findings agree with those of Hootman et al (14), who reported that phorbol esters increased Na$^+$/K$^+$-pump activity in pancreatic acinar cells. In addition, activation of the pump by both DAG and A23187 provide further evidence for direct stimulation of the Na$^+$-K$^+$ pump by PKC. In contrast, Bertorello et al (2) recently reported that PKC-induced phosphorylation of the catalytic subunit of the pump inhibits its activity in preparations obtained from shark rectal gland. Similarly, treatment of Xenopus oocytes with tumor-promoting phorbol esters caused down-regulation of the Na$^+$/K$^+$-pump (6,see also 29). It would thus appear that PKC activation has tissue specific effects, and the conclusion that activation of PKC down-regulates the $Na^+/K^+$-pump may not be valid for all tissues.

The studies on the relation between effects of insulin and phorbol esters on the $Na^+/K^+$-pump further indicate that PKC activation by insulin may be a key early event in insulin-induced increase in $Na^+/K^+$-pump activity, as has been reported for insulin's effect on glucose uptake (11,33; see also 18). Thus, the increase produced by the addition of both phorbol ester and insulin together was not found to be additive, indicating that the two may act via a common mechanism, although other interpretations are also possible. In addition, administration of PMA or PDBu for 20-24 hr, which reportedly down-regulates PKC, consistently blocked effects of insulin and reduced those of phorbol esters on the $Na^+/K^+$-pump. We have also found that insulin-induced stimulation of the $Na^+/K^+$-pump is significantly reduced or eliminated by treatment with staurosporine, a relatively selective PKC blocker. Finally, the insulin-induced increase in binding of PDBu indicates that the hormone translocates PKC from the cytosol to the membrane, an essential aspect of PKC stimulation.

It was reported recently that insulin stimulates the $Na^+/K^+$-pump in rat skeletal muscle in vivo by mobilizing $\alpha2$ isoforms of the catalytic subunit from intracellular stores to the membrane, where the $\alpha1$ isoform normally predominates (16). As mentioned, however, primary cultures of rat skeletal muscle do not express the $\alpha2$-subunit. Our findings do not, however, necessarily contradict and may even complement the results reported by Hundal et al (16). The question as to whether in intact muscle $\alpha2$ subunit isoforms are mobilized in response to an increased demand secondary to phosphorylation and increased activity of already membrane-bound units, or whether mobilization and activation of $\alpha2$ subunits require increased PKC activation, remains to be studied.

The findings in this study, together with the observations that insulin activates (one or several isoforms of) PKC (11,23), and the $\alpha$-subunit of the $Na^+/K^+$ pump is phosphorylated by PKC (2,6,20), while not conclusive, are strongly suggestive that insulin-activation of PKC is a fundamental early signal in stimulation of the $Na^+/K^+$ pump by this hormone in skeletal muscle. Thus, this mechanism may be added to the list of those proposed for insulin-induced stimulation of the $Na^+/K^+$ pump (22). Its relative importance in various tissues remains to be determined.

## ACKNOWLEDGMENTS

This work was supported by the Ben and Effie Raber Neuroscience Research Fund, the Charles Krown Health Sciences Research Fund, and the Otto Meyerhoff Center.

## REFERENCES

1. Azzi, A., D. Boscoboinik, and C. Hensey. The protein kinase C family. *Eur. J. Biochem.* 208: 547-557, 1992.
2. Bertorello, A.M., A. Apeira, S.I. Walaas, A.C. Nairn, and P. Greengard. Phosphorylation of the catalytic subunit of $Na^+/K^+$-ATPase inhibits the activity of the enzyme. *Proc. Natl. Acad. Sci. USA* 88: 11359-11362, 1991.
3. Brodie, C. and S.R. Sampson. Regulation of the sodium-potassium pump in cultured rat skeletal myotubes by intracellular sodium ions. *J. Cell. Physiol.* 140: 131-137, 1989.
4. Brodie, C. and S.R. Sampson. Effects of chronic ouabain-treatment on [$^3$H]-ouabain binding sites and electrogenic component of membrane potential in cultured rat myotubes. *Brain Res.* 347:121-123, 1985.
5. Brodie, C. and S.R. Sampson. Influence of various growth factors and conditions on development of resting membrane potential and its electrogenic component of cultured rat skeletal myotubes. *Int. J. Dev. Neurosci.* 4: 327-337, 1986.

6. Chibalin, A.V., L.A. Vasilets, H. Hennekes, D. Pralong, and K. Geering. Phosphorylation of Na, K-ATPase α-subunits in microsomes and in homogenates of Xenopus oocytes resulting from the stimulation of protein kinase A and protein kinase C. *J. Biol. Chem.* 267: 22378-22384.

7. Clausen, T., and J.A. Flatman. Effects of insulin and epinephrine on Na$^+$/K$^+$ and glucose transport in soleus muscle. *Am. J. Physiol.* 252 (Endocrinol. Metab. 15):E492-E499, 1987.

8. Clausen, T., and O. Hansen. Active Na-K transport and the rate of ouabain binding. The effect of insulin and other stimuli on skeletal muscle and adipocytes. *J. Physiol. Lond.* 270: 415-430, 1977.

9. De Luise, M. A., and M. Harker. Insulin stimulation of Na$^+$/K$^+$-pump in clonal rat osteosarcoma cells. *Diabetes* 37: 33-37, 1988.

10. Erlij, D. and S. Grinstein. The number of sodium ion pumping sites in skeletal muscle and its modification by insulin. *J. Physiol. (Lond.).* 259: 13-31, 1976.

11. Farese, R.V., M.L. Standaert, T. Arnold, B. Yu, T. Ishizuka, J. Hoffman, M. Vila, and D.R. Cooper. The role of protein kinase C in insulin action. *Cellular Signalling.* 4: 133-143, 1992.

12. Gelehter, T.D., P.D. Shreve, and V.M. Dilworth. Insulin regulation of Na/K pump activity in rat hepatoma cells. *Diabetes* 33: 428-434, 1984.

13. Hecht, L.B. and D.S. Straus. Insulin-stimulated protein kinase activity in rat skeletal muscle that phosphorylates ribosomal protein S6. *Bioch. Biophys. Res. Comm.* 152: 1200-1206.

14. Hootman, S.R., M.E. Brown, and J. A. Williams. Phorbol esters and A23187 regulate Na$^+$/K$^+$-pump activity in pancreatic acinar cells. *Am. J. Physiol.* 252 (Gastrointest. Liver Physiol. 15): G499-G505, 1987.

15. Hootman, S.R., and S.A. Ernst. Estimation of Na$^+$/K$^+$-pump numbers and turnover in intact cells with [$^3$H]-ouabain. In: Methods in Enzymology: Biomembranes and Biological Transport, edited by S. Fleischer and B. Fleischer. New York, Academic, 1988.

16. Hundal, H.S., A. Marette, Y. Mitsumoto, T. Ramlal, R. Blostein, and A. Klip. Insulin induces translocation of the α2 and β1 subunits of the Na$^+$/K$^+$-ATPase from intracellular compartments to the plasma membrane in mammalian skeletal muscle. *J. Biochem. Chem.* 267: 5040-5043, 1992.

17. Iannaccone, S.T., K.-X. Li, N. Sperelakis, and D.A. Lathrop. Insulin-induced hyperpolarization in mammalian skeletal muscle. *Am. J. Physiol.* 256(Cell Physiol. 25): C368-C374, 1989.

18. Klip, A. and A.G. Douen. Role of kinases in insulin stimulation of glucose transport. *J. Membrane Biol.* 111: 1-23, 1989.

19. Lotan, I., N. Dascall, Z. Naor, and R. Boton. Modulation of vertebrate brain Na$^+$ and K$^+$ channels by subtypes of protein kinase C. *FEBS Letters* 267: 25-28, 1990.

20. Lowndes, J.M., M. Hokin-Neaverson, and P.J. Bertics. Kinetics of phosphorylation of Na$^+$/K$^+$-ATPase by protein kinase C. *Biochim. Biophys. Acta*, 1052: 143-151, 1990.

21. Martelly, I., Gautron, J., and J. Moraczewski. Protein kinase C activity and phorbol ester binding to rat myogenic cells during growth and differentiation. *Exper. Cell Res.* 183: 92-100, 1989.

22. Messina, J.L., M.L. Standaert, T. Ishizuka, R. Weinstock, and R.V. Farese. Role of protein kinase C in Insulin's regulation of c-fos transcription. *J. Biol. Chem.* 267: 9223-9228, 1992.

23. McGeoch, J.E.M., and G. Guidotti. An insulin-stimulated cation channel in skeletal muscle. *J. Biol. Chem.* 267: 832-841, 1992.

24. Moore, R.D. Effect of insulin upon the sodium pump in frog skeletal muscle. *J. Physiol. Lond.* 232: 23-45, 1973.

25. Omatsu-Kanabe, M., and Kitasato, H. Insulin stimulates the translocation of Na$^+$/K$^+$-dependent ATPase molecules from intracellular stores to the plasma membrane in frog skeletal muscle. *Biochem. J.* 272: 727-733, 1990.

26. Rosic, N.K., M.L. Standaert, and R.J. Pollet. The mechanism of insulin stimulation of Na$^+$/K$^+$-ATPase transport activity in muscle. *J. Biol. Chem.* 260: 6206-62212, 1985.

27. Sampson, S.R., C. Brodie, and S. Alboim. Role of Protein Kinase C in insulin activatin of the Na-K pump in cultured skeletal muscle. *Am. J. Physiol.* 266: C751-C758, 1994.

28. Standaert, M.L., D.J. Buckley, T. Ishizuka, J.M. Hoffman, D.R. Cooper, R.J. Pollet, and R.V. Farese. Protein kinase C inhibitors block insulin and PMA-stimulated hexose transport in isolated rat adipocytes and BC$^3$H-$^1$ myocytes. *Metabolism.* 39: 1170-1179, 1990.

29. Vasilets, L.A. and W. Schwarz. Regulation of endogenous and expressed Na$^+$/K$^+$ pumps in Xenopus oocytes by membrane potential and stimulation of protein kinases. *J. Membrane Biol.* 125: 119-132, 1992.

30. Vigne, P., C. Frelin, and M. Lazdunski. The Na$^+$-dependent regulation of the internal pH in chick skeletal muscle cells. The role of the Na$^+$/H$^+$ exchange system and its dependence on internal pH. *EMBO J.* 3: 1865-1870, 1984.

31. Weil, E., S. Sasson,, and Y. Gutman. Mechanism of insulin-induced activation of Na$^+$/K$^+$-ATP as in isolated rat soleus muscle. *Am. J. Physiol.* 261(Cell Physiol. 30): C224-C230, 1991.

32. Wolitzky, B.A., and D.M. Fambrough. Regulation of the (Na⁺/K⁺)-ATPase in cultured chick skeletal muscle. Regulation of expression by the demand for ion transport. J. *Biol. Chem.* 261: 9990-9999, 1986.

33. Yu, B., M. Standaert, T. Arnold, H. Hernandez, J. Watson, K. Ways, D.R. Cooper, and R.V. Farese. Effects of insulin on diacylglycerol/protein kinase-C signaling and glucose transport in rat skeletal muscles *in vivo* and *in vitro*. *Endocrinology* 130: 3345-3355, 1992.

# 6

# AORTIC AND ABDOMINAL GLOMERA

A. Howe

Physiology Group
Biomedical Sciences Division
King's College London
Campden Hill Road, London W8 7AH, United Kingdom

Firstly, like previous speakers, I would like to pay tribute to John and Hazel Coleridge - both to their seminal research work in the field of cardiovascular receptors, and secondly to them personally. I have known John and Hazel for some 30 years and, over this time, despite the trans-Atlantic gulf between us, we have managed to keep in touch. On more than one occasion, I made visits to work with them in New York and San Francisco -from which our joint publications date. I would like to talk very briefly about some of the work we did together on the thoracic arterial chemoreceptors, the so-called "aortic bodies", by way of introduction to some studies that I carried out subsequently on my own on the "abdominal paraganglia" -putative abdominal chemoreceptors - using the same investigative approach.

I first collaborated with the Coleridges in 1966, when they were at the Royal Free Hospital Medical School in London, and in their early years at New York Medical College. Essentially, John provided the electrophysiological arm of the work and I supplied the histological side. But, in fact, it was Hazel who carried the biggest load, for she did much of the neuro-physiological recordings, as well as sharing with me the labour of scrutinizing the hundreds of serial sections by which we located the minute aortic bodies - not to mention the feeding of the hungry visiting worker (namely myself).

From these studies, we produced the first and to date the only comprehensive survey of the distribution of the thoracic chemoreceptors in the cat and dog, together with a detailed account of their blood supply (Coleridge, Coleridge and Howe, 1967, 1970). We needed to know this in order to be able to stimulate the individual groups of aortic bodies selectively by the close-arterial injection of chemo-excitant drugs, such as cyanide, lobeline and nicotine. I should mention that, at this time, Julius Comroe at the Cardiovascular Research Institute, San Francisco, had surprisingly come to the conclusion that, of the various groups of aortic bodies, only *one* group was in fact functional (namely that situated below the aortic arch - Group 4 in our classification). He believed that all the remaining groups were non-functional. I must say I found this view barely credible, since, from my earlier studies in the 1950's on the cat (Howe, 1956), it was clear to me that (just like the carotid bodies) *all the groups* of aortic bodies that I had identified possessed a rich patent arterial blood supply and venous drainage, and an abundant sensory innervation. Moreover, histologically, they all "looked quite normal" - not in any way redundant or vestigial, as Comroe seems to

*Control of the Cardiovascular and Respiratory Systems in Health and Disease*
Edited by C. T. Kappagoda and M. P. Kaufman, Plenum Press, New York, 1995

57

have believed. With such a combination of electrophysiology and histology, the Coleridges and I were able to show that *all* the thoracic groups were indeed functional chemoreceptors, monitoring the composition of systemic arterial (oxygenated) blood and sending afferent signals up the vagal nerves (Coleridge, Coleridge and Howe, 1967, 1970). In all our studies we found no evidence for the claim made by some workers (Krahl, 1960, 1962; Duke, Green, Heffron and Stubbens, 1963; Hughes, 1965) for the existence of pulmonary arterial chemoreceptors, i.e. chemoreceptors monitoring mixed venous blood - a finding that still holds today (see Orr et al, 1988). Hence, in the light of the earlier classical histological studies of De Castro in Spain in the 1920's, and Muratori in Italy in the 1930's, together with our findings, we can say that the peripheral arterial chemoreceptor system is generally considered to be comprised of 2 sets of so-called "glomus tissue" - the bilateral carotid bodies ("glomera carotica"), which have been found in every mammalian species in which they have been sought, together with the multiple thoracic (aortic) bodies ("glomera aortica"), whose incidence may vary in different species - e.g. relatively plentiful in the cat and dog, sparse or absent in the rat, mouse and guinea pig, and maybe in man (see Kjaergaard, 1973; Böck, 1982). Indeed, most textbooks represent this "classical" view with varying degrees of accuracy.

However, with the subsequent discovery of additional glomus groups in the neck (Matsuura, 1973) and mediastinal region (Easton & Howe, 1983), it is now clear that the arterial chemoreceptor system is more extensive than was previously realised. As depicted in Figure 1 (the most up-to-date inventory of such glomus tissue), this system constitutes an extensive receptor chain, distributed along the course of the 9th and 10th cranial nerves from the carotid sinus region (and possibly even more rostral to the level of nodose and jugular

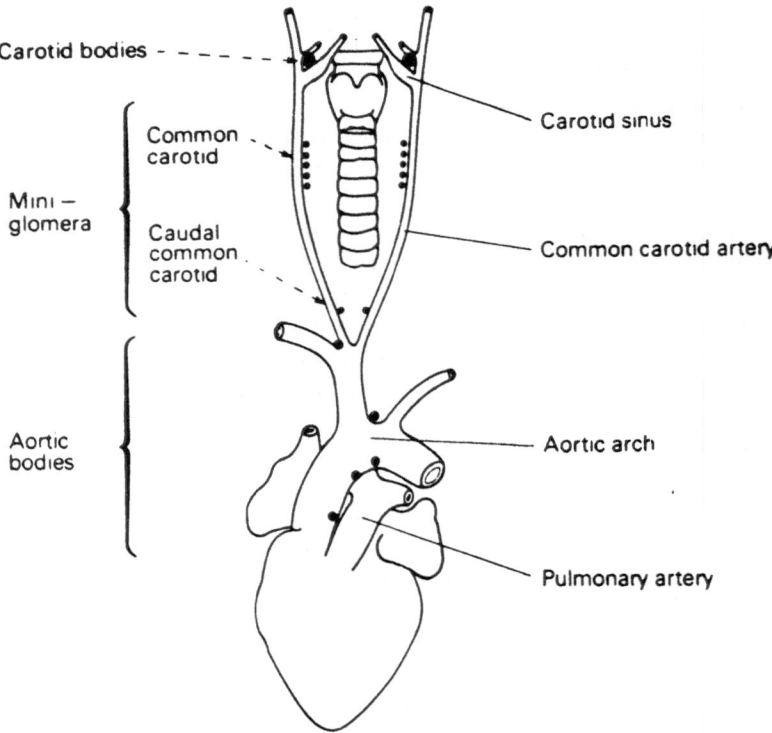

**Figure 1.** Generalized diagram to show the potential distribution of glomus tissue (peripheral arterial chemoreceptor system) in mammals.

ganglia) down to the aortic arch, and including the carotid bodies, the mini-glomera of the cervical and clavicular levels and the thoracic groups (see Easton & Howe, 1983, for details). Whether or not vascular chemoreceptors exist in regions other than these remains to be seen, but minute aggregations of glomus tissue (vagal paraganglia), that bear a striking histological similarity to the classical arterial chemoreceptors, have been described in the abdomen of mice, rats, hamsters and man (Goormaghtigh, 1936; Elliott, 1965). They are closely associated with or embedded in the various divisions of the vagal nerve trunks as they emerge through the diaphragm, course over the surface of the oesophagus and stomach and send branches to various abdominal viscera (Deane, Howe and Morgan, 1975). In fact, some 48 years ago, on very flimsy evidence, Hollinshead (1941, 1946) actually proposed that such abdominal vagal paraganglia (a.v.p.) were vascular chemoreceptors.

Over a number of years, our laboratory has been interested in this paraganglionic system, its similarities with the peripheral arterial chemoreceptor system and with the testing of Hollinshead's hypothesis. Our findings will be summarized under 3 main headings:

1. Histological survey: light and electronmicroscopic observations.
2. Electrophysiological data.
3. Chronic hypoxia: resultant morphological changes.

1. *Histological*: Histologically, the abdominal vagal paraganglia are indistinguishable from the classical arterial chemoreceptors at both the light and electronmicroscopic level, and morphologically distinct from both the S.I.F. cells of the superior cervical ganglia and the chromaffin cells of the adrenal medulla, with which they might conceivably be confused. At the electronmicroscopic level, the same forms of contact between nerve ending and glomus cell are found in the a.v.p. as in the carotid and aortic bodies; the predominant form of junction appears to be afferent in type (Morgan, Pack and Howe, 1976; Howe, Morgan and Pack, 1978).

2. *Electrophysiology*: The abdominal paraganglia are known to be innervated by the vagal nerves. In the anaesthetised rat, electrophysiological techniques revealed the presence of afferent fibres of chemoreceptor type in vagal branches supplying these paraganglia (Andrews, Deane, Howe and Orbach, 1972; Howe, Pack and Wise, 1981). The electroneurograms so recorded closely resembled those elicited from the carotid body and from the mini-glomera (Matsuura, 1973) under various conditions known to provoke arterial chemoreceptor activity, e.g. acute hypoxia.

3. *Chronic Hypoxia*: From studies in man and animals at altitude, and experimentally in hypoxic chambers, it is well established that the carotid bodies become enlarged following chronic hypoxia (Barer, Edwards & Jolly, 1976; Heath & Williams, 1981). A similar study was therefore carried out to see whether the a.v.p. would also show an hypertrophic response under such conditions (Barker, Castro, Howe and Pack, 1984; Howe, Pack and Castro, 1990). Animals were divided into control and experimental groups. The experimental rats were exposed for a period of 21 days in a chamber to an atmosphere of $10\%O_2$ (corresponding to an altitude of 5500m); control animals breathed room air under the same housing conditions. At the end of the experimental period, all animals were anaesthetized and fixed by arterial perfusion in the same way. The carotid bodies and a circumscribed piece of the oesophageal - stomach region (an area in which a.v.p. are regularly located - Deane et al, 1975), were removed and serially sectioned for histology. The size of the carotid bodies and the number and size of a.v.p. were estimated microscopically. In the hypoxic animals, the carotid bodies were significantly larger than in the controls (as were their haematocrits and right ventricular weights - well known indices of chronic hypoxia); they also appeared more vascular than

those of the normoxic controls. There was no significant difference in the number of a.v.p. per animal between control and experimental rats. However, in the hypoxic animals the individual paraganglia were hypertrophied, as compared with the normoxic controls (Barker, Castro, Howe & Pack, 1984; Howe, Pack and Castro, 1990). It is not totally clear which components of the paraganglia - cellular, vascular or stromal - are responsible for this hypertrophy engendered by chronic hypoxia. However, it is established that, like the carotid bodies, the abdominal paraganglia also exhibit an increased vascularity under such conditions of chronic hypoxia.

## CONCLUSIONS

The 3 sets of results, considered collectively, indicate a remarkable similarity between the a.v.p. and the carotid (and aortic) bodies. They may be construed as further support for the view, first advanced by Hollinshead, that the abdominal vagal paraganglia (a.v.p.) belong to the peripheral arterial chemoreceptor system. Whether the a.v.p. are connected with the same systems as the classical chemoreceptors, and are similarly involved in respiratory and cardiovascular control, remains to be seen. Meanwhile, preliminary investigations (Child, Hedley and Howe, 1990) indicate that small doses of cyanide, of the order used to stimulate peripheral arterial chemoreceptors and delivered close-arterially into the vascular bed of the a.v.p., provoke reflex respiratory stimulation. Selective denervation studies show that the sensory pathway for these reflexes is predominantly via the abdominal vagal nerves (and to a lesser extent via the sympathetic splanchnic nerves). It remains to be established whether such reflexes actually emanate from the a.v.p.

From the work of Longhurst and his associates, it is well known that there is a variety of nerve fibres conveying various modalities of afferent information from the abdominal viscera, some of which engender cardiovascular reflexes (see Longhurst,1991). In recalling Hollinshead's original prediction that the a.v.p. may be involved in local (abdominal) vascular control, it is tempting to view the abdominal vagal paraganglia as part of this abdominal "chemosensitive" afferent system.

## REFERENCES

1. Andrews, W.H.H., Deane, B.M., Howe, A. & Orbach, J. (1972) Abdominal chemoreceptors in the rat. *J.Physiol.*, *222*, 84-85P.
2. Barer, G.R., Edwards, C.W. & Jolly, A.T. (1976) Changes in the carotid body and the ventilatory response to hypoxia in chronically hypoxic rats. *Clin.Sci.mol.Med.*, *50*, 311-313.
3. Barker, S.J., Castro, K., Howe, A. & Pack, R.J. (1984) Abdominal vagal paraganglia in the rat: hypertrophic response to chronic hypoxia. *J.Physiol.*, *348*, 74P.
4. Böck, P. (1982) The Paraganglia. Handbuch der mikroskopischen Anatomie des Menschen, Band 6 Teil 8, Springer, Berlin, Heidelberg, New York.
5. Child, B, Hedley, R.J. & Howe, A. (1990) Reflex respiratory responses to abdominal intravascular sodium cyanide in the anaesthetised rat. *J.Physiol.*, *423*, 27P.
6. Coleridge, H., Coleridge, J.C.G. & Howe, A. (1967) A search for pulmonary arterial chemoreceptors in the cat, with a comparison of the blood supply of the aortic bodies in the new-born and adult animal. *Journal of Physiology*, *191*, 353- 374.
7. Coleridge, H., Coleridge, J.C.G. & Howe, A. (1970) Thoracic chemoreceptors in the dog: A histological and electrophysiological study of the location, innervation and blood supply of the aortic bodies. *Circulation Research.*, *26*, 235-247.
8. Deane, B.M., Howe, A. & Morgan, M. (1975) Abdominal vagal paraganglia: distribution and comparison with carotid body, in the rat. *Acta Anatomica.* (Basel), *93*, 19-28.

9. De Castro, F. (1926) Sur la structure et l'innervation de la glande intercarotidienne (Glomus caroticum) de l'homme et des mammifères, et sur un nouveau système d'innervation autonome du nerf glosso-pharyngien. Etudes anatomiques et expérimentales. Trab. Lab. Invest. biol. Univ. Madr. *24*, 365-432.

10. De Castro, F. (1928) Sur la structure et l'innervation du sinus carotidien de l'homme et des mammifères. Nouveaux faits sur l'innervation et la fonction du glomus caroticum. Etudes anatomiques et physiologiques. Trab. Lab. Invest biol. Univ. Madr. *25*, 331-380.

11. Duke, H.N., Green, J.H. Heffron, P.F., & Stubbens, V.W.J. (1963) Pulmonary Chemoreceptors. *Quart. J. Exp. Physiol.*, *48*, 164-175.

12. Easton, J.C. & Howe, A. (1983) The distribution of thoracic glomus tissue (aortic bodies) in the rat. *Cell tiss. Res.*, *232*, 329-356.

13. Elliott, G.B. (1965) Glomus-like bodies on the superior mesenteric artery. *Canad.Med.Assoc.J.*, *92*, 1303-1305.

14. Goormaghtigh, N. (1936) On the existence of abdominal vagal paraganglia in the adult mouse. *Journal of Anatomy* (Lond.), *71*, 77-90.

15. Heath, D. & Williams, D.R. (1981) Man at high altitude. 2nd Ed. London: Churchill Livingstone.

16. Hollinshead, W.H. (1941) Chemoreceptors in the abdomen. *Journal of Comparative Neurology*, *74*, 269-285.

17. Hollinshead, W.H. (1946) The function of the abdominal chemoreceptors of the rat and mouse. *American Journal of Physiology*, *147*, 654-660.

18. Howe, A. (1956) The vasculature of the aortic bodies in the cat. *J.Physiol.*, *134*, 311-318.

19. Howe, A., Morgan, M. & Pack, R.J. (1978) A comparison of the ultrastructure of the abdominal vagal paraganglia and similar tissues in the rat. *J.Physiol.*, *275*, 34-35P.

20. Howe, A., Pack, R.J. & Castro, K. (1990) Hypertrophy of abdominal vagal paraganglia following chronic hypoxia: compared with carotid body. Proceedings of IX International Symposium on Arterial Chemoreception (ISAC) (1988). Springer: New York.

21. Howe, A., Pack, R.J. & Wise, J.C.M. (1981) Arterial chemoreceptor-like activity in the abdominal vagus of the rat. *Journal of Physiology*, *320*, 309-318.

22. Hughes, T. Portal blood supply to glomus tissue and its significance. *Nature*, *205*, 149-150.

23. Kjaergaard, J. (1973) Anatomy of the carotid glomus and carotid glomus-like bodies (non- chromaffin paraganglia). Copenhagen: FADL Forlag.

24. Krahl, V.E. (1960) The Glomus Pulmonale. *Bull. Sch. Med. Maryland*, *45*, 36-38.

25. Krahl, V.E. (1962) The glomus pulmonale: its location and microscopic anatomy, in: CIBA Found[n] Symp. on Pulmonary Structure and Function. *53* - 69.

26. Longhurst, J.C. (1991) Reflex effects from abdominal visceral afferents. In: Tucker, T.H. & Gilmore, J.P. (Eds.), Reflex control of the circulation, Telford Press, Caldwell, N.J. pp551-577.

27. Matsuura, S. (1973) Chemoreceptor properties of glomus tissue found in the carotid region of the cat. *Journal of Physiology*, *235*, 57-73.

28. Morgan, M., Pack, R.J. & Howe, A. (1976) Structure of cells and nerve endings in abdominal vagal paraganglia of the rat. *Cell Tiss. Res.*, *169*, 467-484.

29. Muratori, G. (1933) Contributo istologico allo studio dei riflessi aortici della carotide. Boll. Soc. ital. Biol. sper. *8*, 387-391.

30. Muratori, G. (1935) Connessioni tra tessuto paragangliare e zone recettrici aortiche in vari mammiferi. Monit. Zool. Ital. *45*, 300-310.

31. Orr, J.A., Fedde, R., Shams, H. Roskenbleck. H. & Scheid, P. (1988) Absence of $CO_2$-sensitive venous chemoreceptors in the cat. *Respir.Physiol.*, *73*, 211-224.

# 7

# POSTNATAL DEVELOPMENT OF HYPOGLOSSAL MOTONEURON INTRINSIC PROPERTIES

Albert J. Berger,[*] Douglas A. Bayliss, Mark C. Bellingham,
Masashi Umemiya, and Félix Viana

Department of Physiology and Biophysics
University of Washington
School of Medicine
Seattle, Washington 98195

Motoneurons constitute the final common pathway in the control of the musculature. Among motoneurons innervating muscles of the upper airways are the hypoglossal motoneurons (HMs), which are responsible for control of the tongue musculature, and which play an important role in regulating upper airway patency (Bartlett et al, 1990; Remmers et al, 1978; Wiegand et al, 1991). An understanding of the postnatal development of upper airway motoneuron properties thus may provide new insights into various pathologies including apnea of prematurity (Klesh et al, 1987), and Sudden Infant Death Syndrome (Willinger, 1989).

Although changes in intrinsic electrical properties of motoneurons (input resistance, rheobase, electrical coupling, etc.) as well as sub- and supra-threshold membrane potential responses have been described during postnatal development (Fulton and Walton, 1986; Haddad et al, 1990; Walton and Navarrete, 1991; Núñez-Abades et al, 1993), there is little information on how specific membrane currents change during postnatal development in mammalian motoneurons. The purpose of this chapter is to review our recent work on the postnatal development of HMs. We focus exclusively on data obtained from *in vitro* rat brain stem slice preparations. In these experiments two membrane currents important in the behavior of HMs were studied. The first is the low-voltage-activated (LVA) $Ca^{2+}$ current and the other is the mixed cationic inwardly rectifying current ($I_h$). Both of these currents are inward currents in the physiological range of membrane potentials, and therefore their activation causes membrane potential depolarization. There are important differences in

[*] Please send correspondence to: Albert J. Berger, Ph.D., Department of Physiology and Biophysics, SJ-40, University of Washington School of Medicine, Seattle, Washington 98195 U.S.A. *Telephone:* (206) 543-8196; *FAX:* (206) 685-0619; *E-Mail:* Berger@U.Washington.edu

*Control of the Cardiovascular and Respiratory Systems in Health and Disease*
Edited by C. T. Kappagoda and M. P. Kaufman, Plenum Press, New York, 1995

63

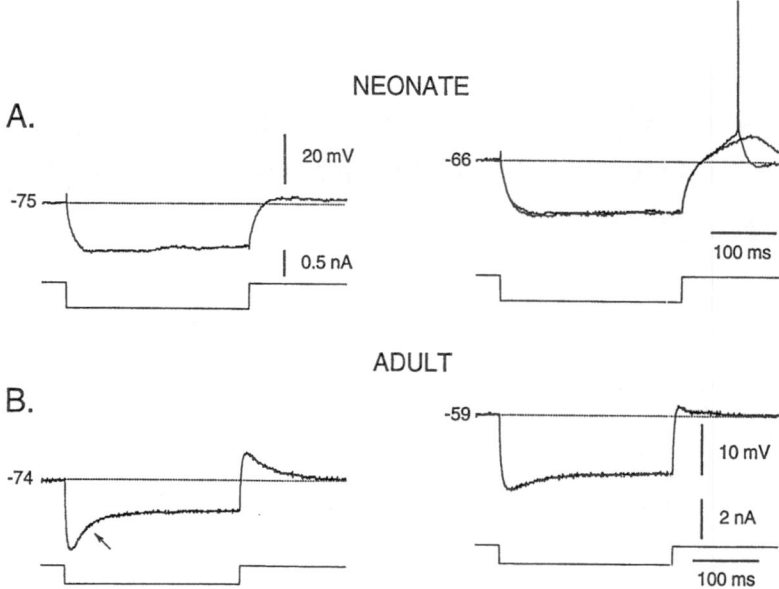

**Figure 1.** Sag and rebound depolarization are age- and voltage-dependent. (A) Voltage response in a neonatal HM at resting potential (right) and at a hyperpolarized potential (left, 0.2 nA negative current injection) in response to a constant 0.5 nA negative current step. Note the absence of sag at both potentials and the disappearance of rebound depolarization when the cell is held hyperpolarized (two-days-old rat). (B) Voltage response in an adult HM at resting potential (left) and at a depolarized potential (right +1.4 nA current injection). Note the degree of sag (arrow) correlated with the size of the rebound response in the adult HM. (From Viana et al, 1994).

these two currents; the LVA $Ca^{2+}$ current is a transient current activated by depolarization, while $I_h$ is a sustained current activated by hyperpolarization.

Figure 1 shows differences in the membrane potential responses of neonatal and adult HMs to injections of negative current pulses, the response are dependent upon the holding membrane potential of the cell. The neonatal HM (Fig. 1A) exhibited a passive exponential-type membrane potential response from both resting and hyperpolarized holding potentials. When the holding potential was sufficiently depolarized (Fig. 1A - right panel) the cell showed a depolarizing overshoot when the negative current pulse was turned off which on occasion led to a rebound spike. As described below, this broad depolarizing overshoot is at least in part due to activation of the LVA $Ca^{2+}$ current present in neonatal HMs. In contrast, adult HMs showed a depolarizing sag in the membrane potential response to negative current pulses (Fig. 1B, arrow). At the end of the negative current pulse there was a transient depolarization. As described below these two responses are due to activation and deactivation of $I_h$ current, respectively. This current is present in much higher density in adult than in neonate HMs (Bayliss et al, 1994).

## A. CHANGES IN CALCIUM CURRENTS WITH POSTNATAL DEVELOPMENT

Calcium currents are important in both the sub- and supra-threshold behavior of neurons. Voltage-activated neuronal $Ca^{2+}$ currents have been divided into two classes. These

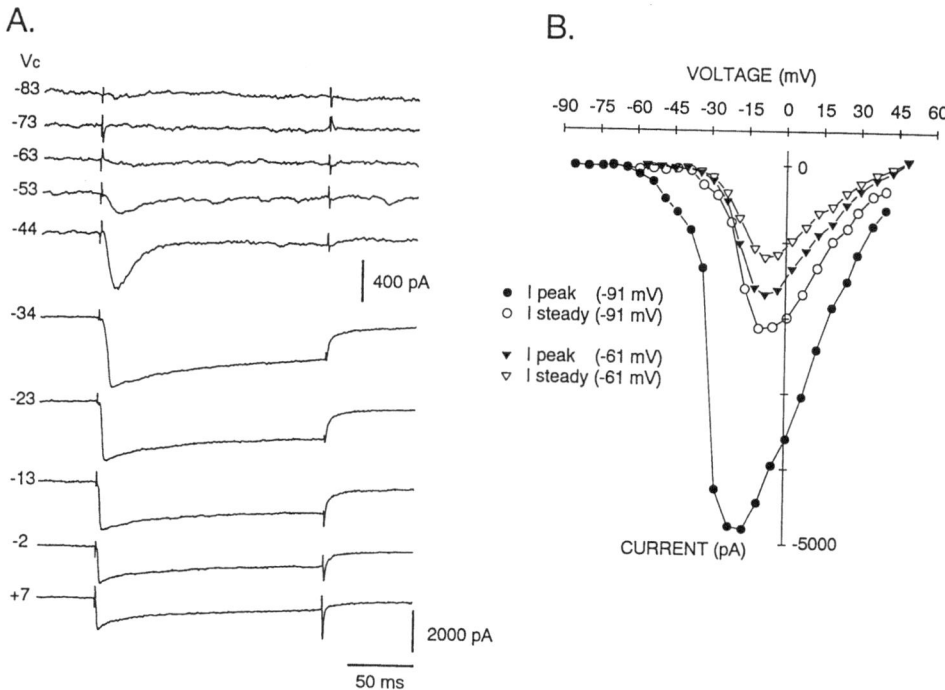

**Figure 2.** Low voltage-activated (LVA) and high voltage-activated (HVA) $Ca^{2+}$ currents in neonatal HMs. (A) Family of $Ca^{2+}$ currents evoked from a holding potential of -93 mV to the command potentials ($V_C$) noted in the left-hand side of the figure. Records have been corrected for linear leak and capacitive currents (three-days-old rat). (B) Peak and "steady-state" current-voltage relationships of a different HM obtained from two different holding potentials. Steady current is measured just before the end of a 200-ms pulse (two-days-old rat). (From Viana et al, 1993).

are LVA and high-voltage-activated (HVA) $Ca^{2+}$ currents (Tsien et al, 1988). The LVA $Ca^{2+}$ current is transiently activated when a cell is depolarized from relatively negative membrane potentials. Hyperpolarization is required to remove inactivation of the current. The LVA current is responsible, at least in part, for the transient depolarizing overshoot of the membrane potential that is observed in neonatal HMs (Fig. 1A - right). Therefore, the LVA $Ca^{2+}$ current can contribute to postinhibitory rebound. Other functions have been described for this current (see below). It has been proposed that the LVA $Ca^{2+}$ current has an important function in determining whether thalamic neurons and inferior olivary neurons fire self-sustaining oscillations (Llinas, 1988; McCormick and Pape, 1990).

Our interest in the LVA $Ca^{2+}$ current stems from earlier studies in which we showed this current to be robust in neonatal HMs (Berger and Viana, 1993; Viana et al, 1993). Figure 2, using whole cell recording of $Ca^{2+}$ currents in a neonatal HMs, shows the presence of the LVA $Ca^{2+}$ current for command voltages of -63 to -44 mV. Command voltages to more depolarized potentials activated a much larger current (note change in current scale); at these potentials the HVA $Ca^{2+}$ current was seen. The functional role of the LVA $Ca^{2+}$ current in neonatal HMs was apparent since from somewhat hyperpolarized membrane potentials (more negative than -70 mV) we found that upon depolarization a subset of HMs fired an initial burst of action potentials (Viana et al, 1993). Burst firing was defined as an initial doublet or triplet of closely spaced spikes that was followed by an augmented afterhyperpolarization. Burst firing was found in 46% of HMs (11 of 24 cells) at postnatal ages 0 to 4 days old, and this frequency declined to 30% (8 of 27 HMs) for postnatal ages 6 to 9 days old, and was absent in all HMs tested ( 8 of 8 HMs) at ages 10 days or older (Viana et al, 1993). The rebound depolarization that underlies the burst firing in the young HMs was $Ca^{2+}$ dependent since substituting the calcium channel blocker $Mn^{2+}$ for $Ca^{2+}$ in the slice-bathing medium strongly reduced the depolarizing response. Further, application of 250 μM of $Ni^{2+}$, which has been shown to block LVA $Ca^{2+}$ currents in other neurons (Ryu and Randic, 1990), caused a substantial reduction in the rebound depolarization that underlies the burst firing (Viana et al, 1993).

In another series of experiments we investigated the postnatal changes that occur in LVA $Ca^{2+}$ current density in HMs (Umemiya and Berger, 1994). In this case, thin slices of the medulla (approx. 130 μm thick) were used to obtain tight-seal whole-cell recordings with patch electrodes from visualized HMs. We used a pharmacological approach to isolate the LVA $Ca^{2+}$ current. To do this, we applied $Ca^{2+}$ channel blockers to block the three components of the HVA $Ca^{2+}$ current present in these cells; ω-CgTx was used to block N-type HVA $Ca^{2+}$ channels, ω-Aga-IVA to block P-type HVA $Ca^{2+}$ channels and nimodipine to block L-type HVA $Ca^{2+}$ channels. We assessed the $Ca^{2+}$ current amplitude in the presence and absence of these three blockers. The inward $Ca^{2+}$ current was measured by stepping the cell's membrane potential to 0 mV from a holding potential of -70 mV. $Ba^{2+}$ was used as a charge carrier in place of $Ca^{2+}$ to lessen $Ca^{2+}$ current run-down. The dominant component of this resistant $Ca^{2+}$ current in the presence of the three HVA channel blockers is the LVA $Ca^{2+}$ current. Based on the argument put forth above we would predict that the LVA $Ca^{2+}$ current should decrease with postnatal age. To test this hypothesis we compared the percentage of the total $Ca^{2+}$ current that was resistant to the three HVA channel blockers in 3 to 6 day old rat HMs. In order to normalize for cell size, we also estimated cell capacitance. As shown in Fig. 3

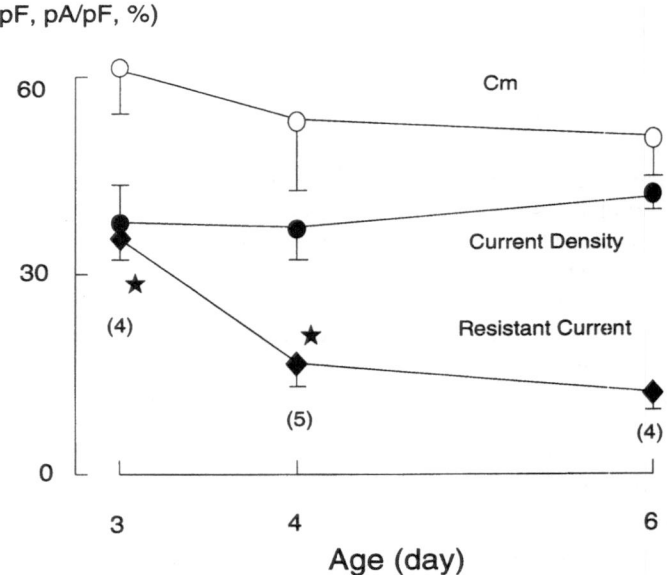

**Figure 3.** Developmental changes in membrane capacitance (Cm, open circles), total $Ca^{2+}$ current density (filled circles), and percentage of resistant $Ca^{2+}$ current (filled diamonds). Resistant current was measured after application of all three HVA $Ca^{2+}$ channel blockers. Numbers in parentheses are the number of HMs at each postnatal age. Stars indicate significant differences in the percentage of resistant current between 3- and 4-days-old rats. Currents were measured by stepping to a 0 mV test potential from a holding potential of -70 mV. Values and error bars are means ± SEs. (From Umemiya and Berger, 1994).

(open circles) we found that there was no significant change in capacitance over this time period for the population of HMs studied. These results indicate that HMs exhibit little change in size during this period, since capacitance is proportional to neuronal surface area. These results are consistent with other studies showing a little change in HM size in this postnatal period (Viana et al., 1994; Núñez-Abades and Cameron, in press). During this period we found that the total $Ca^{2+}$ current density, expressed as picoamps per picofarad, did not change, yet the percentage of resistant current (presumably reflecting the LVA $Ca^{2+}$) significantly decreased. At 3 days old it was on average 35.6%, falling to 16.5% at 4 days old and at 6 days old it was 9.5%. Thus the density of the LVA $Ca^{2+}$ current declines over this postnatal period. This change in the LVA $Ca^{2+}$ current density may be responsible for the very different firing behaviors exhibited by young compared with older HMs.

# B. CHANGES IN $I_h$ WITH POSTNATAL DEVELOPMENT

A striking feature of adult HMs is the large inward rectification that manifests itself as a depolarizing sag in the membrane potential response to hyperpolarizing current pulses (Figs. 1B and 6B). Although the characteristics of this response and the current that underlies it have been extensively described in the literature, until recently little was known about its postnatal development (McCormick and Prince, 1987). Some preliminary results indicated that in adult HMs a depolarizing sag in response to hyperpolarizing current pulse occurred, but this type of response was absent or nearly so in neonatal HMs (Haddad et al, 1990; Núñez-Abades et al, 1993; Viana et al, 1994). The purpose of our recent study was to use single-electrode voltage and current clamp recordings to investigate the postnatal changes in $I_h$ that occur and to correlate these with the amount of sag present in HMs (Bayliss et al, 1994).

These experiments were performed in thick (400-μm) transverse slices taken from the rat medulla. HMs were impaled with conventional intracellular microelectrodes and studied under current- and voltage-clamp recording conditions.

The amplitude of $I_h$ during the early postnatal period (<P9) was compared with the current in adult rat HMs. Figure 4 shows typical examples of the currents elicited by hyperpolarizing voltage steps applied to a typical neonatal (Fig 4A) and to an adult HM (Fig. 4B). It is apparent that the size of the time dependent current, that is the difference between the instantaneous and the steady state current, was much greater in adult than in neonatal HMs. These differences reflect much greater $I_h$ in adult than in neonate HMs. Figure 4C shows the average current-voltage relationship for $I_h$ in neonate (filled circles, n = 7 cells) and adult HMs (filled triangles, n = 20 cells). This figure clearly shows that $I_h$ is much larger in adult HMs. In fact at the half activation voltage of $I_h$ in adult HMs, -84 mV, the amplitude of the $I_h$ current in adult HMs was 10-times larger than that in neonatal Hms.

To compare the voltage dependency of the current we next scaled the current amplitude of neonatal HMs by a factor of 10. Figure 4D shows the results of such scaling and indicates that the adult and scaled neonatal current-voltage relationships were virtually identical. These results provide evidence that the smaller $I_h$ current in neonate HMs was not the result of a shift in the voltage dependency of this current. Additional experimental data revealed that the average reversal potential of $I_h$ in adult and neonatal HMs was not significantly different (-39 versus -31 mV, respectively). Thus we concluded that the approximate 10-fold increase in $I_h$ in adult HMs reflects an increase in $I_h$ current density.

The consequences of the increase in $I_h$ with postnatal development on the subthreshold membrane potential responses were also investigated. To do this we injected both positive and negative rectangular current pulses in neonate (Fig. 5A) and adult (Fig. 5B), and HM membrane potential responses were recorded. As evident from Fig. 5A and B the

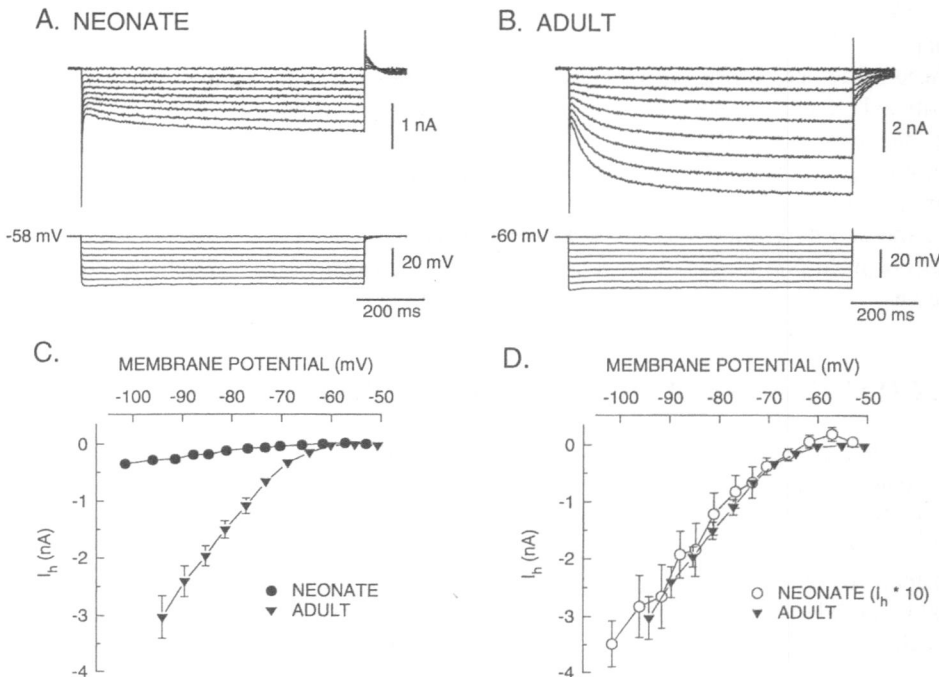

**Figure 4.** Postnatal change in $I_h$ amplitude of rat HMs. Currents (top traces) were evoked by hyperpolarizing voltage steps (bottom traces) in a neonatal HM (six-days-old rat in A) and in an adult HM (28-days-old rat in B). The instantaneous current jumps (reflecting input conductance) and the time-dependent current ($I_h$) were both substantially smaller in the neonatal HM (note the difference in current calibration). (C) Current-voltage relationships for $I_h$ were obtained from neonatal ($\leq$ eight-days-old, n = 7 HMs) and from adult HMs (>21-days-old, n = 20 HMs) and averaged. It is clear that the mean amplitude of $I_h$ in neonatal HMs (filled circles) is substantially smaller than that in adult HMs (filled triangles). (D) Mean current ($\pm$ SE) from neonatal HMs was scaled by a factor of 10 and the calculated values plotted as a function of membrane potential (open circles). For comparison, the unscaled values from adult HMs are also shown. Note that the shape of the relationships are nearly identical, suggesting that the voltage dependence of $I_h$ in neonatal and adult HMs was similar. (From Bayliss et al, 1994).

voltage response is markedly different between neonate and adult HMs. Specifically, in response to negative current pulses adult HMs exhibit a voltage response that has an early peak followed by a large depolarizing sag (arrows). At the end of the current pulse there is a depolarizing overshoot (arrowhead). The sag and overshoot are due to activation and deactivation of $I_h$, respectively. In adult HMs, in response to positive current pulses, the membrane potentials show hyperpolarizing relaxations which are due to deactivation of $I_h$. In contrast, in neonate HMs the sag and overshoot seen in the adults are virtually absent and only become apparent with pulses that result in large membrane potential changes. In the neonate, in response to large positive current pulses, a depolarizing transient was observed (Fig. 5A, asterisk). Since the voltage excursion needed to activate the depolarizing transient was quite large compared with that needed in the adult, the depolarization in the neonate probably reflects activation of the LVA $Ca^{2+}$ current (see above). Figure 5C shows the current-voltage relationships, and indicates that the input resistance measured at either the peak (filled circles and triangles) or the steady state (open circles and triangles), was much higher in neonatal HMs. The larger divergence of the peak and steady state I-V relationships

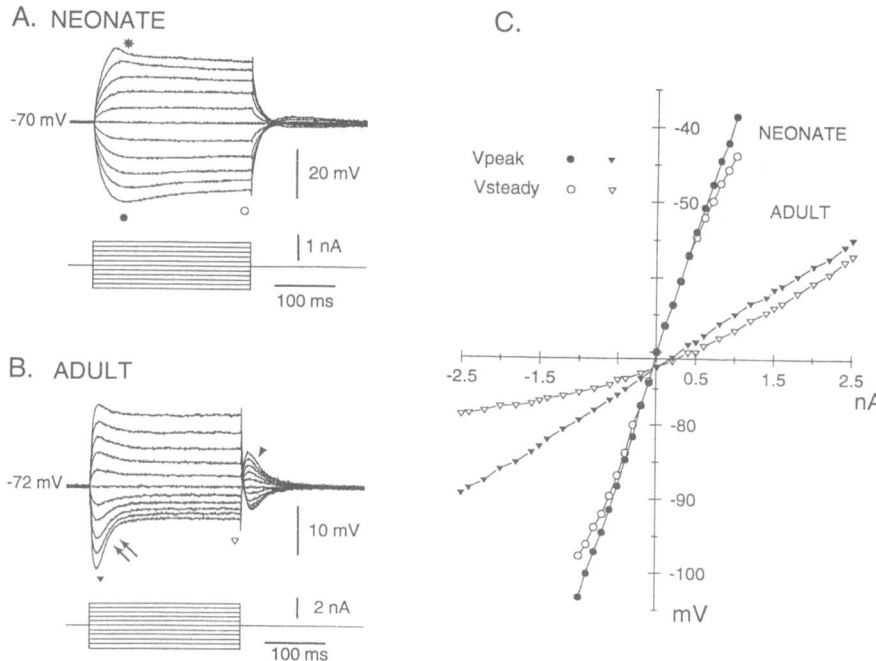

**Figure 5.** Postnatal changes in input resistance and sag in rat HMs. The voltage responses (top traces) to rectangular hyperpolarizing and depolarizing current pulses (bottom traces) were recorded in a neonate (two-days-old rat in A) and an adult HM (20-days-old rat in B). Note the difference in current and voltage calibrations. The depolarizing sag (arrows), which reflects activation of $I_h$, is much more pronounced in adult HM. The depolarizing rebound potential (arrowhead) is also more prominent in the adult HM. Also note hyperpolarizing relaxations during depolarizing current steps are evident even with the smallest steps in the adult HM but that the depolarizing transient in the neonatal HM (asterisk) only occurs at substantially more depolarized potentials. (C) Peak (filled circles and triangles) and steady-state (open circles and triangles) voltage were plotted as a function of injected current for neonatal (circles) and adult (triangles) HMs. The increased slope of the data from the neonatal HM reflects its higher input resistance (lower input conductance). (From Bayliss et al, 1994).

in the adult versus the neonate reflects the large sag in the adult. These data show that the postnatal developmental increase in $I_h$ causes a markedly different membrane potential response of adult compared with neonatal Hms.

## C. SUMMARY

This review has provided evidence that marked changes are occurring in ionic currents present in upper airway motoneurons during the early postnatal period. Our results have shown that the density of the LVA $Ca^{2+}$ current decreases during this period, and this probably reflects a reduced expression of the $Ca^{2+}$ channel responsible for this current, the so-called T-type channel. These results help to explain the changes in burst firing behavior of HMs during the early postnatal period. We have shown that the fraction of HMs exhibiting burst firing behavior was the greatest among HMs just at or after birth, and disappeared by 10 days of age (Viana et al , 1993). The LVA $Ca^{2+}$ current contributes to this firing behavior.

In contrast to the reduction in the LVA $Ca^{2+}$ current density with postnatal development, there is an apparent increase in $I_h$ current density during this period. The increase in $I_h$ provides a basis for a number of differences in the electrophysiological properties of adult versus neonate HMs. These include a striking depolarizing sag and overshoot during and immediately after application of hyperpolarizing current pulses in adult HMs. It is of interest that rebound depolarization following hyperpolarization can be observed in neonatal HMs even though there is little $I_h$ present. This response probably reflects the activation of a LVA $Ca^{2+}$ current.

Other differences in neonate versus adult HMs also are in part probably due to differences in $I_h$ current density. Since $I_h$ is active at normal resting membrane potential (approximately -70 mV), $I_h$ may contribute to the lower input resistance of adult compared with neonatal HMs (Haddad et al, 1990; Núñez-Abades et al, 1993; Viana et al, 1994), and the lower apparent membrane resistivity of older HMs (Viana et al 1994). The larger $I_h$ in the adult may be a factor in the shorter spike afterhyperpolarization observed in adult versus neonatal HMs (Viana, et al, 1994). This may be a consequence of the greater amount of $I_h$ activated during the afterhyperpolarization in adult HMs. The larger $I_h$ in adult HMs may also contribute to differences in how synaptic inputs are integrated. For example, inhibitory inputs which hyperpolarize the membrane potential may have their effect lessened due to $I_h$ activation with hyperpolarization. Thus in adult HMs $I_h$ may weaken prolonged or strong hyperpolarizations that occur in response to inhibitory synaptic inputs, while depolarizing responses arising from excitatory synaptic inputs may not be compromised. In contrast, neonatal HMs, which lack a substantial $I_h$ current, do not have the stabilizing influence upon membrane potential that is due to $I_h$. Therefore, these cells may be more susceptible to such inhibitions.

In conclusion, this chapter has described the changes that take place in two ionic currents during postnatal development, and how they contribute to distinct subthreshold and firing properties of neonatal and adult motoneurons.

## ACKNOWLEDGMENTS

This work was supported by a Javits Neuroscience Investigator Award NS-14857 and HL-49657 from the National Institutes of Health to A.J.B. D.A.B. was supported by a Parker B. Francis Fellowship.

## REFERENCES

Bartlett, D., Jr., Leiter, J.C. and Knuth, S.L. (1990). Control and actions of the genioglossus muscle. In: *Sleep and Respiration*, F.G. Issa, P.M. Suratt and J.E. Remmers, eds. New York: Wiley-Liss, Inc., pp. 99-108.

Bayliss, D.A., Viana, F., Bellingham, M.C. and Berger, A.J. (1994). Characteristics and postnatal development of a hyperpolarization-activated inward current ($I_h$) in rat hypoglossal motoneurons *in vitro*. *Journal of Neurophysiology* **71**, 119-128.

Berger, A.J. and Viana, F. (1993). Properties of brainstem neurons: calcium currents in hypoglossal motoneurons. In: *Respiratory Control: Central and Peripheral Mechanisms*, D.F. Speck, M.S. Dekin, W.R. Revelette and D.T. Frazier, eds. Lexington: University Press of Kentucky, pp. 6-11.

Fulton, B.P. and Walton, K. (1986). Electrophysiological properties of neonatal rat motoneurones studied *in vitro*. *Journal of Physiology (Lond.)* **370**, 651-678.

Haddad, G.G., Donnelly, D.F. and Getting, P.A. (1990). Biophysical properties of hypoglossal neurons in vitro: intracellular studies in adult and neonatal rats. *Journal of Applied Physiology* **69**, 1509-1517.

Klesh, K.W., Brozanski, B.S. and Guthrie, R.D. (1987). Apnea of prematurity: current theories of pathogenesis and treatment. In: *Neonatal Intensive Care*, R.D. Guthrie, ed. New York: Churchill Livingston, Inc., pp. 91-122.

Llinàs, R.R. (1988). The intrinsic electrophysiological properties of mammalian neurons: Insights into central nervous system function. *Science* **242**, 1654-1664.

McCormick, D.A. and Pape, H.-C. (1990). Properties of a hyperpolarization-activated cation current and its role in rhythmic oscillation in thalamic relay neurones. *Journal of Physiology (Lond.)* **431**, 291-318.

McCormick, D.A. and Prince, D.A. (1987). Postnatal development of electrophysiological properties of rat cerebral cortical pyramidal neurones. *Journal of Physiology (Lond.)* **393**, 743-762.

Núñez-Abades, P.A., Spielmann, J.M., Barrionuevo, G. and Cameron, W.E. (1993). In vitro electrophysiology of developing genioglossal motoneurons in the rat. *Journal of Neurophysiology* **70**, 1401-1411.

Núñez-Abades, P.A. and Cameron, W.E. (in press). Morphology of developing rat genioglossal motoneurons studied *in vitro*: relative changes in diameter and surface area of somata and dendrites. *Journal of Comparative Neurology.*

Remmers, J.E., de Groot, W.J. Sauerland, E.K. and Anch, A.M. (1978). Pathogenesis of upper airway occlusion during sleep. *Journal of Applied Physiology* **44**, 931-938.

Ryu, P.D. and Randic, M. (1990). Low- and high-voltage-activated calcium currents in rat spinal dorsal horn neurons. *Journal of Neurophysiology* **63**, 273-285.

Tsien, R.W., Lipscombe, D., Madison, D.V. Bley, K.R. and Fox, A.P. (1988) Multiple types of neuronal calcium channels and their selective modulation. *Trends in Neuroscience* **11**, 431-438.

Umemiya, M. and Berger, A.J. (1994). Properties and function of low- and high-voltage-activated $Ca^{2+}$ channels in hypoglossal motoneurons. *Journal of Neuroscience* **14**, 5652-5660.

Viana, F., Bayliss, D.A. and Berger, A.J. (1993). Calcium conductances and their role in the firing behavior of neonatal rat hypoglossal motoneurons. *Journal of Neurophysiology* **69**, 2137-2149.

Viana, F., Bayliss, D. A. and Berger, A. J. (1994). Postnatal changes in rat hypoglossal motoneuron membrane properties. *Neuroscience* **59**, 131-148.

Walton, K. and Navarrete, R. (1991). Postnatal changes in motoneurone electrotonic coupling studied in the *in vitro* rat lumbar spinal cord. *Journal of Physiology (Lond.)* **433**, 283-305.

Wiegand, L., Zwillich, C.W., Wiegand, D. and White, D.P. (1991). Changes in upper airway muscle activation and ventilation during phasic REM sleep in normal men. *Journal of Applied Physiology* **71**, 488-497.

Willinger, M. (1989). SIDS a challenge. *Journal of NIH Research* **1**, 73-80.

# 8

# CENTRAL NERVOUS MECHANISMS RESPONSIBLE FOR CARDIO-RESPIRATORY HOMEOSTASIS

K. M. Spyer[*]

Department of Physiology
The Royal Free Hospital School of Medicine
Rowland Hill Street
London NW3 2PF

There is a body of data that indicates that the generation of respiratory activity involves the interplay of a limited number of distinct classes of respiratory neurones in the ventrolateral medulla [1]. Similarly the cardiovascular system is regulated by the activity of premotor and preganglionic neurones localised in an equivalent region of the CNS [2]. Cardio-respiratory homeostasis depends on the integration of the activity of the two systems and it could be envisaged that this was achieved by the action of two independent control systems that are loosely dependent on one another. Studies over the last decade have drawn attention, however, to the fact that the two systems are tightly coupled being sensitive to the same reflex inputs and participating in the expression of behavioral activities in a totally co-ordinated manner [3]. This has led to a recognition that rather than operating through independent control systems the CNS acts to maintain cardio-respiratory homeostasis by means of a single group of 'cardio-respiratory' neurones located in the ventrolateral medulla which regulate the activity of the two systems. This is achieved by generating complementary patterns of discharge of bulbospinal neurones that control respiratory motoneuronal and sympathetic preganglionic neuronal firing and through brainstem interneurones influencing the activity of vagal motoneurones that supply the heart, lungs and airways (and subsidiary muscles of respiration). This notion is not novel being firmly based in the 19th century literature but the experimental justification is only now emerging [4]. This results from an extensive electrophysiological analysis of the properties of the various motoneurones and interneurones in these networks, the nature of the reflex inputs impinging on them and the action of regions of the forebrain in generating changes in their activity.

This report will endeavour to summarise some recent studies indicating that a major factor in this co-ordination process is the modulation of reflex inputs at the level of the nucleus tractus solitarii (NTS) [2]. This process involves the integration of reflex inputs,

[*] Telephone: 0171 431 0009; fax: 0171 433 1921: email: spyer@rfhsm.ac.uk

*Control of the Cardiovascular and Respiratory Systems in Health and Disease*
Edited by C. T. Kappagoda and M. P. Kaufman, Plenum Press, New York, 1995

their modulation by descending inputs from regions of the CNS concerned with cardio-respiratory control and the consequent patterning of the efferent projections from the NTS that impinge on neurones of the ventrolateral medulla and elsewhere.

In the context of the influences that pulmonary, cardiac, baroreceptor and chemoreceptor afferents exert on the cardiovascular and respiratory systems, the Coleridges [reviewed in 5, 6] (and others notably Michael Daly [7]) have provided some of the most important data. The Coleridges have described the detailed receptor properties of many of these afferents and together these contributions have been remarkable in demonstrating that receptors in the respiratory system affect cardiovascular activity, and conversely receptors localised in the cardiovascular system influence respiratory activity. However it is now apparent that many of the distinctive features of the responses to activating different classes of afferent can be attributed to their differential actions on respiration [5,7].

The central nervous processing of reflex inputs is as yet poorly understood other than a well established recognition that the nucleus tractus solitarii (NTS) is the primary site of interaction and control [2]. In regard of the interplay between respiratory and cardiovascular control most information in the literature concerns the regulation of the vagal supply to the heart [8] and the present report will concentrate on the role of the NTS and the synaptic mechanisms that determine the level of activity of the cardiac vagal preganglionic neurones (CVM).

## METHODS

The experiments that form the basis of this review have all been undertaken in the anaesthetized and paralysed cat. Full experimental details and protocols are provided elsewhere [9,10,11]. The studies have involved the use of intracellular and extracellular recording techniques in order to study the response characteristics of NTS (and CVM) on stimulating a range of afferent inputs and activation of the hypothalamic defence area (HDA).

## RESULTS

### Arterial Baroreceptor Input

There is an extensive literature that details the importance of the baroreceptor input in maintaining the level of activity of CVM and so determining the level of heart rate. Indeed this input provides the primary excitatory drive to CVM (see McAllen & Spyer, 1978 [12]) The effectiveness of this input is, however, powerfully modulated by respiratory activity. In particular there is a direct synaptic control exerted by the brainstem respiratory network that elicits a Cl- dependent hyperpolarisation of CVM during inspiration; they exhibit maximal activity during stage I of expiration (post-inspiration) with a variable degree of excitability being displayed during stage II expiration [13]. Further studies have indicated that lung stretch afferents also exert an inhibitory control of CVM discharge resulting in tachycardia as a consequence of a reduction in vagal efferent discharge frequency [7, 14]. A direct inhibitory action of lung inflation (or of vagal afferent stimulation) was not revealed, however, in an intracellular recording study [13], nor were NTS neurones that were excited by baroreceptor afferents, and are presumed to relay onto CVM, seen to be affected by lung inflation [9]. Accordingly the mode of action of these afferents in slowing the heart remains to be elucidated. As lung inflation inhibits inspiratory neuronal activity the main action may be via a disinhibition of CVM but an independent action is also indicated by other indirect observations (see [3] for discussion).

The activity of CVM is also markedly affected by stimulation in the HDA. Heart_rate rises during such stimulation as a result of sympatho-excitation and a fall in CVM discharge [2, 8]. The latter is produced in two ways. First, those NTS neurones that are excited by baroreceptor stimulation are powerfully inhibited on HDA stimulation. This is mediated by GABA acting at GABAa receptors located on these neurones. The source of the GABA input appears to be a group of GABA-containing neurones localised within the NTS that are

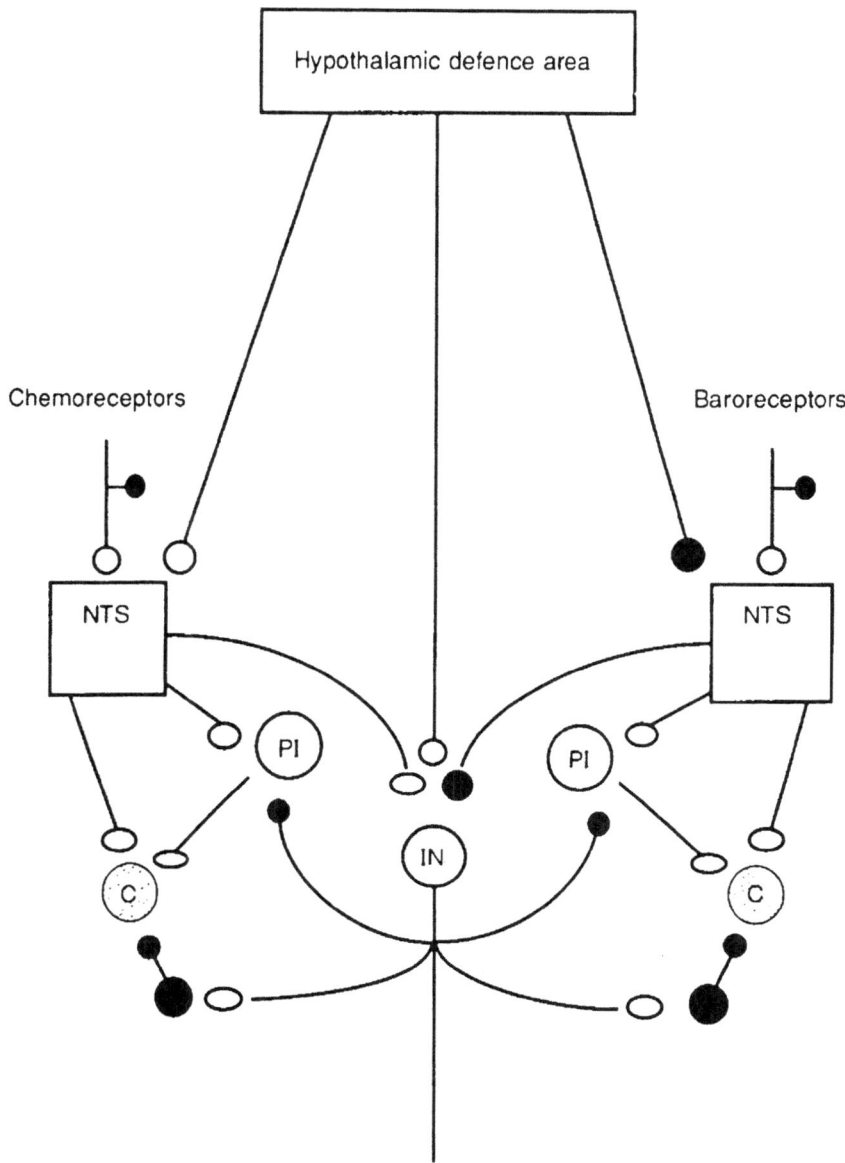

**Figure 1.** Schematic diagram of the pathways and synaptic interactions whereby the hypothalamic defence area affects the arterial chemoreceptor (left side) and the arterial baroreceptor (right side) reflex control of the vagal outflow to the heart. Excitatory connections, —O; inhibitory connections, —O; although these are not necessarily direct monosynaptic action. C, cardiac vagal motoneurones in the NA; IN, inspiratory neurones; NTS, nucleus tractus solitarii; PI, post-inspiratory (or stage 1 expiratory) neurones. (Reproduced from Spyer, K.M., 1994 with permission).

excited on HDA stimulation [2]. This effectively disfacilitates CVM but they also receive a direct GABA innervation which is activated on HDA stimulation. Secondly since HDA stimulation also induces an increase in inspiratory drive there will be an enhanced inhibitory input impinging on these neurones also. Together these two actions of HDA stimulation ensure that as minute volume increases, there is a parallel increase in cardiac output. Such a pattern of influences will follow any increase in inspiratory drive [see Figure 1].

## Arterial Chemoreceptor Input

Until recently relatively little was known of the characteristics of the NTS neurones that mediate the arterial chemoreceptor reflex.The consensus is now that the majority of those NTS neurones that are monosynaptically activated by SN and chemoreceptor inputs are localised close to the obex in the commissural and medial subnuclei of the NTS [15, 16]. These neurones do not show a respiratory rhythm in their discharge nor are they modulated by inputs associated with lung inflation (see also baroreceptor activated neurones as described above). They are also unaffected by baroreceptor stimulation although other NTS neurones that are excited on chemoreceptor stimulation are often inhibited by activating the baroreceptors. In some earlier extracellular recording experiments it was reported that a small

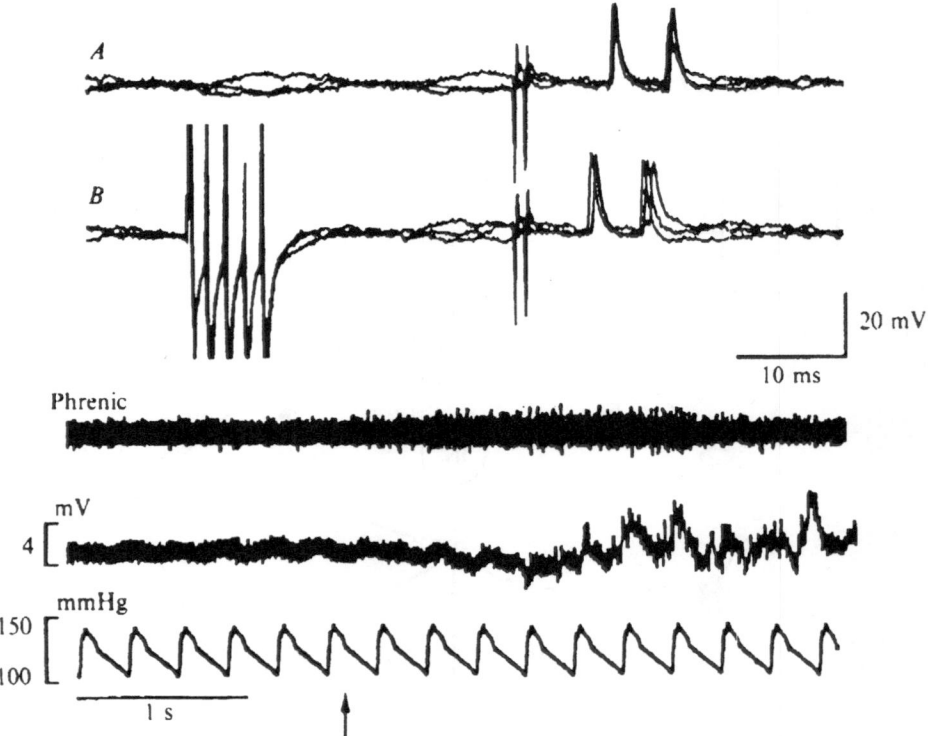

**Figure 2.** Response of an NTS neurone (membrane potential -40 mV) to stimulation of SN, HDA and arterial chemoreceptors. Upper records, responses to electrical stimulation of SN (two pulses separated by 1 ms) in the absence of (A) and after (B) a conditioning stimulus to the HDA (five pulses, 500 Hz) that precedes the SN stimulus by 30 ms. Lower traces show response to an intracarotid injection of $CO_2$-saturated saline (arrow). Traces from above: phrenic nerve activity, membrane potential of the neurone, arterial blood pressure. At this stage the neurone was not generating action potentials. (Reproduced from Silva-Carvalho, L., Dawid-Milner, M.S., Goldsmith, G.E. & Spyer, K.M. 1993, with permission).

proportion of neurones in the vicinity of the NTS showed excitatory responses to both baroreceptor and chemoreceptor stimulation [17]. Such neurones have not been identified in intracellular recordings.

Evidence has also accumulated of a facilitatory interaction between the chemoreceptor reflex and the defence response. Hilton & Joels (1965) [18] first showed in the cat that chemoreceptor stimulation could elicit a defence response, and studies in the rat and cat by Marshall [19] have sustained this observation. Similarly low intensity stimulation in the HDA can facilitate the effects of chemoreceptor activation. This effect has been shown to involve both a convergence of excitatory input at the level of the NTS and a true facilitatory action [11] whereby the former neurones feed an excitatory input onto neurones that are themselves unaffected by either input when delivered separately (see Figure 2). Some of the neurones receiving epsps on stimulation of both HDA and SN (and chemoreceptors) were also inhibited on specific baroreceptor activation.

## Laryngeal Afferent Inputs

Natural stimulation of laryngeal mechanoreceptors and electrical stimulation of the SLN lead to a triad of cardio-respiratory responses [7]. These are a bradycardia (both through vagal activation and sympatho-inhibition), a fall in blood pressure and apnoea. These responses are suppressed when there is a simultaneous activation of the HDA [20]. When the two stimuli are adjusted so that when the HDA is stimulated at an intensity just above threshold and the SLN is stimulated submaximally the effects of the HDA stimulus suppress the action of SLN stimulation. As such this action of the HDA resembles its influence on the baroreceptor input (see above). Since there is evidence that many NTS neurones that are excited by SN and baroreceptor stimulation are also excited by SLN stimulation [9,10,20] the possibility that the action of the HDA might be mediated by a GABAergic mechanism involving a common pool of NTS neurones was investigated in a recent series of investigations.

In these experiments neurones that showed evidence of a convergence of SN and SLN inputs were tested for their responses to a range of natural stimuli - arterial baroreceptor and chemoreceptor, laryngeal mechanoreceptor and chemosensory endings. There was a significant population of NTS neurone that received excitatory inputs from both arterial baroreceptors and laryngeal mechanoreceptors [20]. These neurones were also inhibited on HDA stimulation, the ipsps having the same characteristics as those that have been shown previously to be mediated by GABA acting at GABAa receptors in the case of baroreceptor activated neurones. Further indications were obtained that some of those NTS neurones that were excited by the arterial chemoreceptors were often excited by activation of chemosensitive inputs from the larynx.

## DISCUSSION

These data indicate that the NTS plays an essential role in cardio-respiratory integration. The patterns of convergence of different classes of afferent appear to have clear functional consequences. Neurones within the NTS do not appear to be coded for individual reflex inputs in general - some second order neurones may display only limited indications of convergence and in consequence it is likely that the code is determined by the efferent connections of each class of functionally defined NTS neurone [2].

We have obtained evidence for the presence of NTS neurones that are excited by arterial chemoreceptor activation and are inhibited by baroreceptor inputs. These would be expected to provide an excitatory input to sympathetic premotor neurones of the ventrolateral medulla as well as to inspiratory neurones. Notably they were also shown to be excited on HDA stimulation. They would not however be expected to affect CVM since these are excited by both baroreceptor and chemoreceptor inputs [2]. Conversely those NTS neurones that were excited by both baroreceptors and laryngeal mechanoreceptors were inhibited by HDA stimulation. These would contribute to the excitatory control of CVM and the inhibitory control of sympathetic activity. Such neurones were never seen to excited by chemoreceptor inputs which is consistent with the sympatho-excitatory role of the arterial chemoreceptors, and so may also contribute an inhibitory influence on inspiratory activity whilst exciting post-inspiration.

The apparent segregation of baroreceptor and chemoreceptor evoked effects within the NTS contrasts with other evidence in the literature but is consistent with their reciprocal actions on efferent sympathetic and inspiratory control. Their synergistic actions on CVM is probably achieved through independent pathways or by means of a class of NTS neurone that is concerned exclusively with the control of CVM activity or post-inspiration, or both.

These observations provide provocative insights into the integration of the reflex control of circulation and ventilation. They further imply that the pattern of afferent input emerging from the peripheral receptors in the cardiovascular and respiratory systems that have been so exquisitely described by John and Hazel Coleridge may have considerable importance in determining the pattern of activity in central circuits, and particularly those originating in the NTS, that are sensitive not only to these varied inputs but also to descending inputs derived from areas of the CNS that are concerned with affective behaviour and perhaps other functions. The implications for behavioral influences on cardio-respiratory homeostasis remain to be fully resolved.

## ACKNOWLEDGMENTS

The original studies in this review were funded by grants from the British Heart Foundation and Wellcome Trust. These studies were performed in collaboration with L. Silva-Carvalho and S. Dawid-Milner whose contributions are gratefully acknowledged.

## REFERENCES

1. Richter, D.W. (1982). Generation and maintenance of the respiratory rhythm. *Journal of Experimental Biology*, **100**, 93-107.
2. Spyer, K.M. (1994). Central nervous mechanisms contributing to cardiovascular control. *Journal of Physiology*, **474.1**, 1-19.
3. Richter, D.W. & Spyer, K.M. (1990). Cardiorespiratory control. In: *Central Regulation of Autonomic Functions*, ed, Loewy, A.D. & Spyer, K.M. pp 180-207, Oxford University Press, New York.
4. Richter, D.W., Spyer, K.M., Gilbey, M.P., Lawson, E.E. Bainton, C.R. & Wilhelm, Z. (1991). On the existence of a common cardiorespiratory network. In *Cardiorespiratory and Motor Co-ordination,* ed, Koepchen, H.-P. & Houpaniemi, T. pp. 118-139. Springer-Verlag, Berlin.
5. Coleridge H.M., Coleridge J.C.G. & Jordan, D. (1991). Integration of ventilatory and cardiovascular control systems. In *The Lung: Scientific Foundations,* ed. R.G. Crystal, J.B. West, P.J. Barnes, N.S. Cherniack, E. R. Weibel. **2**, 1405-18. New York, Raven.
6. Coleridge, H.M and Coleridge, J.C.G. (1994) Pulmonary Reflexes: Neural Mechanisms of Pulmonary Defense. *Annual Review of Physiology*, **56**,69-91
7. Daly, M de B.(1985). Interactions between respiration and circulation. In *Handbook of Physiology*, section II, *The Respiratory System*, pp 529-594. American Physiological Society, Bethesda, MD, USA.
8. Spyer, K.M. (1990) The central nervous organisation of reflex circulatory control. *Central Regulation of Autonomic Functions*, ed. Loewy, A.D. & Spyer, K.M. pp 168-188. Oxford University Press, New York.

9. Mifflin, S.W., Spyer, K.M. & Withington-Wray, D.J. (1988a) Baroreceptor inputs to the nucleus tractus solitarius in the car: postsynaptic actions and the influence of respiration. *Journal of Physiology,* 399,349-367.

10. Mifflin, S.W., Spyer, K.M. & Withington-Wray, D.J. (1988b). Baroreceptor inputs to the nucleus tractus solitarius in the cat; modulation by the hypothalamus. *Journal of Physiology,* **399**, 369-387.

11. Silva-Carvalho, L., Dawid-Milner, M.S., Goldsmith, G.E. & Spyer, K.M. (1993). Hypothalamic-evoked effects in cat nucleus tractus solitarius facilitating chemoreceptor reflexes. *Experimental Physiology,* **78**, 425-428.

12. McAllen, R.M. & Spyer, K.M. (1978b) The baroreceptor input to cardiac vagal motoneurones. *Journal of Physiology,* **282**, 365-374.

13. Gilbey, M.P., Jordan, D., Richter, D.W. & Spyer, K.M. (1984). Synaptic mechanisms involved in the inspiratory modulation of vagal cardio-inhibitory neurones in the cat. *Journal of Physiology,* **356**, 65-78

14. Davidson, N.S., Goldner, S. & McClosky, D.I. (1976). Respiratory modulation of baroreceptor and chemoreceptor reflexes affecting heart rate and cardiac vagal efferent nerve activity. *Journal of Physiology,* **259**, 523-530.

15. Mifflin, S.W. (1992). The arterial chemoreceptor input to nucleus tractus solitarius. *American Journal of Physiology,* **263**, R368-375.

16. Spyer, K.M., Izzo, P.N., Lin, R.J., Paton, J.F.R., Silva-Carvalho, L. F. & Richter, D.W. (1990). ~The central nervous organisation of the carotid body chemoreceptor reflex. In *Chemoreceptors and Chemoreceptor Reflexes,* ed. Acker, H., Trzebski, A & O'Regan, R.G., pp 317-321. Plenum Press, New York.

17. Lipski, J., McAllen, R.M.& Trzebski, A. (1976). Carotid baroreceptor and chemoreceptor inputs onto single medullary neurons. *Brain Research,* **107**, 132-136.

18. Hilton, S.M. & Joels, N. (1965). Facilitation of chemoreceptor reflexes during the defence reaction. *Journal of Physiology,* **176**, 20-22P

19. Marshall, J. (1987). Contribution to overall cardiovascular control made by the chemoreceptor-induced alerting defence response. In: *Neurobiology of the Cardiorespiratory System.* Edited E.W. Taylor, Published Manchester University Press, pp 221-249

20. Dawid-Milner, M.S., Silva-Carvalho, L., Goldsmith, G.E. & Spyer, K.M. (1994). Hypothalamic modulation of laryngeal reflexes in the anaesthetized cat. *Journal of Physiology,* **475.P,** 112P.

# 9

# NEUROENDOCRINE REGULATION OF VASCULAR CAPACITANCE

F. Karim

Department of Physiology
Leeds University
The Worsley Medical and Dental Building
Leeds, LS2 9NQ, United Kingdom

## INTRODUCTION

Vascular capacitance of an organ or region is defined as the volume of blood contained in it at a particular distending pressure. Since the veins contain about 70% of the total blood volume, these are called the capacitance vessels (Green, 1950). A change in the diameter of the veins can cause a significant change in venous return and thus cardiac filling and cardiac output. The walls of the veins, contain a contractile (smooth muscle) and a viscoelastic element, and are very thin. Therefore, even a slight change in the transmural pressure can cause a great change in their diameters. Although the dominant influence on the veins and venous return of mechanical factors such as respiration, muscular exercise and posture has been well recognised for a very long time (see Brecher 1956, Karim 1987), the neurohumoral regulation of the venomotor tone and vascular capacitance has been the subject of great interest for the past thirty years (Shepherd & Vanhoutte, 1975, Rothe, 1983, Hainsworth, 1986). Generally, the veins (particularly those in the splanchnic bed) have rich sympathetic innervation, and numerous investigations have provided evidence that the individual veins or selected venous beds are under nervous control and react to the vasomotor centre activity in the same manner as do the arteries (see Folkow & Mellander 1964). Several investigations have clearly demonstrated that the capacitance vessels are more sensitive to changes in the sympathetic nerve activity than are the resistance vessels both in the hind limb of anaesthetised cats (Mellander, 1960) and dogs (Hainsworth et al., 1975, Karim et al., 1980) and in the abdomen of anaesthetized dogs (Karim & Hainsworth, 1976). The regulation of the capacitance vessels has recently been extensively reviewd (Rothe, 1983, Hainsworth, 1986). The purpose of this short chapter is to briefly evaluate the current evidence for the neuroendocrine regulation of vascular capacitance in different regions of the body and the application of the knowledge in normal and abnormal conditions. However, before these are discussed it is necessary to give a brief account of the methods for assessing the functions of vascular capacitance and capcitance vessels.

*Control of the Cardiovascular and Respiratory Systems in Health and Disease*
Edited by C. T. Kappagoda and M. P. Kaufman, Plenum Press, New York, 1995

# METHODS OF STUDY

One of the main difficulties to assess the venous or capacitance responses is to find a suitable method. Changes in venous tone or vascular capacitance in an area of the circulation can be determined either by (1) recording changes in distending pressure at a constant volume (e.g. Karim & Ali, 1969) or (2) by recording changes in volume at a constant distending pressure . The latter technique is superior for several counts as explained below.

In order to make an accurate and quantitative assessment of resistance and capacitance responses in an area, three essential criteria must be satisfied:

1. Firstly, the area must be vascularly and if possible humorally isolated in order to avoid changes in pressure and blood volume occuring through collaterals from changes in another part of the circulation.
2. Secondly, the area must be perfused at a constant flow. The advantages of a constant flow perfusion are: (a) any change in perfusion pressure will then give a quantitative change in vascular resistance, (b) since the flow is constant any change in concentrations of vasoactive local metabolites will be minimised, (c) the constant flow prevents elastic recoil and passive distension, and (d) it also prevents absorption and filtration of tissue fluid.
3. Thirdly, the venous outflow pressure must be held constant to prevent changes in venous diameter and the volume passively. Obviously, the above factors if not attended, can add considerable errors in the assessment of capacitance responses from active constriction of the veins due primarily to neurohumoral factors.

The above criteria were used in anaesthetised and artificially ventilated dogs to determine effects on abdominal and hind limb vascular capacitance of direct stimulation of sympathetic nerves and alteration of sympathetic activity from baroreceptor and chemoreceptor reflexes (Karim & Hainsworth, 1976, Hainsworth & Karim 1976, Hainsworth, Karim, McGregor & Wood, (1983a.c). The similar principle of recording volume changes at a constant pressure has recently been used to determine effects of cardiovascular reflexes on the total body capacitance (Karim & Majid, 1991, Karim & Al-Obaid, 1993). This method of measurement of volume shift into an extracorporial reservoir has been used by many investigators for studying effects of drugs and cardiovascular reflexes (e.g Ross et al 1961; Braunwald et al 1963; Rothe, 1976). However the volume shift method, though surgically less traumatic, is the result of both active venoconstriction and the passive fluid shift (Rothe, 1976).

# ABDOMINAL VASCULAR CAPACITANCE MAKES A
# SIGNIFICANT CONTRIBUTION TO BARORECEPTOR REFLEX

It has been well established that the abdominal vascular bed acts as a potential blood reservoir and can play an important role by mobilising its content for circulatory homeostasis under various conditions such as changes of posture and haemorrhages (Rothe, 1976, Hainsworth & Karim 1976, Hoka, Bosnjack, Siker, Luo & Kampine 1988). The following investigation would provide a clear evidence for a vital role of this circulatory bed.

The abdominal circulation was vascularly isolated without opening the abdominal cavity (see Fig. 1 and Karim & Hainsworth, 1976 for details). The blood flow through the spinal canal was prevented by packing surgical gauze into the canal. The hind limbs were excluded from the abdominal circulation by strong ligatures at the upper ends of the limbs. Blood was pumped at a constant flow from the proximal to the distal aorta to perfuse the

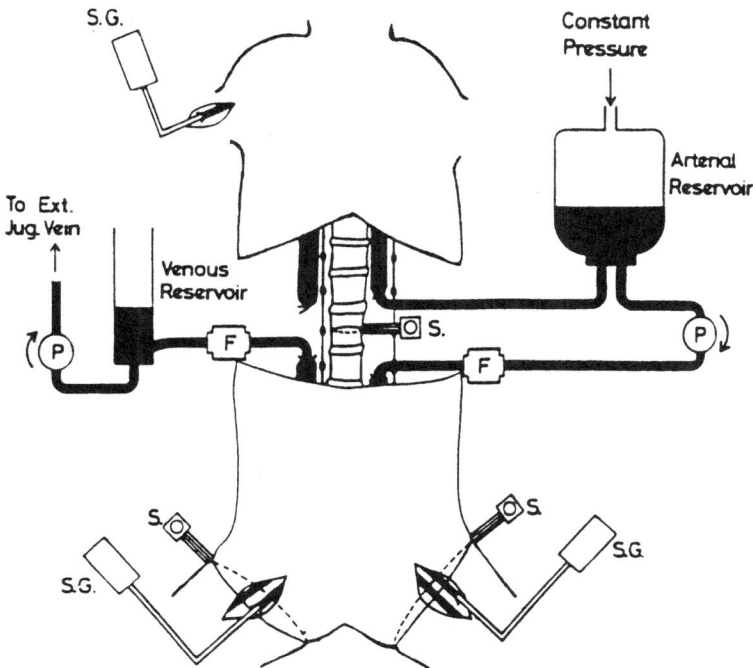

**Figure 1.** Diagram of abdominal preparation. Blood from descending thoracic aorta collected and pumped at a constant flow into the distal end of the aorta above the diaphragm. Blood from inferior vena cava collected in a reservoir and returned to external jugular vein. SG, strain gauge; P, roller pump; F, electromagnetic-flow-meter cannulating probe; S, snare-nylon cord tightened with ratchet (modified from Karim & Hainsworth, 1976).

abdominal circulation (Fig. 1.) The perfusion pressure was recorded from a cannula passed through a femoral artery so that the tip of the cannula lay at the upper part of the abdominal aorta. The blood draining the area was collected into a reservoir through a wide cannula inserted into the inferior vena cava above the diaphragm. This blood was then pumped into the upper part of the circulation through an external jugular vein at a rate to keep the inferior vena caval (IVC) pressure constant. The IVC pressure was measured by a cannula passed through a femoral vein so that the tip of the cannula lay at the level of the diaphragm. The inflow and outflow were measured by using two electromagnetic flow meters. The sympathetic trunks (splanchnic nerves) were crushed or tied (see Fig. 2) and stimulated electrically (10-15 V, 2 ms duration and different frequencies). Direct stimulation of sympathetic nerves at 1 to 20 Hz or lowering pressures in vascularly isolated carotid sinuses from about 250 down to about 60 mm Hg, resulted in a large transient increase in outflow within lower range and a sustained increase in the perfusion pressure within higher range of sympathetic efferent nerve activity. As the inflow and inferior vena caval pressure were kept constant the transient increase in outflow indicated a reduction in abdominal vascular capacitance, the volume expelled was calculated by integrating the outflow change with time using a planimeter (See Fig. 3).

Stimulation of both spanchnic nerves at 20 Hz increased vascular resistance by 135% and reduced capacitance by about 150 ml (7.20 ml Kg$^{-1}$). The unloading of carotid barorecep-tors by lowering carotid pressure from 250 to 60 mm Hg caused average capacitance and resistance responses of about 115 ml and 70% respectively (Fig. 5). Shoukas & Sagawa (1973) observed in dogs a change of total capacitance of 7.5 ml Kg$^{-1}$, when pressure in the isolated

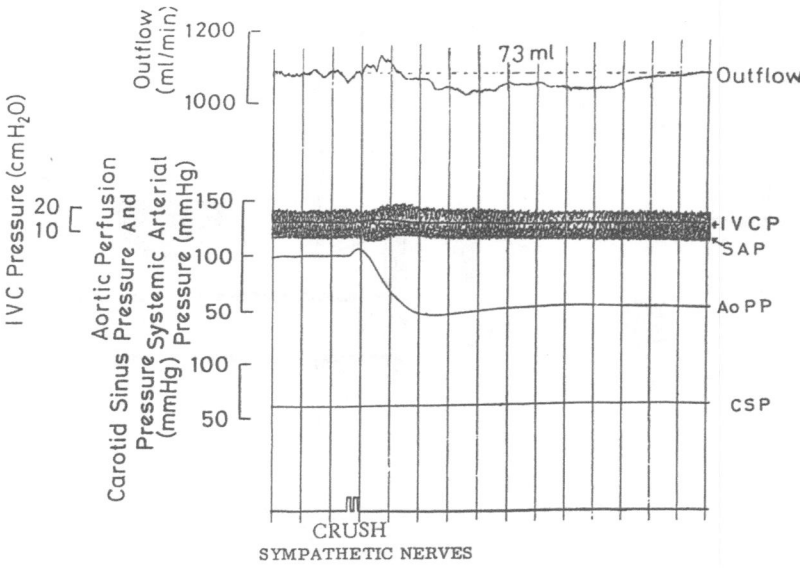

**Figure 2.** An example of the effect of crushing both sympathetic trunks above the diaphragm on abdominal vascular capacitance and perfusion pressure at a constant inflow and at a constant outflow pressure in anaesthetized and artificially ventilated dogs (abdomen was vascularly isolated as described previously by Karim and Hainsworth, 1976, also see Fig. 1.). Note that crushing (tying) both sympathetic trunk was followed by a reduction of outflow by 73 ml indicating a retention of this amount of blood in the abdominal vascular bed since inflow was held constant by a roller pump. There was a concomitant reduction of aortic perfusion pressure (AOPP) suggesting a reduction of abdominal vascular resistance by that amount. The isolated carotid sinus pressure was held constantly low at about 65mmHg. The results indicates that abdominal capacitance element has ongoing sympathetic activity to keep the capacitance smaller, particularly at a lower than normal carotid sinus pressure. Crushing the limb sympathetic nerve in an identical hind limb preparation does not produce any change in capacitance or venous tone. However, hind limb vascular resistance decreases. From Karim & Hainsworth (unpublished).

carotid sinuses changed from 75 to 200 mmHg. However, in their more recent study (Shoukas & Brunner, 1980), they reported a significantly greater response (12.35 ml Kg$^{-1}$ obtained by changing carotid pressure from 50 - 200 mmHg, which is similar with our result (13.7 ml Kg$^{-1}$) obtained by changing carotid pressure from 70 - 200 mmHg by using a simpler canine preparation (Karim & Al-Obaidi, 1993, also see Fig 6). Fig 4A shows the results of stimulation of peripheral ends of both sympathetic trunks with different frequencies. Fig 4B shows the effects of lowering the pressure from vascularly isolated carotid sinuses in vagotomised and paralysed dogs. It is interesting to observe that both direct and reflex activation of the sympathetic nerves produced a greater effect on the vascular capacitance than on the vascular resistance at lower frequencies. In 13 dogs comparision of the reflex reponses with that from direct stimulation of the sympathetic nerves with 10 V., 2 ms and at different frequencies showed that the reflex capacitance responses to lowering of pressure in carotid sinuses over the whole baroreceptor sensitivity range were similar to that obtained from direct stimulation of the sympathetic nerves at 2 Hz, whereas it was necessary to stimulate the nerves at 5-10 Hz to obtain the resistance responses similar to that occurred reflexly. In 11 dogs when the carotid sinus pressure (CSP) was decreased in small steps the vascular capacitance response appeared first without any change in the vascular resistance. On the other hand at the lower end of CSP a step decrease in CSP resulted in a large change in the vascular resistance without any significant change in the vascular capacitance (Fig 4B). When the total changes in the

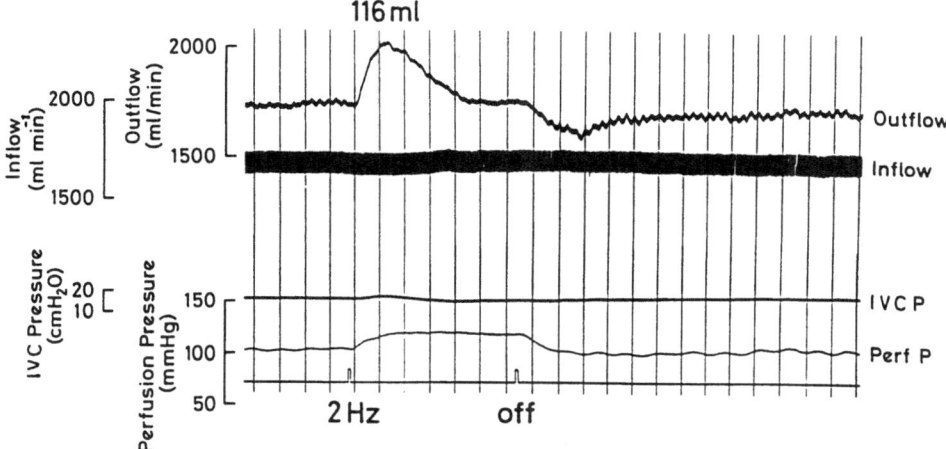

**Figure 3.** Responses to bilateral stimulation of splanchnic nerves with 10 V, 2 ms and 2 Hz in a dog weighing 27 Kg. Records of inferior vena caval outflow and aortic inflow, inferior vana caval pressure (IVCP), abdominal aortic perfusion pressure (Perf P), and datum line. Since inflow and inferior vena caval pressure were held constant, changes in outflow were due to changes in vascular capacitance and changes in perfusion pressure were due to changes in vascular resistance. The volume (116 ml) expelled from the abdominal vascular bed was calculated by integrating changes in outflow with respect to time. Modified from Karim & Hainsworth, 1976.

capacitance and perfusion pressure were expressed as 100% and other changes at each step change in CSP as per cent of the maximum responses the capacitance responses occurred at a significantly higher CSP than for resistance responses (Fig 4B).

The impulse frequency going to the capacitance and resistance vessels at a particular CSP was determined as described by Hainsworth and Karim (1976). The results show that at each step change in carotid sinus pressure the stimulus frequency required to induce the resistance response was the same as that required to induce the capacitance response. This indicated that the veins or capacitance vessels of the abdomen responded near maximally at a low frequency of stimulation. This was consistent with our earlier study in which near maximal capacitance responses to direct electrical stimulation of the sympathetic trunk at a stimulation frequency of 2 Hz were seen (Fig 4A). It was therefore clear that capacitance changes occur at significantly higher carotid sinus pressures than that for resistance responses and this was due to the fact that capacitance elements responded near maximally at low frequency of stimulation.

Similar capacitance and resistance responses from changing pressures to aortic arch baroreceptors were seen at low and normal carotid pressure in 15 anaesthetised dogs. However, the effects at a high carotid sinus pressure i.e. high carotid baroreceptor inputs, could be greatly reduced showing an interaction between aortic and carotid baroreceptors in these responses, (Karim et al., 1978). All the reflex responses were abolished when the sympathetic nerves to the abdominal circulation were sectioned or when the vagus nerves were cooled to 5 °C.

## HIND LIMB VASCULAR CAPACITANCE DOES NOT MAKE A SIGNIFICANT CONTRIBUTION TO BAROREFLEX

The hind limb was vascularly isolated by tying the muscles of the upper end of the limb and the responses of the whole hind limb vascular capacitance and resistance were studied using the method similar to that as described for the study of the abdominal circulation (Karim & Hainsworth, 1976, Hainsworth et al., 1983a,b). During constant flow perfusion, both changes in pressures in the vascularly isolated carotid sinuses and direct stimulation of efferent sympathetic nerves resulted in large changes in vascular resistance without significant changes in vascular capacitance (Fig 5). During constant-pressure perfusion, both reflex and direct stimulation resulted in not only significant changes in resistance but also in capacitance which were much larger than those obtained during constant-flow perfusion. Similar changes in volume were obtained when the flow rate was manually changed by altering the pump speed suggesting that the effect on the volume change was secondary to flow change and not related to the change in the sympathetic activity (Hainsworth et al. 1983c.) Our finding was similar to that reported earlier (Lesh & Rothe, 1969).

Although unloading of carotid and aortic baroreceptors or direct sympathetic stimulation did not cause any significant change in the limb vascular capacitance (blood volume shift), the tone of the superficial veins showed a large change in response to *direct* stimulation of the lumbar sympathetic trunks (Hainsworth et al. 1975; Karim et al., 1980) and a small change in response to unloading of carotid and aortic baroreceptors even when the sensitivity of the veins to adrenergic stimulation was increased by cooling the perfusate to 31°C. Browse et al (1966) also observed a lacks significant role of the limb veins in the baroreflex. Therefore the majority of the sympathetic neurones supplying the superficial veins do not seem to be connected to the baroreceptor afferents and lack basal tone at normal temperature. Since the muscle veins have very scanty sympathetic nerve supply (Fuxe & Sedvall, 1965; Chen et al., 1993) the neural regulation of these veins does not seem to be very important.

# ADRENOCEPTORS ARE INVOLVED IN THE VASCULAR CAPACITANCE RESPONSES TO UNLOADING OF BARORECEPTORS

The role of beta adrenergic receptors in vascular capacitance responses is still unknown (Imai et al 1978, Rutlen et al 1980). Recently, a simple method of measurement of changes in total body vascular capacitance from bilateral carotid occlusion was made in anaesthetised pigs in order to determine the receptor mechanism for vascular capacitance response at a constant systemic pressure (see Fig 6 for details). Bilateral carotid occlusion always resulted in a shift of blood volume into a graduated extra-corporeal reservoir, about 80 ml from a 25 kg pig, (Fig. 7). The syetemic pressure

---

Figure 4. (A) Abdominal vascular capacitance and resistance responses to stimulation of splanchnic nerves at supramaximal intensity and different frequencies. In each dog changes in capacitance and resistance produced by stimulation at 20 Hz were expressed as 100% and responses at other frequencies were calculated as percentages of these responses (at 20 Hz). Values shown are means ± 1 SE. Probabilities calculated for paired comparisons: * at both 1 and 2 Hz, P<001. Note that at 2 Hz stimulation, the response of the capacitance vessels was abut 70% of the maximal, whereas that of the resistance vessels was only abut 25% of the maximal. (From Karim & Hainsworth, 1976). (B) Responses of abdominal resistance (open circles) and capacitance (closed circles) vessels in anasthetised dogs, to changes in pressure in the isolated carotid sinuses, expressed as percentages of the maximal responses induced when carotid sinus pressure was decreased to 60 mmHg . Note that at the same carotid sinus pressures capacitance responses were significantly greater than the corresponding resistance respinses (*P<0.005, **, P<0.01; ***, P<0.005: paired t tests). (Modified from Hainsworth & Karim, 1976).

was held constant and the right atrial pressure did not change. Since a large proportion of this volume shift occurred within a minute of occlusion a considerable part of this volume shift was likely due to a reduction of the total vascular capacitance from reflex vasoconstriction and the remaining part is due to passive elastic recoil of the venous wall and fluid absorption at capillaries (Mellander, 1960; Brooksby & Donald,

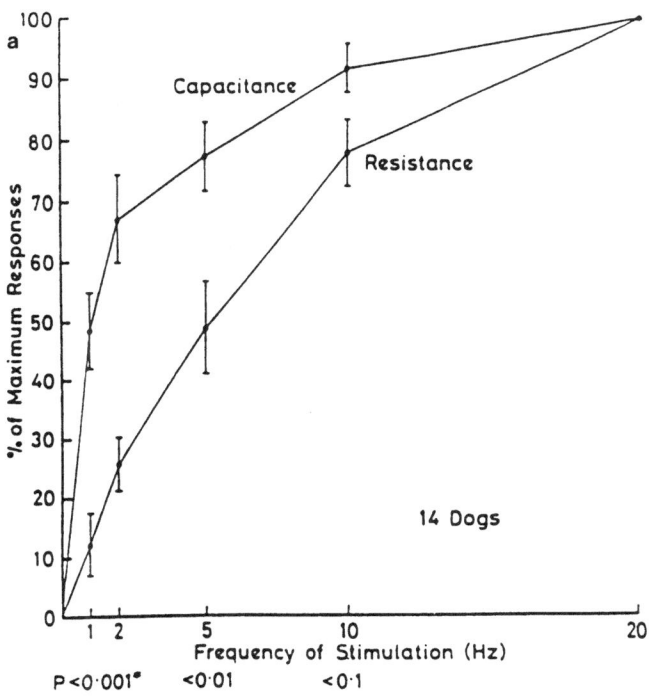

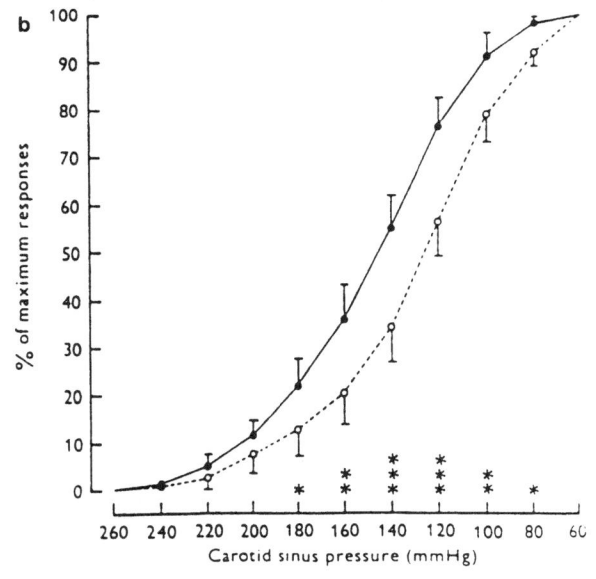

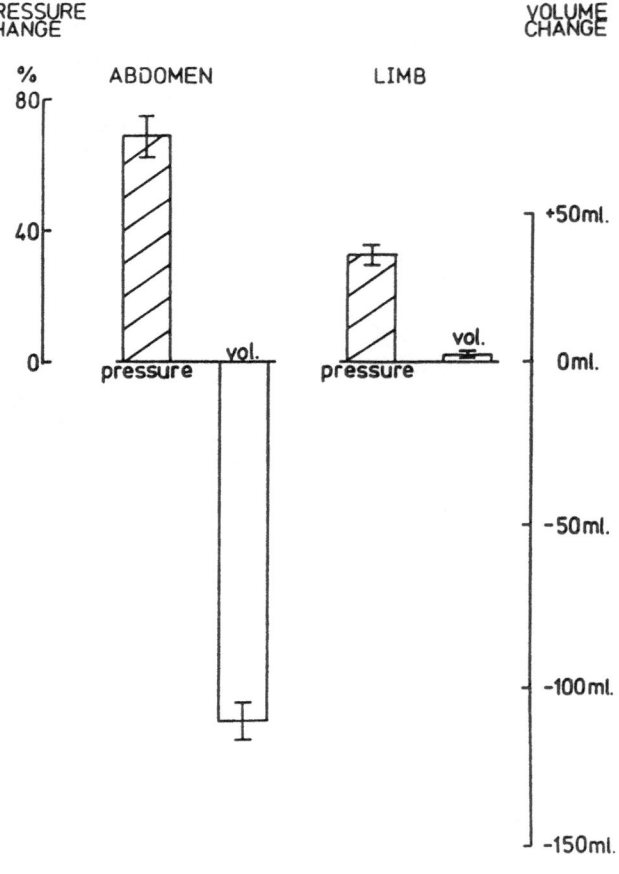

**Figure 5.** Comparison of the responses of abdominal and hind limb vascular capacitance (open bar) and resistance (hatched bar) elements to unloading of carotid baroreceptors in anaesthetized and artificially ventilated dogs (mean ± SE N= 10). Both areas were vascularly isolated and perfused with blood at a constant flow in two separate series. Pressures in the vascularly isolated carotid sinuses were decreased from above the supramaximal down to the threshold level for these baroreceptors. Note that a decrease in carotid sinus pressure caused a significant decrease in abdominal vascular capacitance (about 110 ml) and no change in limb vascular capacitance, whereas the perfusion pressures (vascular resistance) increased significantly in both vascular beds. From Karim, Hainsworth and Wood (unpublished).

1972; Drees & Rothe, 1984). The responses did not change after cholinergic blockade with atropine. However, after beta-adrenergic receptor blockade by atenolol the responses were significantly enhanced and after alpha adrenergic blockade the responses were abolished almost completely (Fig. 7, also see Muller - Ruchholtz et al. 1977a, b: Bennett et al 1984). These results suggested that the vascular capacitance responses to unloading of baroreceptors were mediated via alpha adrenergic receptors, the action of which can be enhanced when the vasodilator effect of beta-adrenoreceptors was removed by beta blockers (Abboud et al. 1968). On the other hand, as expected, administration of epinephrine can greatly attenuate baroreflex- induce capacitance responses (Shoukas & Brunner, 1980). The results of these simple experiments, therefore, explain why the patients suffering from syncope benefits from beta-blocker therapy (Sra et al. 1992). The alpha receptor induced constriction of capacitance elements following a change of position of the beta adrenergic blocked patient from lying to standing can occur unopposed by beta receptors. Thus a greater amount of blood would be mobilised towards the heart to maintain an adequate cardiac filling and output for blood pressure and cerebral perfusion. Whether these patients have greater than normal beta-receptor population in their capacitance element remains to be seen. Also, it remains to be seen whether the Starling's mechanism for the maintenance of cardiac output becomes dominant under beta blocker therapy when the heart rate cannot increase following a change in position of the body from lying to

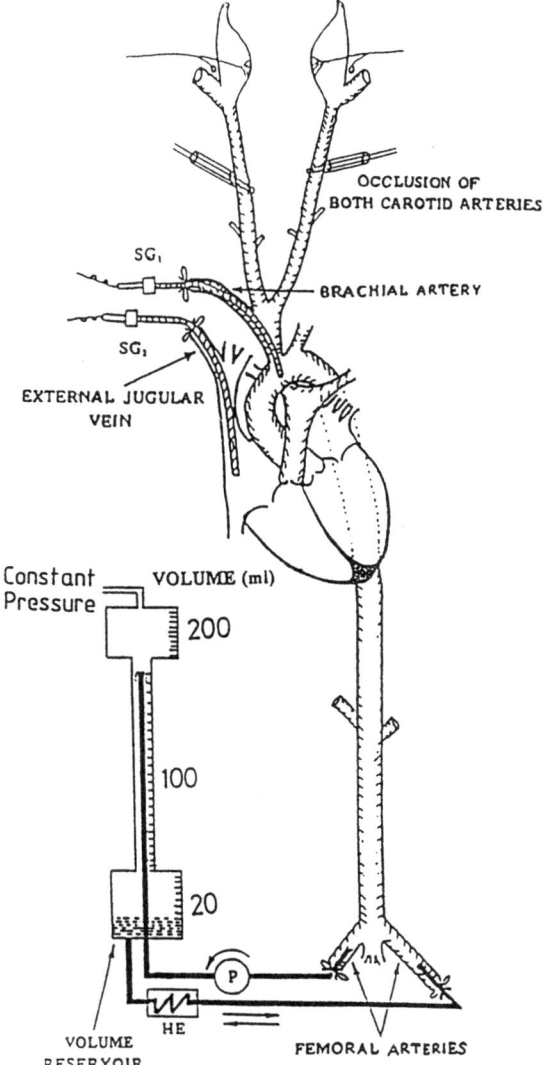

**Figure 6.** Diagram of the preparation. Both common carotid arteries are exposed in the neck. The arterial blood pressure is recorded by a strain gauge manometer (SG,) attached to a cannula placed in the aorta via the right brachial artery and held constant by means of a pressure control device consisting of a Starling resistance and compressed air system (see Karim & Al-Obaidi, 1993 for details). Blood is pumped from the right femoral artery using a roller pump (P) into a calibrated volume reservoir primed with 20-30ml dextran/saline and returned to the animal via a heat exchanger (HE) and a cannula connected to the left femoral artery. A desired pressure is initially set in the system by the pressure control device and maintained constant throughout the experiment.

standing. However, persons with postural hypotension show some inpairement of reflex vasomotor activity (Page et al. 1955).

## THE SIGNIFICANCE OF THE RESPONSES TO BARORECEPTOR LOADING AND UNLOADING

The sympathetic activity in normal humans and animals is expected to be low (Wallin et al. 1973, Hainsworth et al. 1975) and the abdominal veins contain over 20% of the total blood volume making it a potential blood reservoir (Greenway & Innes, 1980), the capacity of which can be significantly altered by tuning sympathetic activity at this lower range (Fig 4B). Changes in sympathetic activity to this extent are likely to occur in ordinary circulatory stresses such as postural changes and during deep breathing. The normal respiratory movements which are accompanied by arterial pressure swings at the most sensitive range

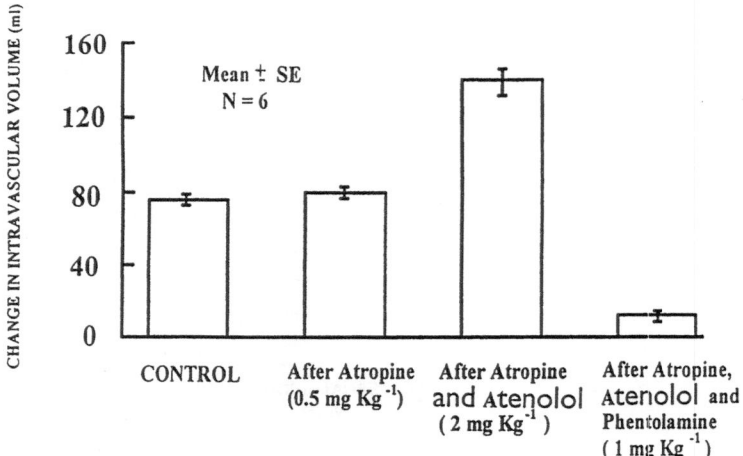

**Figure 7.** Changes in intravascular volume following bilateral carotid occlusion before and after cholinergic and adrenergic receptor blockade at a constant pressure in anaesthetized pigs. The carotid occlusion was maintained for 3-4 minutes. Note that carotid occlusion decreased the total vascular capacitance by shifting about 80 ml of blood into the extracorporeal reservoir: the responses were augmented after administration of atenolol and abolished after phentotamine. From Karim & Al-Obaidi (unpublished).

of baroreceptors may use this mechanism for venous return in addition to their mechanical effects: during inspiration intrathoracic pressure decreases and intra abdominal pressure increases constituting abdominothoracic pump for venous return (Brecher 1956, Abel & Walddhousen 1968). However, under clinical conditions of high sympathetic nerve activity the neurohumoral component of this mechanism for venous return would be negligible, and the maintenance of arterial blood pressure would depend largely on the *mechanical factors* acting on the venous side and the *neurohumoral factors* acting largely on the arterial side. The mechanical factors include (1) respiratory movements; (2) reduction of blood flow due to constriction of precapillary resistance vessels leading to passive elastic recoil of the thin wall capacitance vessels and (3) fluid shifts into the intravascular space due to a fall in capillary pressure from a reduction in blood flow. The lack of response in the superficial vein could be related to thermoregulation e.g. if the body is exposed to cold it can increase arterial blood pressure (cold pressor response), which in turn can stimulate baroreceptors resulting in an inhibition of the sympathetic activity to superficial veins. This mechanism would have caused heat loss. The absence of capacitance response in the muscle bed could be related to an attempt of conservation of intravascular volume during hypotension, e.g. if the veins would constrict strongly in response to hypotension (unloading of baroreceptors), an increase in post to pre-capillary resistance ratio would have caused a loss of fluid from the capillaries by filtration and thus the condition could be made worse. Therefore, the limb veins should not play any role in the baroreceptor reflex.

## STIMULATION OF PERIPHERAL CHEMORECEPTORS PRODUCE SIGNIFICANT CHANGES IN VASCULAR CAPACITANCE

Stimulation of carotid chemoreceptors with venous blood resulted in a greater reduction of abdominal vascular capacitance at high carotid pressure than at lower range (Hainsworth, Karim, McGregor & Wood 1983a). When carotid sinus pressure was held

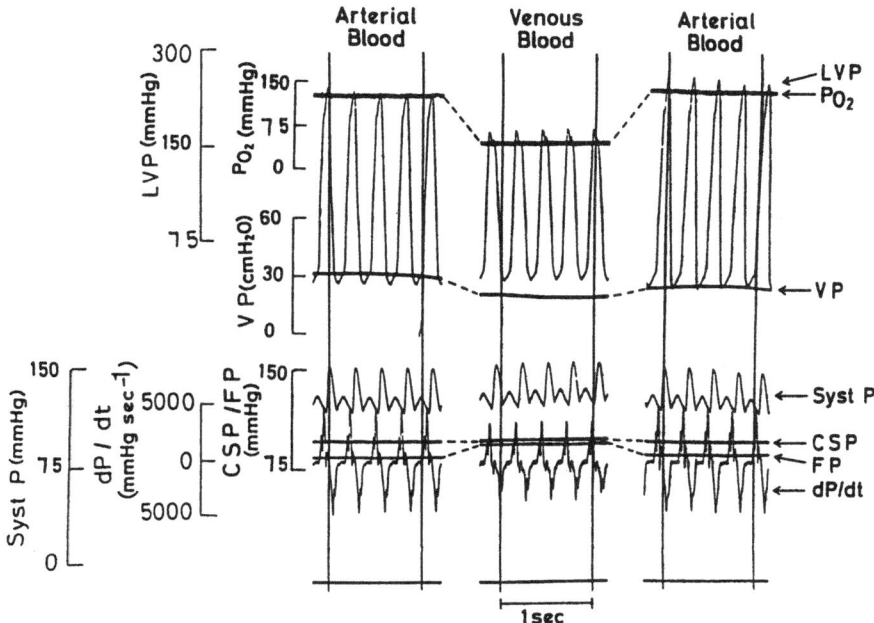

**Figure 8.** Steady state responses of dP/dt max, perfused femoral artery (FP) and superficial metatarsal vein (V) to stimulatiojn of carotid body chemoreceptors with venous blood. Records taken during arterial perfusion of the chemoreceptors (left panel), two minutes after venous perfusion of the receptors (middle panel) and 2 minutes after re-establishing the arterial perfusion (right panel). Note that the chemoreceptor stimulation resulted in a decrease in dP/dt max accompanied by constriction of the perfused artery and dilation of the vein. LVP, left ventricular pressure; $P_{O_2}$, oxygen tension of perfusate; HR, heart rate; syst P, systemic arterial pressure. From Sofola, Karim & Hainsworth (unpublished).

below 130 mmHg, chemoreceptor stimulation increased perfusion pressure in isolated canine abdominal vascular bed (prepared as described above) by 55 mm Hg and decreased capacitance by 0.6 ml/kg and when carotid sinus pressure was held above 130 mmHg chemoreceptor stimulation increased perfusion pressure by about 30 mmHg and decreased vascular capacitance by 1.1. ml/kg. The responses at high and low carotid sinus pressure are significantly different. These results are consistent with our earlier observation that the capacitance element is more sensitive to sympathetic stimulation, i.e. when carotid sinus pressure was high sympathetic activity was very low probably about 0.5 -1 impulse sec[-1] and a small increase in sympathetic activity can result in relatively large capacitance responses and very little resistance responses. At the other end of the carotid sinus pressure a high sympathetic activity would result in near maximal constriction of capacitance vessels, although further constriction of resistance vessels could still occur. However, more recently, Karim and Al-Obaidi (1993) reported in anaesthetised dogs an interaction between carotid baroreceptors and chemoreceptors in the responses of total vascular capacitance: a reduction of total vascular capacitance of about 160 ml (7.5 ml Kg[-1]) at low carotid sinus pressure (CSP, about 70 mmHg) and about 65ml (3.1 ml (Kg body weight)[-1]) at high CSP (about 195 mmHg) in response to carotid chemoreceptor stimulation under a constant mean systemic arterial blood pressure (about 90 mmHg). Under the same experiemental condition changing carotid sinus pressure from about 70 mmHg to about 200 mmHg resulted in a very large capacitance response (about 300 ml., 13.7 ml (Kg body wt)[-1]) showing the dominance of baroreceptor influence in regulating vascular capacitance. Stimulation of aortic chemoreceptors, either by venous blood or by injection of sodium cyanide into the aortic arch, also results in an increase in abdominal vascular resistance and a decrease in abdominal vascular capacitance (Hainsworth et al 1983b).

The superficial venous tone decreases when the peripheral chemoreceptors are stimulated Shepherd & Vanhoutte 1975; Britton & Donald, 1982; Sofola, Hainsworth & Karim, 1978, Unpublished Observation (Fig.8). The significance of this response of the superficial vein is not known ( Pelletier & Shepherd, 1972). Also, the effect of stimulation of aortic chemoreceptors on the limb veins is not very clear.

# EFFECTS OF STIMULATION OF CARDIOPULMONARY RECEPTORS ON VASCULAR CAPACITANCE ARE NOT CLEAR

The influence of the cardiopulmonary receptors on the capacitance vessels has not yet been fully explored (Abboud et al 1979; Rothe, 1983). In anaesthetised dogs inflation of the lungs with moderate pressure up to 20 cm $H_2O$) has been shown to cause a reflex arterial constriction without any significant effect on the capacitance vessels of the limb and abdomen (Greenwood et al., 1977; Karim et al. 1987). However, in anaesthetised rabbits positive pressure breathing with positive end expiratory pressure above 19 mmHg has been shown to cause constriction of inferior vena cava (Stinnet et al., 1983). Also, in conscious man, a deep breath causes an increase in pressure in the vein of an occluded limb (Samuloff et al., 1966, Epstein et al 1968; Browse & Hardwick, 1969), but a deep breath is not a specific stimulus. Localised stimulation of left atrial receptors does not produce a significant reflex change in the total body vascular capacitance, whereas a change in isolated carotid sinuses produce a very large change under the same condition (Karim & Majid, 1991). This is consistent with our earlier study in which stimulation of left atrial receptors did not cause any change in the activities of the splenic and lumbar sympathetic efferents, whereas the cardiac sympathic activity increased and renal nerve activity decreased (Karim et al 1972). The absence of capacitance responses to stimulation of the low pressure receptors is contrary to the general belief that the reflex control of the capacitance vessels from these receptors should be more important than from high pressure receptors, because blood volume and vascular capacitance interact to influence cardiac filling pressure (Rothe, 1983). Tutt et al. (1988) observed a small increase in adbominal vascular capacitance in response to stimulation of left ventricular receptors in anaesthetized dogs. Distension of left ventricle has been shown to cause an increase in vascular capacitance in the splanchnic and a decrease in the extra splanchnic areas showing the importance of abdominal vascular capacitance in the cardiovascular homestasis (Hoka et al 1988).

# CONCLUSION

Abdominal capacitance vessels (veins) play a very important role via baro and chemoreceptor reflexes during circulatory stresses such as postural changes and haemorrhages in the blood pressure homeostasis. The limb capacitance vessels (muscle and superficial veins are not significantly influenced by baroreceptors. Therefore, these receptors do not interfere with the thermoregulatory mechanism and absorption of fluid from limb muscles into capillaries after haemorrhage. The beta-adrenergic blockers can potentiate the capacitance responses to baroreceptor inhibition due to hypotension during haemorrhage and syncope (fainting). Therefore, the patients suffering from postural hypotension are likely to benefit from these types of drugs.

# REFERENCES

1. Abel F.L. Waldhausen JA (1968) Effects of anaesthesia and artificial ventilation on caval flow and cardiac output. J Appl Physiol 25: 479-484.
2. Abboud FM, Schmid PG, Eckstein JW (1968) Vascular reponses after alpha-adrenergic blockade. I. Responses of capacitance and resistance vessels to norepinephrine in man. J Clin Invest 47:1-9.
3. Abboud FM, Eckberg DL, Johannsen UJ, Mark AL (1979) Carotid and cardiopulmonary baroreceptor control of splanchnic and forearm vascular resistance during venous pooling in man. J Physiol 286: 173-184.
4. Bennett TD, Wyss CR, Scher AM (1984) Changes in vascular capacity in awake dogs in response to carotid sinus occlusion and administration of catecholamines. Circ Res 55: 440-453.
5. Braunwald E, Ross J Jr, Kahler RL, Gaffney TE, Goldblatt A, Mason DT (1963) Reflex control of the systemic venous bed. Effects on venous tone of vasoactive drugs, and of baroreceptor and chemoreceptor stimulation . Circ Res 12: 539-550
6. Brecher GA (1956). Venous return. New York: Grune & Stratton.
7. Britten SL, Donald DE (1982) . Response of large hindlimb veins of dog to aortic arch stimulation. Am J Physiol 242: H1050-H1055.
8. Brooksby GA, Donald DE (1972). Release of blood from the splanchnic circulation in dogs. Circ Res 31: 105-118.
9. Browse NL, Hardwick PJ, (1969). The deep-breath-vasoconstriction reflex. Clin Sci 37:125-135.
10. Browse NL, Donald DE, Shepherd JT (1966) Role of the veins in the carotid sinus reflex. Am J Physiol 210: 95-102.
11. Chen, H.J., Ta Chi (1994) The sypathetic efferent innervation of the cutaneus and muscle veins in cats. A comparative study using retrograde localization with Horseradish Peroxidase JANS 46: 189-197.
12. Drees JA, Rothe CF (1974) Reflex vasoconstriction and capacity vessels pressure-volume relationships in dogs. Circ Res 34:360-373.
13. Epstein SE, Beiser GD, Stampfer M, Braunwald E (1968) Role of the venous system in baroreceptor-mediated reflexes in man. J Clin Invest 47: 139-152.
14. Folkow B, Mellander S (1964) Veins and venous tone. Am Heart J 68:397-408.
15. Fuxe K, Sedvall G (1965) The distribution of adrenergic nerve fibres to the blood vessels in skeletal muscle. Acta Physiol Scand 64:75-86.
16. Green HD (1950) In: Glasser O (ed) Medical Physics, vol 2. Year Book, Circulation: Physical Principles, Chicago, p231.
17. Greenway CV, Innes IR (1980) Effects of splanchnic nerve stimulation on cardiac preload, afterload and cardiac output in cats Circ. Res 46: 181-189
18. Greenwood PV, Hainsworth R, Karim F, Morrison GW, Sofola OA (1977) Peripheral vascular responses from lung inflation. J Physiol 273:55-56P.
19. Hainsworth R, (1986) Vascular capacitance: its control and importance. Rev. Physiol Biochem Pharmacol 105: 101-173.
20. Hainsworth R, Karim F, Stoker JB (1975) The influence of aortic baroreceptor on venous tone in the perfused hindlimb of the dog. J Physiol. 244: 337-351.
21. Hainsworth R, Karim F (1976) Responses of abdominal vascular capacitance. in the anaesthetized dog to changes in carotid sinus pressure. J Physiol (Lond) 262: 659-677.
22. Hainsworth R, Karim F, McGregor KH, Wood LM (1983a) Responses of abdominal vascular resistance and capacitance to stimulation of carotid chemoreceptors in anaesthetized dogs J Physiol 334: 409-419.
23. Hainsworth R, Karim F, McGregor AH, Rankin AJ (1983b) Effects of stimulation of aortic chemoreceptors on abdominal vascular resistance and capacitance in anaesthetised dogs. J Physiol (Lond) 334:421-431.
24. Hainsworth R, Karim F MacGregor KH, Wood LM, (1983c) Hind-limb vascular-capacitance responses in anesthetised dogs. J Physiol (Lond) 337:417-428.
25. Hoka, S, Bosnjack, ZJ, Siker, D, Luo, R, Kampine, JP (1988) Dynamic changes in venous outflow by baroreflex and left ventricular distension 254: R212-R221.
26. Imai Y, Satoh K, Taira N (1978) Role of pheripheral vasculature in changes in venous return caused by isoproterenol, norperephrine, and methoxamine in anaesthetised dogs. Circ Res 43:553-561.
27. Karim F, Ali, H. (1969) Effect of electrical stimulation of the carotid sinus on venometer tone of the superior vena cava in dogs. Life Science 8: 791-799.
28. Karim F, Hainsworth R (1976) Responses of abdominal vascular capacitance to stimulation of splanchnic nerves. Am J Physiol 231:434-440.

29. Karim F, Kidd C, Malpus CM, Penna PE (1972) The effects of stimulation of the left atrial receptors on sympathetic efferent nerve activity. J physiol 227:243-260.

30. Karim F, Hainsworth R, Pandey RP (1978) Reflex responses of abdomunal vascular capacitance from aortic baroreceptors in dogs. Am J Physiol 235: H488-H493.

31. Karim F Araneda G, Hainsworth R (1980) The influence of perfusate temperature on the responses of a superfucial vein in the carotid baroreceptor reflex in dogs. Pfleugers Archiv 383: 79-85.

32. Karim F, (1987) Mechanical effects of respirationon on heart function (ACP, Applied Cardiopulmonary Pathophysiology 1: 109-126.

33. Karim F. Morrison & Hainsworth (1987) Vascular responses in the hind-limb to lung inflation anaesthetized dogs ACP (Applied Cardiopulmonary Pathophysiology 1:127-136.

34. Karim F, Majid DSA , (1991) Inhibition of atrial receptor-induced renal responses by stimulation of carotid baroreceptors in anaesthetized dogs. J Physiol. 432, 509-520.

35. Karim F, Al-Obaidi, M (1993) Modification of carotid chemoreceptor -induced changes in renal haemodynamics and function by carotid baroreflex in dogs. J. Physiol 466: 599-610.

36. Lesh TA, Rothe CF (1969) Sympathetic and haemodynamic effects on capacitance vessels in dog skeletal muscle. Am J Physiol 217:819-827.

37. Mellander S (1960) Comparative studies on the adrenergic neuro-hormonal control of resistance and capacitance blood vessels in the cat. Acta Physiol Scand 50 (Suppl 176):1-86.

38. Muller-Ruchholtz ER, Losch HM, Gund E, Lochner W (1977a). Effect of alpha adrenergic receptor stimulation on integrated systemic venous bed. Pfleugers Arch 370:241-246.

39. Muller-Ruchholtz ER, Losch HM, Gund E, Lovhner W (1977b) Effect of beta adrenergic receptor stimulation on integrated systemic venous bed. Pfleugers Archiv. 370:247-251.

40. Page EG, Hickson JB, Sieker HO, McIntosh HE, Pryor WW (1955) Reflex venomotor activity in normal persons and in patients with postural hypotension. Circulation 11:262-270.

41. Pelletier CL, Shepherd JT (1972) Venous responses to stimulation of carotid chemoreceptors by hypoxia and hypercapnia. Am J Physiol 223:97-103.

42. Ross J Jr, Frahm CJ, Baunwald E (1961) The influence of carotid baroreceptors and vasoactive drugs on systemic vascular volume and venous distensibility. Circ Res 9:75-82.

43. Rothe CF (1976) Reflex vascular capacity reduction in the dog. Circ Res 39: 705-710.

44. Rothe CF (1983) Reflex control of veins and vascular capacitance. Physiological Review 63:1281-1342.

45. Rutlen DL, Supple EN, Powell WJ Jr (1981) Beta adrenergic regulation of total systemic intravascular volume in the dog. Circ Res 48:112-120.

46. Shepherd JT, Vanhoutte P (1975) Veins and their control. Saunders Philadelphia

47. Samuloff SL, Bevegard BS, Shepherd JT (1966) Temporary arrest of circulation to a limb for the study of venomotor reactions in man J. Appl. Physiol. 21: 341-346.

48. Shoukas AA, Brunner MC (1980) Epinephrine and the carotid sinus baroreceptor reflex influence on capacitance and reflex properties of the total systemic vascular bed of the dog. Circ Res 47:249-257.

49. Shoukas AA, Sagawa K (1973) Control of total systemic vascular capacity by the carotid sinus baroreceptor reflex. Circ Res 33: 22-33.

50. Sra JS, Murthy VS, Jazayeri MR, Shen Y, Toup PJ, Avital B, Akhtar, M (1992) Use of intravenous esmold to predict efficacy of oral beta adrenergic blocker therapy in patients with neurocardiogenic syncope. JACC 19: 402-408.

51. Stinnett HO, Magnusson MR, Gourley MF (1983) Vein segment diameter and pressure responses during lung hyperinflation in the rabbit, Proc 29th Int., Physiol Congr Abstr 322, 21 Sydney

52. Tutt SM, McGregor KH, Hainsworth R (1988) Reflex vascular responses to changes in left ventricular pressure in anaesthetized dogs. Quat. J. Exp. Physiol. 73: 425-437.

53. Wallin BG, Delius W, Hagbath K-E (1973) Comparison of sympathetic nerve activity in normotensive and hypertensive subjects. Circ. Res. 33: 9-21.

# 10

# THE MICROPHYSIOLOGY OF LUNG LIQUID CLEARANCE

Jahar Bhattacharya [*]

Departments of Medicine and Physiology
College of Physicians and Surgeons
Columbia University
St. Luke's-Roosevelt Hospital Center
1000 10th Avenue, New York, NY 10019

Pulmonary edema results from failure at essentially two levels of regulation in the lung microvascular bed, namely that of of liquid production by microvascular filtration and that of filtrate clearance from the interstitium. Of these, the regulation of filtration has attracted by far the greatest amount of interest. The reasons are not difficult to understand. In the sixties and seventies the burgeoning interest in the pulmonary circulation produced a number of elegant techniques for quantification of the determinants of lung microvascular filtration. To take some examples, the applications of the gravimetric, the lymph sampling and the indicator dilution techniques to the lung, advanced the quantification of the lung microvascular barrier and furthered the understanding of microvascular filtrate composition (reviewed in 27).

This progress notwithstanding, investigators were well aware that an understanding of microvascular filtration mechanisms could not be divested from that of perimicrovascular hydrodynamics. This reasoning was forced by first, the Starling equation which pointed to the hydrostatic and protein osmotic pressures of the perimicrovascular interstitium as being respectively, negative and positive determinants of filtration. Hence these interstitial forces had to be evaluated. Second, it was known from the thirties that the alveolar septum is devoid of lymphatics. Therefore it was necessary to postulate a flow of the alveolar capillary filtrate across the substantial interstitial distance (~200 μm) that separates septal from juxtaseptal interstitium, in order that the filtrate reach and be removed by the first collecting lymphatics. The regulation of this flow required clarification.

Although the need to understand post-filtration hydrodynamics fueled the inception of lung interstitial research, it also sprouted some of the most intense controversies that

---

[*] Phone: (212) 523-7310; Fax: (212) 523-7442

*Control of the Cardiovascular and Respiratory Systems in Health and Dis*ease
Edited by C. T. Kappagoda and M. P. Kaufman, Plenum Press, New York, 1995

continue to polarize thinking in the field to this day.[*] Basically, universal agreement is lacking on almost on every issue in interstitial hydrodynamics - the theory and value of the lung interstitial pressure, the nature of interstitial liquid flow, the sieving properties of the interstitium and the importance of interstitial charge, to name a few areas of controversy.

These problems and controversies are at least partly attributable to a failure in the literature, to distinguish the liquid exchange from the structural compartment of the lung interstitium. The liquid exchange compartment consists of the perimicrovascular space surrounding microvessels in the alveolar septum and those in the first few juxtaseptal generations. The interstitial thickness is <2 μm, hence for dynamic studies, it is amenable essentially, only to microscopy based experimental approaches. The structural compartment is the loose fascia that tethers large blood vessels (diameter >1 mm), the large bronchi and the interlobular septa to the parenchyma. Although the two interstitial compartments are physically contiguous, the structural compartment does not normally participate in the lung's liquid exchange because the filtrate is siphoned off by lymphatics in the liquid exchange compartment.

The difficulty is that investigators have regularly obtained data in the structural interstitium which is more accessible to standard experimental probes (20), and have applied their interpretations to the liquid exchange interstitium. We may call this the leap-of-faith approach because the two interstitia are assumed to be similar, even when they are clearly not![†]

An egregious example of this practice is seen in the interstitial pressure literature. Traditional measurements of the lung interstitial pressure were obtained using probes of mm dimensions (capsules, wicks, hypodermic needles) that could only be wedged into the structural fascia (reviewed in 29). These measurements were then applied for the interpretation of filtration mechanisms in the microconfines of the liquid exchange interstitium. Another example is the evolving effort to implicate matrix biochemistry in the regulation of liquid exchange. Although the hypothesis itself is potentially intriguing, the bulk of the biochemical data appears to be relevant to the structural and not the liquid exchange interstitium, the matrix chemistry of which remains poorly characterized.

We decided that an approach that might resolve some of these problems would be to focus directly on the perimicrovascular interstitium of the subpleural microcirculation, by means of optical and micropuncture methods. The subpleural region gives optical access to lung microvessels that are mostly venules of diameters in the 10-30 μm range, as shown for example in Figure 1. The sketch in Figure 2 though more impressionistic, summarizes the functional anatomy of the liquid exchange pathway in a system of subpleural vessels which in most small laboratory mammals, consists purely of pulmonary and not bronchial vessels.

---

[*] One of the memorable controversies of the seventies was the Guyton Wiederhielm lung interstitial pressure controversy. The exact value of the lung interstitial pressure was (and is!) at issue. Guyton's view that the pressure had to be highly subatmospheric because it was generated by lymphatic suction, was opposed by Wiederhielm who believed that the pressure was attributable to the swelling pressure of the interstitial matrix and was subatmospheric by only a few cmH$_2$O. As a postdoctoral fellow newly grappling with these questions, I remember attending the 1979 debate at which the main protagonists faced off on their positions. I was struck by the ascerbetic tone of the exchanges and especially by the fact that neither speaker appeared to give the other much credence. At one point, one pleaded with the other to stop rejecting their papers! The Guyton and Wiederhielm theories have each sprung forth a different worldview of lung interstitial hydrodynamics. Echoes of the 1979 debate are still audible.

[†] An important difference between the structural and liquid exchange compartments of the lung interstitium is the matrix chemistry, which I will not discuss in detail. Because of its considerably greater mechanical stresses, the structural interstitium is richer in matrix constituents such as collagen and glycosaminoglycans (30, 31) and may thereby, possess hydrodynamic properties that are different from those of the liquid exchange compartment.

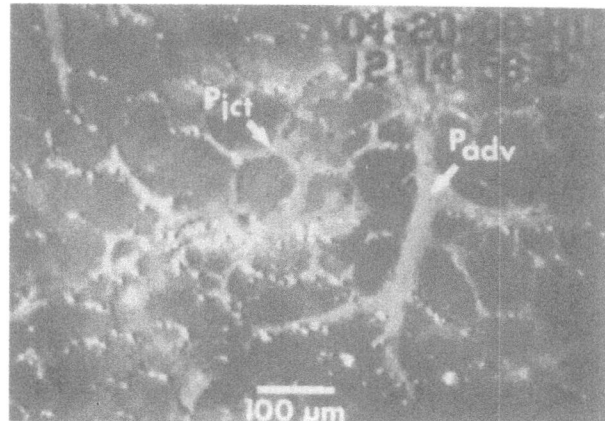

**Figure 1.** The subpleural microcirculation. A venule is shown to the right of center. Pjct and Padv denote sites of interstitial pressure measurement respectively, in the septal interstitium at the alveolar junctions and in juxtaseptal interstitium in the venular adventitia. (*Reprinted from J. Appl. Physiol. 66:2600-2605, 1989*).

The investing interstitium of these microvessels is accessible by micropuncture at essentially two locations, namely in the compact space at the introitus of the alveolar septum and in the loose, lymphatic-containing interstitium surrounding vessels that are 2-3 orders of branching separated from the septal capillaries (Figure 2). We may call these respectively, the septal and the juxtaseptal interstitia. Therefore the liquid exchange interstitium can be

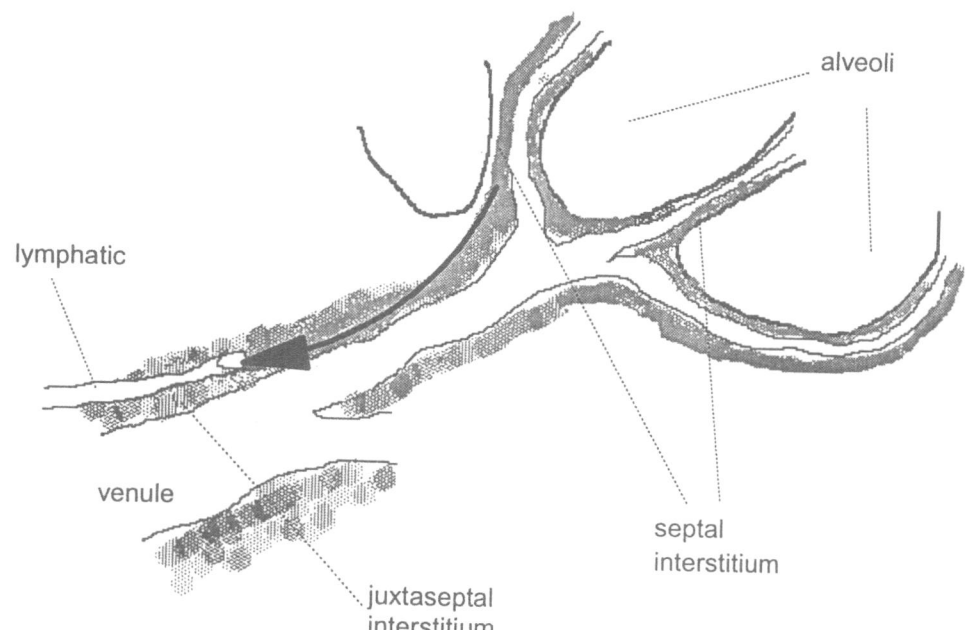

**Figure 2.** Schematic of the liquid flow pathway (arrow) in lung interstitium. The septal interstitium invests capillaries that lie in the space between alveolar walls. Lymphatics lie in the juxtaseptal interstitium that invests vessels outside the septum. Density of grey shading denotes presumed compactness of interstitium at different regions. Interstitial liquid flows from the septal to the juxtaseptal interstitium, en route to the collecting lymphatic.

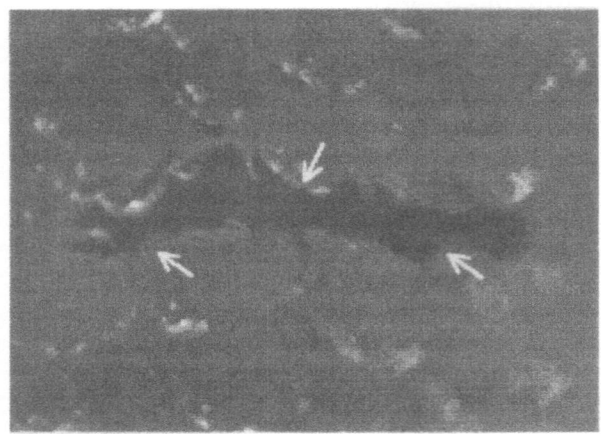

**Figure 3.** Oil filled subpleural venule. The black oil was injected to demarcate the sites of compression by surrounding alveoli (arrows).

studied along the septum-lymphatic pathway that is taken by the filtrate, after its formation in the septal capillaries.

Though reductionist, the single microvessel approach for interstitial studies has several advantages. The alveolar pressure surrounds the interstitium of both septal and juxtaseptal microvessels and therefore, alveolar compression determines interstitial pressures at these locations (Figure 3 ). Hence, the important relationship between the alveolar pressure and lung interstitial hydodynamics, can be directly determined. The increasing sophistication of optical methods promises the possibility of unique structure-function studies of capillary and interstitial transport in the single microvessel. We have already developed methods for the quantification of filtration across these single microvessels (22,23,24) and are in the process of imaging the vessels for quantification of transport chemistry. It may be possible ultimately, to combine fluorescence data that allow interpretation of interstitial flow and structural chemistry, with interstitial pressure data obtained by the micropuncture-servonull technique.

Our present objective was to determine the postulated interstitial pressure gradient that may drive interstitial liquid flow from filtration sites at the alveolar junctions to collecting lymphatics in the venular adventitia (27). In this brief review I have limited the discussion mainly to our pressure data obtained in the past ten years, as also to our recent data on liquid flow in the venular adventitia. Numerous other reviews of lung interstitial biology are available. Those by Taylor and Parker (29) and Lai-Fook (12), provide exhaustive coverage of pertinent literature. The brief reviews by Turino (30) and by Weibel and Crystal (31) give interesting insights into the matrix chemistry and cell biology of the lung interstitium. The lively discussion in Staub's 1974 review provides a provocative view of lung interstitial hydrodynamics (27). These reviews should be consulted for a more complete understanding of the current thinking on the regulation of lung interstitial liquid clearance.

## MATRIX SWELLING

According to the classical view, lymphatics completely remove the microvascular filtrate in the normal lung. In high filtration states a point is reached at which filtration exceeds the lymphatic removal capacity, hence liquid fails to be cleared by lymphatics and it accumulates first in the interstitium surrounding large vessels (perivascular cuffs) and then in alveolar sacs (27). This popular hypothesis for pulmonary edema, which we may call the

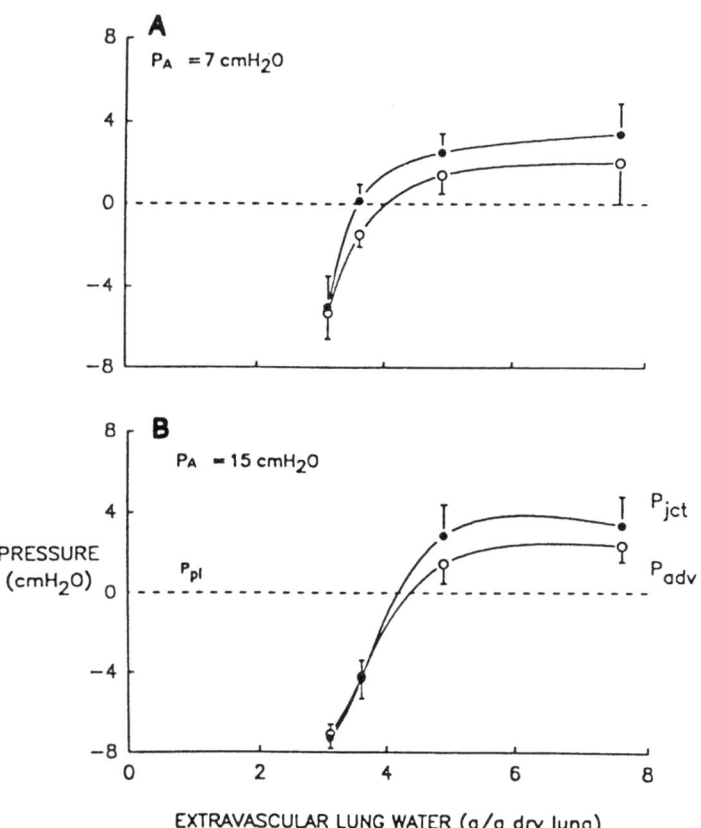

**Figure 4.** Pressure-hydration plot for lung perimicrovascular interstitium. PA = alveolar pressure. Ppl = pleural pressure. Pjct = pressure in septal interstitium at alveolar wall junctions. Padv = juxtaseptal adventitial pressure. Pjct and Padv were obtained as paired measurements at each level of lung water. Baseline Pjct, Padv values are second from left in each panel. Lung water was reduced below baseline by lung dehydration. Note, increase of PA abolished the Pjct-Padv gradient at baseline, but not in edema. (*Reprinted from* Glucksberg, M.R. and J. Bhattacharya. Effect of dehydration on interstitial pressures in the isolated dog lung. J. Appl. Physiol. 67:839-845, 1989).

the theory of lymphatic failure, fails to explain two classical findings. These are first, that during pressure-induced increases of filtration, lung water accumulates in the interstitium well before the maximum lymphatic capacity is reached (7). Second, focal increases of microvascular permeability which locally increase filtration, cause focal edema even though the increase of filtration is clearly unable to overwhelm the lung's lymphatic clearance capacity. Evidently, other factors are also important.

In recent years there has been a major revision of the "old" view that the peri-microvascular interstitium is essentially, a biologically inert conduit for microvascular filtrate and it is now understood that matrix constituents of the interstitium may play a critical role in ordering interstitial liquid flow. Considerable interest has focused on hyaluronan (14) which is abundantly present in lung interstitium and the content of which correlates with lung water content both under normal and diseased conditions (2,18). The hydration dependent swelling of hyaluronan and other matrix glycosaminoglycans explains the interstitial expansion in edema, classically seen as cuff formation around large pulmonary vessels (17). However it should be pointed out that presence of hyaluronan or the other glycosaminoglycans remains unconfirmed in the liquid exchange interstitium of the normal lung.

Nevertheless, the swelling property of the interstitium is significant because during the onset of edema, it imparts a hyperbolic profile to the interstitial pressure-hydration curve (Figure 4) and essentially regulates the responses of the perimicrovascular interstitial pressure.[*]

The transmicrovascular force balance attributable to the vascular-perimicrovascular pressure difference is important for regulation of not only filtration as defined in the Starling equation, but also of vascular diameter. Hence a sharp increase of perimicrovascular pressure in edema would not only have the beneficial effect of reducing filtration, but it could also cause the flow reducing effects of vascular compression. As it happens, perimicrovascular pressure increases very little, by only 2-3 cmH$_2$O, even when edema is extremely severe (4,8,11). This pressure increase is limited by the interstitial swelling.

One may view interstitial swelling as a "loosening up" of the matrix because as the interstitium becomes edematous, interstitial compliance increases dramatically. Figure 4 exemplifies the differential pressure responses in the non-edematous and edematous ranges of the pressure-hydration plot. An increment of lung water by 1 g/g of dry lung increases interstitial pressure by ~4 cmH$_2$O in the non-edematous range (3 to 4 g/g dry), but by <0.5 cmH$_2$O in the edematous range (5 to 6 g/g dry). These effects occur even during lung

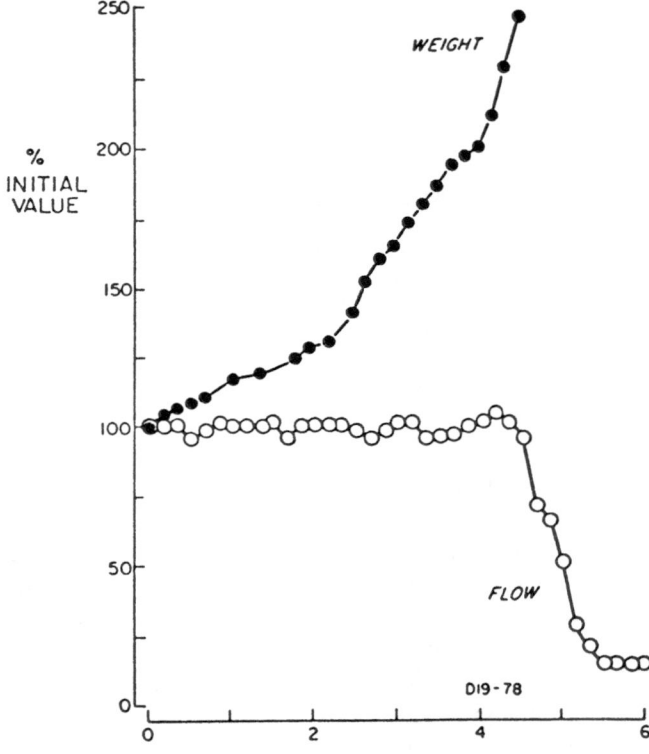

**Figure 5.** Effect of edema on lung blood flow. Increase of weight quantifies extent of edema. Note that blood flow remained at baseline levels even when lung weight had increased by 100%. (*Reprinted from* Bhattacharya, J., Nakahara, K. and N.C. Staub. The effect of edema on pulmonary blood flow in the isolated perfused dog lung lobe. J. Appl. Physiol. 48:444-449, 1980).

[*] A significance of the swelling that I have excluded from present consideration, is that swellinginduced geometrical distortions may activate interstitial sensory afferents (25). This important regulation is well discussed in the review by the Coleridges (6).

expansion and they protect against the vascular compressive effects of increased interstitial pressure.

Our data for lung perfusion in edema corroborate this conclusion. In Figure 5 which illustrates the perfusion history of a lung during progressive edema, it is worth noting that even when the edema was sufficiently severe that the lung weight doubled, there was no reduction of organ blood flow[*] (5). The doubling of lung weight signifies the doubling of the lung water content, and it is indicative of a highly advanced clinical stage of pulmonary edema. The inability of such advanced edema to diminish blood flow, implies that the increase of interstitial pressure in edema is not sufficient to compress blood vessels. We may interpret that an intrinsic matrix property, namely swelling, blunts increases of interstitial pressure and protects vascular patency in edema. Therefore clinically, decreases of pulmonary blood flow associated with pulmonary edema should be attributed not to compression by interstitial pressure, but rather to other complications of edema such as for example, hypoxic pulmonary vascoconstriction.

## THE INTERSTITIAL PRESSURE GRADIENT

The recognition that the interstitial matrix actively participitates in the regulation of interstitial pressure in edema, suggests that the matrix regulates interstitial liquid flow under normal conditions. The pressure gradient that drives this flow is predicted to occur along the flow pathway from the septal interstitium at the alveolar wall junctions, to the lymphatics of the juxtaseptal interstitium (Figure 2). A major candidate for this pressure gradient is the interstitial pressure difference that we determined by the micropuncture-servonull technique (3,4,8). These measurements revealed interstitial pressure was ~1.5 $cmH_2O$ higher in the septal than in the juxtaseptal interstitium.

An important feature of the septal-juxtaseptal pressure gradient is its filtration independence. In several experimental series, the gradient was well maintained in the absence of microvascular filtration (3,4,8), indicating that it is not determined by the filtration-induced interstitial liquid flow. The gradient may result from the difference of matrix swelling pressures at the septal and juxtaseptal locations. The swelling pressure of a gel compresses liquid held in the interstices of the gel. For interstitium, the gel interstices are liquid flow channels that presumably collapse in the absence of filtration. However during filtration as liquid continues to be added to the interstitium, liquid pressure probably builds in the flow channels until channel pressure slightly exceeds the surrounding gel pressure. Under these conditions, channel liquid pressure must vary directly with the surrounding swelling pressure.[†] Hence, it may be proposed that a gradient of swelling pressures attributable to regional variations in matrix properties, generates a gradient of liquid pressures that drives interstitial liquid flow.

The higher swelling pressure of the septal interstitium, reflects its higher water imbibation properties than that of the juxtaseptal interstitium. This impression is strengthened by the fact that a global increase of interstitial water widens the septal-juxtaseptal pressure gradient (4). When water is globally removed as by lung dehydration, the pressure gradient is abolished (Figure 4) (8). The reason why swelling pressures should be different at the septal as opposed to the juxtaseptal interstitia remains unclear, although it may reflect differences of interstitial gel structure at the two locations.

---

[*] The blood flow decrease in Figure 5 occurred after edema had increased to well over agonal levels. At such massive accumulations of lung water, the weight of the alveolar liquid directly compresses vessels.

[†] This is analogous to the "waterfall" or "Starling resistor" hypothesis of blood flow regulation in lung.

The most important regulatory effects on the septal-juxtaseptal pressure gradient are seen during lung expansion, or following increases of vascular pressure. An increase of the transpulmonary pressure from 5 to 15 $cmH_2O$ increases lung volume from 50 to 90% TLC (total lung capacity) and concomitantly, abolishes the pressure gradient (Figure 4). By contrast, increase of vascular pressure from 5 to 15 $cmH_2O$ almost doubles the pressure gradient (9). Given that the pressure gradient drives interstitial liquid flow, these responses may be of significance in the regulation of interstitial liquid clearance. The effect of lung expansion suggests that an increase of lung volume, as for example in inspiration, impedes interstitial liquid flow because the pressure gradient decreases or disappears. Hence low transpulmonary pressures such as those occuring at end-expiration, are likely to be the most conducive for liquid flow out of alveolar junctions. These considerations are likely to be the physiological basis for establishing positive end-expiratory pressures and lung volumes during therapy for pulmonary edema.

The effect of vascular pressure on the other hand, suggests that a safety mechanism exists for increasing the interstitital pressure gradient and driving out liquid from the alveolar junctions during pressure-induced increases of filtration. An interesting aspect of the vascular pressure effect was that although the gradient increased when vascular pressure increased from 5 to 15 $cmH_2O$, the gradient increased no further when vascular pressure was further raised to 25 $cmH_2O$ (9). It is possible therefore, that the development of edema at high vascular pressures is at least partly attributable to the fact, that the increase of the septal-juxtaseptal interstitial pressure gradient fails to keep pace with the increase of filtration. Hence, excess filtrate fails to be cleared from the alveolar junctions. This explanation as different from the lymphatic failure hypothesis, attributes the development of pulmonary edema to failure of liquid clearance in the prelymphatic interstitium.

## INTERSTITIAL CHARGE

Although it has long been known that the lung interstitium carries negative charges, the importance of negative charge as a determinant of interstitial liquid flow is being understood only from recent experiments. In one approach, Parker and colleagues determined the lung interstitial distribution volume of two positively charged tracers, LDH5 and LDH1 (19,21). LDH5, the more positively charged tracer, occupied a larger interstitial volume which is consistent with its expected distribution behaviour in a negatively charged matrix. In a different approach involving flow measurements across the interstitial compartments of a lung slice (13,28), Lai-Fook and colleagues showed that interstitial transit of positively charged molecules was less impeded than that of negatively charged molecules. Hyaluronan has been identified as the cause of the anionic impedance, because hyaluronidase treatment improves anion flow (17,28). The overall conclusion to be derived from these studies is that the lung interstitium carries a global net negative charge that excludes plasma anions from the interstitial matrix.

The ability of interstitial charge to determine the interstitial flow of charged proteins, is of potential significance in the development of edema, especially if we consider that the charge content may decrease or increase under different conditions. In pressure or sepsis induced increases of filtration, loss of hyaluronan in lung lymph (15,16) may decrease negative charge in the lung interstitium. Conversely, upregulation of hyaluronan synthesis as part of the repair process after injury (1,10,18), may increase interstitial negative charge. Decrease of negative charge may reduce the interstitial anionic barrier and facilitate interstitial entry of anionic plasma proteins such as albumin, that filter across the microvascular membrane. If interstitial volume remains unchanged, increased interstitial albumin content may increase the interstitial protein osmotic pressure which is a pro-filtration force, and may

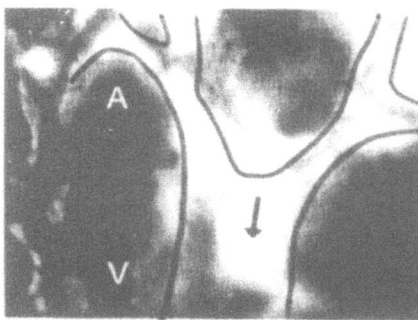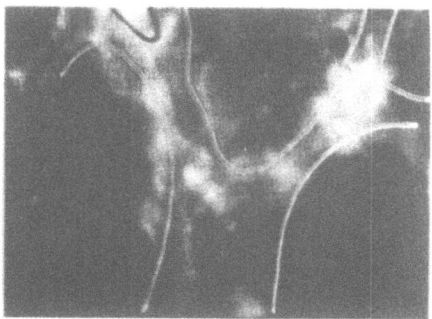

**Figure 6.** Images of microinjected liquid in lung periseptal interstitium. A system of branching venular capillaries consisting of septal and juxtaseptal vessels, is outlined. 20 nl of fluorescent albumin was injected at the bifurcation of the main juxtaseptal venule. Arrow shows direction of blood flow. _Left panel:_ recorded immediately after injection. Interstitial spread of fluorescence is denoted by white areas. Alveoli (not visible) occur in the nonfluorescent, black areas. Note that fluorescence spread both in the direction of the arrow, into the adventitia of the venule (V), as well as in the opposite direction, into the interstitium surrounding septal vessels juxtaposed to alveoli (A). _Right panel_ recorded 1h after injection, shows retention of dye at 'A' but not at 'V'.

thereby promote edema. Increase of negative charge may also predispose to edema because anion loading may promote a Donnan based swelling of the matrix gel amounting to an increase of the interstitial water content. The extent to which these considerations apply to the liquid exchange interstitium remains unresolved.

## PERIMICROVASCULAR FLOW

A major obstacle to the understanding of perimicrovascular hydrodynamics has been the difficulty of directly assessing interstitial liquid flow at this location. To address this problem, we have begun quantification of perimicrovascular flows by the microscopic imaging of microinjected liquid in the rat lung (32). A 20 nl aliquot of fluorescent albumin microinjected at the bifurcation point of venules of the first post-capillary generation, tracked longitudinally along the perimicrovascular space both peripherally towards septal capillaries, as also centrally in the hilar direction, along the venular cuff (Figure 6). An important difference in the behaviour of the flows was that the liquid which flowed into the septal interstitium remained uncleared for the duration of our experiment (~4 h) (Figure 6), whereas the liquid that flowed along the venular cuff was completely removed by a lymphatic, in minutes (Figures 6 and 7).

It needs to be mentioned that the lymphatic uptake of the microinjected liquid was entirely non-pulsatile and our impression is that the liquid flowed passively into the lymphatic. This absence of pulsatility, which is consistent with our previous non-pulsatile pressure recordings at this location, appears to be inconsistent with Guyton's view that pulsatile lymphatic suction draws interstitial liquid into initial lymphatics. However the suction theory may still be applicable to our findings with the modification, that the pulsatile suction generated in lymphatic segments downstream from the initial collecting segment, induces a steady negative pressure in the lumen of the initial segment. This luminal pressure needs to be slightly more negative than the surrounding adventitial pressure in order to provide the adventitial-lymphatic pressure gradient required for generating flow into the initial lymphatic. The critical step of liquid removal from the lung interstitium is then,

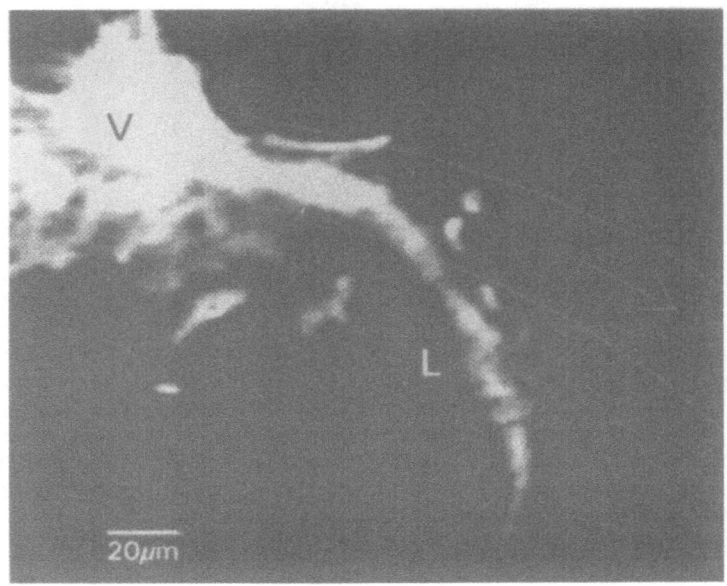

**Figure 7.** Removal of microinjected interstitial liquid by perivenular lymphatic. 'V', venular interstitium. L, lymphatic. Arrow denotes direction of flow in venule. Microinjection site (not shown) was ~100 μm upstream of 'V'. Note absence of fluorescent staining in postlymphatic adventitia. (*Reprinted from* Ying X., R. Qiao, S. Ishikawa and J. Bhattacharya. Removal of albumin microinjected in rat lung perimicrovascular space. J. Appl. Physiol. *In press*).

attributable to this pressure difference that is likely to be very small, probably of the order of 0.1 cmH$_2$O, because we were unable to detect it by the servonull method (our unpublished observations).

　　Although for lymphatic uptake, pressure must be higher outside than inside the initial lymphatic, this pressure difference does not cause lymphatic compression. Lymphatic patency is probably protected by the tethering of lymphatics to surrounding tissue. This is analogous to the situation in juxtaseptal lung microvessels, in which tissue tethering maintains microvascular patency even at luminal pressures well below surrounding pressure (26). In addition, a structural interstitial-lymphatic coupling probably directs interstitial liquid flow into lymphatics. This is indicated in our experiments in which microinjected liquid appeared always, to follow a track directed towards the lumen of the collecting lymphatic and showed no tendency to flow past the lymphatic lumen (Figure 6). Loss of this structural coupling in edema, may deteriorate the lung's liquid clearance mechanism.

　　An important conclusion to be drawn from these imaging studies is that interstitial liquid clearance properties differ between septal and juxtaseptal locations. The juxtaseptal clearance was efficient in that the liquid overload was completely removed from the microinjection site even though it had to flow distances of 100-150 μm before reaching a lymphatic. We have shown that this liquid flow was convective (32) and therefore, it could have been driven by the interstitial pressure gradient that we had determined in earlier experiments. By contrast, the liquid overload in the septal interstitium persisted indicating that interstitial hydraulic conductivity was considerably lower at the septal location. The reason for this is not clear, although it may be attributable to the greater compactness of the septal interstitium and possibly, to the local matrix chemistry. Whatever the reason, it seems that as compared with the extraseptal interstitium, a filtration overload in the septal intersti-

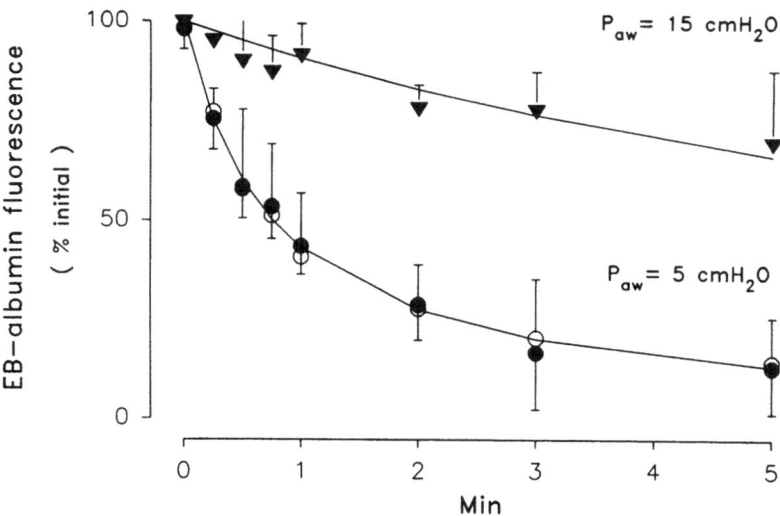

**Figure 8.** Quantification of perivenular liquid flow in nonedematous lung. Paw, airway pressure. Rate of fluorescence decay denotes rate of liquid clearance. Lung expansion inhibited the clearance. (*Reprinted from* Ying X., R. Qiao, S. Ishikawa and J. Bhattacharya. Removal of albumin microinjected in rat lung perimicrovascular space. J. Appl. Physiol. *In press*).

tium is likely to be less efficiently cleared and may be the more potent cause for pulmonary edema.

A puzzling pattern of liquid clearance is encountered when edema results from a focal increase of microvascular permeability. Such permeability increases follow lung microvascular injury and typically, localized regions of the lung develop edema that clear poorly. The interstitial mechanisms that inhibit liquid clearance under these conditons remain inadequately understood.

In our imaging experiments, the injection of nanoliter volumes into a space of micrometer dimensions models focal edema. The fact that this was cleared with a half-time of >1 min (Figure 8), indicates that an appreciable delay occurs in the perimicrovascular clearance of liquid overloads. This delay could contribute substantially to interstitial liquid retention, especially when excess liquid volumes are being continuously added to the interstitium under high filtration conditions. By our conservative estimate the perimicrovascular hydraulic conductivity is at least 10 times higher than the hydraulic conductivity of the microvascular membrane (32). Thus although the high interstitial hydraulic conductivity ensures that the interstitium poses no significant hindrance to filtrate flow under normal conditions, flow hindrance may become appreciable in the presence of a large filtrate load.

Some confusion exists in the literature on whether the hydraulic conductance of interstitial flow channels in the capillary-lymphatic pathway changes in edema. According to a classical proposal, edema increases the conductance by recruiting new channels. Although channel recruitment clearly occurs when edema liquid spills over into the perivascular space that is normally not the filtrate's pathway, such recruitment in the standard capillary-lymphatic pathway remains unconfirmed. Our recent findings indicate that flows of injected interstitial fluid were similar in non-edematous and moderately edematous lungs (Figure 9). Hence we find no evidence for edema-induced channel recruitment in the perimicrovascular space, and we do not believe that perimicrovascular liquid removal is significantly enhanced in edema.

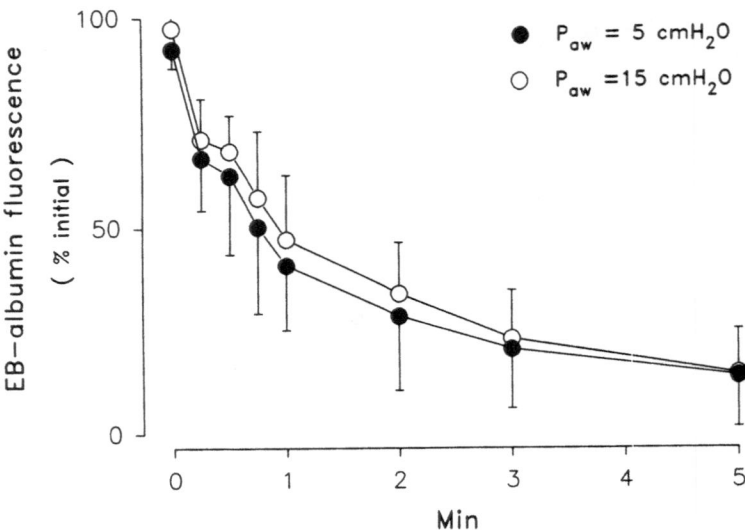

**Figure 9.** Perivenular flow in edema. Symbols same as Figure 8. Data obtained in the presence of moderate edema (extravascular lung water 40% higher than normal). Note lung expansion failed to inhibit liquid removal rate. (*Reprinted from* Ying X., R. Qiao, S. Ishikawa and J. Bhattacharya. Removal of albumin microinjected in rat lung perimicrovascular space. J. Appl. Physiol. *In press*).

Lung expansion which imposes global, if regionally differential stresses on the lung interstitium, needs to be counted amongst factors that may well affect the geometry of interstitial flow channels. It is well known that lung expansion secondary to increase of alveolar pressure, increases diameters of large vessels ("extra-alveolar vessels") but decreases diameters of septal capillaries ("alveolar vessels"). These effects are attributable respectively, to enhancement of the outward interstial stress on large vessels and to direct compression of septal vessels. However in the physiological range of microvascular pressures, juxtaseptal microvessels do not significantly change diameter during increased alveolar pressure (26). Hence at this location, outward and compressive stresses are in balance during lung expansion.

We may conclude from these considerations that lung expansion does not compress the juxtaseptal interstitium and that thereby, it does not compress juxtaseptal lymphatic channels. However it is significant that in our experiments (32), increased alveolar pressure markedly inhibited the perimicrovascular flow of microinjected liquid in the non-edematous lung (Figure 8). Interestingly, this inhibition did not occur in the presence of moderate edema.

These findings when interpreted in the light of our previous interstitial pressure data for lung expansion (4), reveal a possible pressure-flow relation for the liquid exchange interstitium. When we compare the interstitial effects of lung expansion at different levels of lung hydration, we find that perimicrovascular flow was inhibited and the septal-juxtaseptal pressure gradient was abolished in the nonedematous lung (Figures 4 and 8). However both the flow and the pressure gradient persisted when lung expansion was instituted in the presence of edema (Figures 4 and 9). These findings allow us to draw the causal interpretation, that the existence of perimicrovascular flow predicates the existence of the interstitial pressure gradient. This correlative evidence is the strongest yet, that the longitudinal interstitial pressure gradient determined by micropuncture, is in fact the force that determines interstitial liquid clearance.

# CONCLUSION

The new understanding derived from our microphysiological studies of the liquid exchange interstitium, is that factors intrinsic to the perimicrovascular matrix regulate clearance of the lung microvascular filtrate. The importance of this concept must be understood in the context of the previous, and indeed prevailing view, that lung liquid clearance is determined entirely by the efficacy of lymphatic clearance and that the interstitium plays only a nominal role in this process. The old view is clearly incorrect because we now have firm evidence that regional differences in the structure of the interstitium result in differences in local swelling pressure. This in turn, establishes an interstitial pressure gradient that we quantified previously and that we now show, qualifies well as the critical determinant of perimicrovascular liquid removal. To the extent that the gradient is a function of matrix properties that govern interstitial swelling pressures along the flow pathway, interstitial structural considerations must be taken into account when interpreting mechanisms of interstitial liquid removal.

We must therefore, dismantle the view that the liquid exchange interstitium is more or less homogenous in structure and function. Even within the microregions of the perialveolar interstitium, we find that clearance rates of overloaded liquid are considerably different as indicated in the markedly slower septal than extraseptal clearance. This raises the potential concern that the generation of pulmonary edema following augmented filtration rates, may be partly attributable to failure of liquid clearance mechanisms in regions of low interstitial hydraulic conductivity. We have suggestive evidence that the septal interstitium may be one such region and we speculate that the glycosaminoglycan rich interstitium in regions of lung repair, may be another.

One of the challenges to future research is clearly that posed by the microphysiology of the periseptal interstitium. Not only the hydrodynamics, but also the chemistry and the hydrated anatomy of this region needs to be well understood. The application of modern research methods to this critical liquid exchange region of the lung will provide important clues on liquid removal mechanisms that operate normally and that fail during the formation of pulmonary edema.

# ACKNOWLEDGMENTS

I am grateful to Dr. Norman C. Staub and Dr. Matthew Glucksberg for numerous discussions on the lung interstitium. I thank Dr. G.M. Turino for critically reviewing the manuscript. Our research was supported by grants from the National Heart, Lung and Blood Institute, the American Heart Association and the New York Lung Association.

# REFERENCES

1. Anderson-Bray, B., P.M. Sampson, M. Osman, A. Giandomenico and G.M. Turino. Early changes in lung tissue hyaluronan (hyaluronic acid) and hyaluronidase in bleomycin-induced alveolitis in hamsters. Am. Rev. Respir. Dis. 143:284-288, 1991.
2. Bhattacharya, J., T. Cruz, S. Bhattacharya and B. Bray. Hyaluronan affects extravascular water in lungs of unanesthetized rabbits. J. Appl. Physiol. 66:2595-2599, 1989.
3. Bhattacharya, J., M.A. Gropper, and J.M. Shepard. Lung expansion and the perialveolar interstitial pressure gradient. J. Appl. Physiol. 66:2600-2605, 1989.
4. Bhattacharya, J., Gropper, M.A. and N.C. Staub. Interstitial fluid pressure gradient measured by micropuncture in excised dog lung. J. Appl. Physiol. 56:271-277, 1984.

5. Bhattacharya, J., Nakahara, K. and N.C. Staub. The effect of edema on pulmonary blood flow in the isolated perfused dog lung lobe. J. Appl. Physiol. 48:444-449, 1980.

6. Coleridge, J.C.G. and H.M. Coleridge. Afferent vagal C fibre innervation of the lungs and airways and its functional significance. Rev. Physiol. Biochem. Pharmacol. 99:2-110, 1984.

7. Erdmann, A.J., T.R. Vaughan, K.L. Brigham, W.C. Woolverrton and N.C. Staub. Effect of increased vascular pressure on lung fluid balance in unanesthetized sheep. Circ. Res. 37:271-284, 1975.

8. Glucksberg, M.R. and J. Bhattacharya. Effect of dehydration on interstitial pressures in the isolated dog lung. J. Appl. Physiol. 67:839-845, 1989.

9. Glucksberg, M.R. and J. Bhattacharya. Effect of vascular pressure on interstitial pressures in the isolated dog lung. J. Appl. Physiol. 75:268-272, 1993.

10. Hallgren, R., T. Samuelsson, T.C. Laurent and J. Modig. Accumulation of hyaluronan (hyaluronic acid) in the lung in adult respiratory distress syndrome. Am. Rev. Respir. Dis. 139:682-687, 1989.

11. Lai-Fook, S.J. Perivascular interstitial fluid pressure measured by micropipettes in isolated dog lung. J. Appl. Physiol. 52:9-15, 1982.

12. Lai-Fook, S.J. Mechanical factors in lung liquid distribution. Ann. Rev. Physiol. 55:155-180, 1993.

13. Lai-Fook, S.J. and L.V. Brown. Effect of electric charge on hydraulic conductivity of pulmonary interstitium. J. Appl. Physiol. 70:1928-1932, 1991.

14. Laurent, T.C., J. Fraser and E. Robert. Hyaluronan. Faseb J. 6:2397-2404, 1992.

15. Lebel, L., L. Smith, B. Risberg, B. Gerdin and T.C. Laurent. Effect of increased hydrostatic pressure on lymphatic elimination of hyaluronan from sheep lung. J. Appl. Physiol. 64:1327-1332, 1988.

16. Lebel, L., L. Smith, B. Risberg, T.C. Laurent and B. Gerdin. Increased lymphatic elimination of interstitial hyaluronan during E. Coli sepsis in sheep. Am. J. Physiol. 256:H1524-H1531, 1989.

17. Li, J., S.J. Lai-Fook, and R.L. Conhaim. Effect of hyaluronidase on interstitial cuff and pressure response in liquid-inflated rabbit lung. J. Appl. Physiol. 72:1261-1269, 1992.

18. Nettelbladt, O., A. Tengblad, and R. Hallgren. Lung accumulation of hyaluronan parallels pulmonary edema in experimental alveolitis. Am. J. Physiol. 257:L379-L384, 1989.

19. Parker, J.C., S. Gilchrist and J.T. Cartledge. Plasma-lymph exchange and interstitial distribution volume of charge macromolecules in the lung. J. Appl. Physiol. 59:1128-1136, 1985.

20. Parker, J.C. and A.E. Taylor. Comparison of capsular and intra-alveolar fluid pressure in the lung. J. Appl. Physiol. 52:1444-1452, 1982.

21. Parker, J.C. Transport and distribution of charged macromolecules in lung. Adv. Microcirc. 13:150-159, 1987.

22. Qiao, Ren-Li and J. Bhattacharya. Segmental barrier properties of the pulmonary microvascular bed. J. Appl. Physiol 64:2562-2567, 1991.

23. Qiao, Ren-Li, R. Sadurski and J. Bhattacharya. Hydraulic conductivity of ischemic pulmonary venules. Am. J. Physiol.: Lung Cell. Mol. Physiol. 264:382-386, 1993.

24. Qiao, Ren-Li, X. Ying and J. Bhattacharya. Hyperoncotic albumin decreases the endothelial barrier in rat lung. Am. J. Physiol. 265:H198-H204, 1993.

25. Roberts, A.M., J. Bhattacharya, H.D. Schultz, H.M. Coleridge. Effect of lung congestion and edema on pulmonary vagal afferent activity in dogs. Circulation Res. 58:512-522, 1986.

26. Sadurski R., X. Ying, H. Tsukada, S. Bhattacharya and J. Bhattacharya. Diameters of juxta-capillary venules determined by an oil-drop method in rat lung. J. Appl. Physiol. 77:718-725, 1994.

27. Staub, N.C. Pulmonary edema. Physiol. Rev. 54:678-811, 1974.

28. Tajaddini, A. L.V. Brown and S.J. Lai-Fook. Effect of hydration on lung interstitial permeability response to albumin and hyaluronidase. J. Appl. Physiol. 76:578-583, 1994.

29. Taylor, A.E. and J.C. Parker. Pulmonary interstitial spaces and lymphatics. In Handbook of Physiology. Eds. S.R. Geiger. Bethesda, Md. Am. Physiol. Soc. 1984. Vol IV. pp 167-230.

30. Turino, G.M. The lung parenchyma - a dynamic matrix. Am. Rev. Respir. Dis. 132:1324-1334, 1985.

31. Weibel, E.R. and R.G. Crystal. Structural organization of the pulmonary interstitium. In "THE LUNG: Scientific Foundations. Eds. R.G. Crystal, J.B. West eta al. Raven Press, Ltd., New York. 1991. pp 369-380.

32. Ying X., R. Qiao, S. Ishikawa and J. Bhattacharya. Removal of albumin microinjected in rat lung perimicrovascular space. J. Appl. Physiol. (In press).

# 11

# CARDIAC RECEPTOR ACTIVITY IN HEART FAILURE[*]

## Implications for the Control of Sympathetic Nervous Outflow

Irving H. Zucker,[†] Wei Wang, and Harold D. Schultz

Department of Physiology and Biophysics
University of Nebraska College of Medicine
600 S. 42nd Street
Omaha, Nebraska 68198-4575

## INTRODUCTION

The role of cardiac receptors in reflex control of the circulation has been of interest to physiologists and anatomists for many years. The early anatomical work of Nonidez (41) clearly identified receptors in the atria with unencapsulated end organs and myelinated fibers. Subsequently, many investigators have detailed the anatomical structure of a variety of receptors in the atria and ventricles (31). While the early work of Bainbridge (2) implicated atrial receptors in the tachycardia which ensues during atrial distension or volume expansion in the dog, the importance of atrial receptors in reflex control of the circulation and in fluid balance homeostasis was solidified by a variety of prominent investigators, some of whom contributed to this symposium. An outstanding review of the atrial receptor physiology and anatomy has been provided by Linden and Kappagoda in their 1982 monograph (31). While controversy still remains as to the importance of atrial stretch receptors in humans (11), the majority of investigators feel that these receptors are ideally located to function as volume receptors and play some role in both fluid balance and sympathetic control, especially to the kidney (13-15,18,26,27,29,32).

Ventricular receptors also play an important role in cardiovascular regulation. Once again some of the early physiological studies of ventricular receptor discharge properties were done by many of the contributors to this symposium. While the early reflex studies of Von Bezold and Hirt (53) and Jarisch and Richter (25) supported the view that these afferents are capable of initiating bradycardia and depressor reflexes, it was not until the work of the

[*] Many of the studies reported in this review were supported by NIH grants # HL33359 and HL38690.

[†] Address all Correspondence to: Irving H. Zucker, Ph.D., Department of Physiology and Biophysics, University of Nebraska College of Medicine, 600 S. 42nd Street, Omaha, Ne. 68198-4575, (402) 559 7161, FAX: (402) 559 4438, EMAIL: IZUCKER@NETSERV.UNMC.EDU

*Control of the Cardiovascular and Respiratory Systems in Health and Disease*
Edited by C. T. Kappagoda and M. P. Kaufman, Plenum Press, New York, 1995

**109**

Coleridges (8) that a more comprehensive understanding of these receptors was elucidated. While the normal physiological function of cardiac receptors still is not clear, we do know that they inhibit sympathetic outflow when stimulated. It was the work of the Coleridges which provided, to a large extent, the impetus for the studies that were carried out on the role of cardiac receptors in abnormal reflex regulation in heart failure in my laboratory in collaboration with Dr. Joseph P. Gilmore, Dr. Wei Wang, Dr. Marian Brändle and Dr. Harold D. Schultz. I am grateful for the concise and comprehensive studies done by John and Hazel Coleridge over the years. These studies paved the way for investigations of abnormal cardiovascular reflex function and its influence on the circulation.

## ATRIAL RECEPTORS

Since the early work of Henry and Gauer and their colleagues (19,20), it has been generally assumed that vagal afferents located in the low pressure side of the circulation, primarily in the atria, play an important role in fluid balance. The regulation of vasopressin secretion (27), renal sympathetic nerve activity (26,58), renin release (4) and sodium excretion (13,36) have been attributed to changes in the discharge of atrial stretch receptors. If this is indeed the case, then it seems reasonable to hypothesize that a reduction in the sensitivity of these afferents could lead to a depression in renal function to the extent that salt and water retention would ensue. Since resetting and desensitization of arterial baroreceptors has been shown to occur in hypertensive states (9,28), we postulated that a similar phenomenon would exist for atrial receptors in the setting of chronic heart failure. Greenberg et al. (17) showed a depression in right atrial receptor sensitivity in dogs with low output heart failure due to pulmonary artery stenosis and tricuspid avulsion. Anjuchovskij et al. (1) presented similar results in cats and suggested that this was mediated by a reduction in atrial compliance in heart failure. Because the study of Greenberg et al. (17) used an intervention which necessitated thoracotomy with resultant adhesions, we were concerned that the healing process would alter atrial compliance. Studies carried out in this laboratory were undertaken in which the discharge properties of left atrial receptors were evaluated in dogs with an abdominal aorta to vena caval fistula (AV fistula) which resulted in a state of high cardiac output heart failure (56). After approximately 8 weeks these dogs had grossly elevated left ventricular end diastolic pressure and dilated hearts. Table 1 presents the hemodynamics of these dogs compared to sham animals. Pressure-discharge curves were constructed by slow volume expansion and hemorrhage so that the same atrial pressure range was covered in the dogs with heart failure and in the sham operated controls. Figure 1 shows the relationship between left atrial receptor discharge and left atrial pressure taken at the peak of the 'v' wave. As can be seen, the dogs with AV fistulas had a higher threshold, lower slope and lower peak discharge than the sham operated controls. In previous studies we have shown that *acute* heart failure has no effect on atrial discharge sensitivity (12). The mechanism for this depression of discharge sensitivity may be related to structural or functional abnormalities of the receptors themselves or to altered atrial compliance. In order to address the compliance issue, we measured the diameter of the left atrial appendage in four dogs (2 sham and 2 heart failure) using piezoelectric crystals. As can be seen from figure 2, the slope of the pressure diameter relationship is steeper in the heart failure animals indicative of a less compliant chamber. It is also obvious that the resting diameter is significantly increased in the animals with AV fistulas.

Using the methylene blue staining technique, we examined the gross morphology of atrial unencapsulated endings in the left atrial endocardium in dogs with and without high output heart failure. Figure 3 shows representative photomicrographs of receptors from dogs with AV fistulas (right) and normal dogs (left). Normal afferent endings showed the charac-

**Table 1.** Hemodynamics of dogs with an AV fistula and of sham dogs

|  | CHF | SHAM |  |
|---|---|---|---|
| Body Wt. (kg) | 20.0±0.96* (11)+ | 20.6±0.73 (8) | NS |
| Heart Wt. (g) | 212.7±10.8 (11) | 152.0±7.2 (8) | P<.0005 |
| Ht. Wt./bod.wt.(g/kg) | 10.8±0.66 (11) | 7.4±0.35 (8) | P<.0025 |
| MAP (mm Hg) | 98.3±6.8 (11) | 121.4±7.2 (8) | P<.025 |
| LVEDP (mm Hg) | 24.2±8.6 (7) | 4.8±.93 (6) | P<.025 |
| dp/dtmax (mm Hg/s) | 2150.8±206.6 (7) | 1920.7±130.8 (6) | NS |
| dp/dtmax/LVEDP(s) | 90.5±16.3 (6) | 659.5±282.4 (6) | P<.05 |
| CO (l/min) before AV fistula or sham | 2.80±.81 (5) | 2.53±.31 (2) | NS |
| CO(l/min) after AV fistula or sham | 4.37±.83 (5) | 1.92±.26 (2) | P<.0025 |

*± 1 SEM, + numbers in parenthesis refer to the number of animals on which each measurement was made. MAP= mean arterial pressure; LVEDP= left ventricular end diastolic pressure; CO=cardiac output. From ref. 56.

teristic morphology of an arborizing end organ and myelinated fiber. In contrast, the endings from dogs with heart failure exhibited a morphology characteristic of being in various stages of degeneration although the myelinated fiber was still clearly visible. Of course, one can never be sure that the endings observed are representative of the population as a whole. In this regard, we did not observe any normally appearing endings in the dogs with chronic AV fistula's. In a

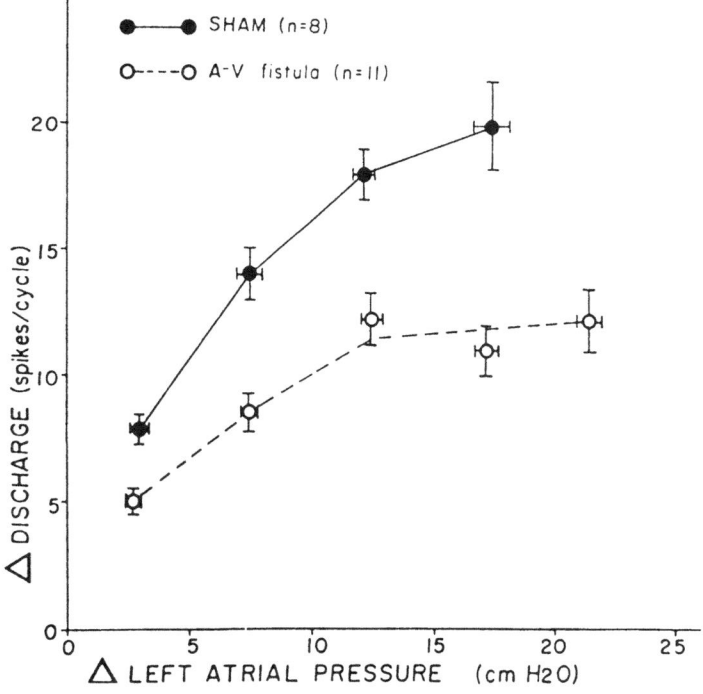

**Figure 1.** The relationship between the change in discharge in spikes/cardiac cycle and the change in left atrial peak v wave pressure. ●, Sham operated dogs; ○, heart failure dogs. Vertical and horizontal bars are ±1SEM. The data was accumulated on 16 receptors from 11 heart failure dogs and 14 receptors from 8 sham dogs. (reproduced with permission from ref. 56).

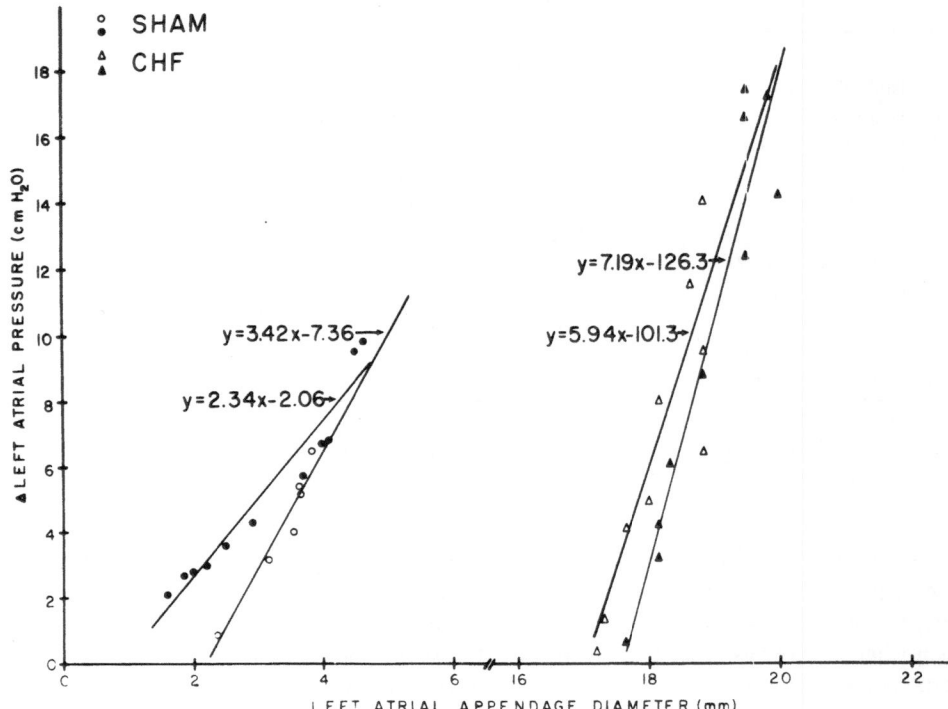

**Figure 2.** The relationship between the change in left atrial end diastolic (peak v wave) pressure and the diameter of the left atrial appendage. (circles represent two separate sham-operated dogs; triangles represent two separate heart failure dogs). (reproduced with permission from ref. 56).

subsequent study, we closed the shunt after the dogs had developed clinical and hemodynamic signs of high output failure (57). The shunt was closed and the dogs were allowed to recover for a period of approximately 8 weeks. While the receptor discharge-pressure relationship had returned to normal, the morphology of the receptors still appeared abnormal; however, we saw many normally appearing endings which was in distinct contrast to the dogs which exhibited symptoms of overt heart failure. We concluded from these studies that chronic atrial dilation reduced the sensitivity of atrial stretch receptors by two possible mechanisms, a change in compliance of the atria and morphological alteration in the receptor endings. Other mechanisms (such as ionic) may also be operative.

In addition to the electrophysiological changes that were documented for atrial receptors in heart failure, the reflexes mediated by these receptors were also investigated (58,62). Inflation of a balloon in the left atrium normally results in a diuresis and natriuresis in dogs (19). However, this diuretic reflex is abolished in dogs with heart failure (62). Interestingly, even though the diuretic response was abolished it was still possible to see a suppression of vasopressin secretion during balloon inflation in dogs with heart failure. Of course, the level to which vasopressin fell was far in excess of that necessary for evoking a diuretic response. Another reflex which has been attributed to the stimulation of atrial receptors is a suppression of renal sympathetic nerve activity (26). We investigated the ability of dogs with high output failure to suppress renal nerve activity during balloon inflation (58). Figure 4 shows that there was a reduction in the ability to suppress renal nerve activity in dogs with heart failure. In fact, renal nerve activity actually increased as a result of arterial baroreceptor unloading.

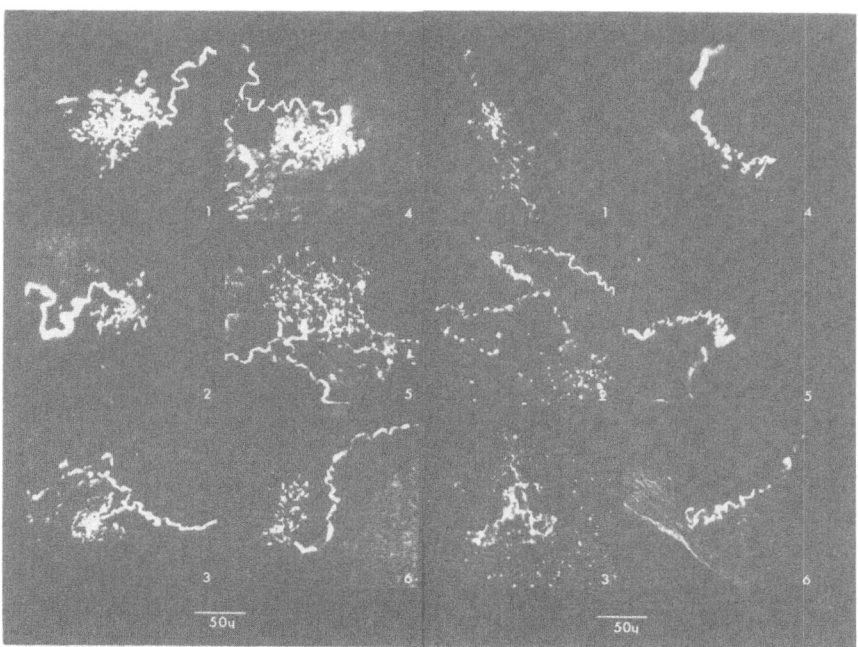

**Figure 3.** Light micrographs of left atrial receptors from sham-operated dogs (left panels) and from dogs with heart failure (right panels). (reproduced with permission from ref. 56).

In summary, the lack of atrial stretch receptor modulation of sympathetic nerve activity and renal function in heart failure most likely contributes to salt and water retention with an eventual exacerbation of the heart failure state.

## VENTRICULAR RECEPTORS: VAGAL

Because ventricular receptors with vagal afferents exert potent sympathoinhibitory effects (34,48,55) these afferents are naturally one reflexogenic area that is of interest in understanding the mechanisms of sympathoexcitation in heart failure. Two main types of

**Figure 4.** The mean data showing suppression of renal sympathetic nerve activity (RSNA) derived from normal and AV fistula dogs in response to left atrial balloon inflation. Open bars depict data from the normal dogs. Closed bars show data from AV fistula dogs. LAP = left atrial pressure. *= significantly different from zero. (reproduced with permission from ref. 58).

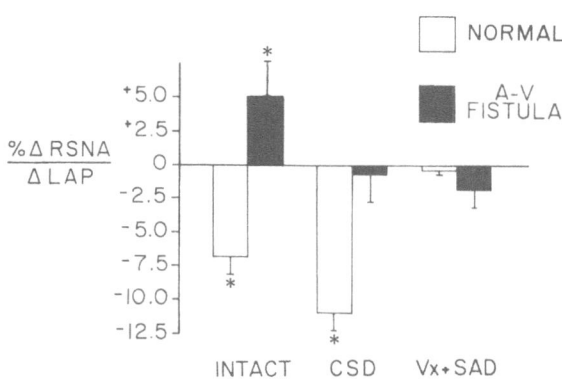

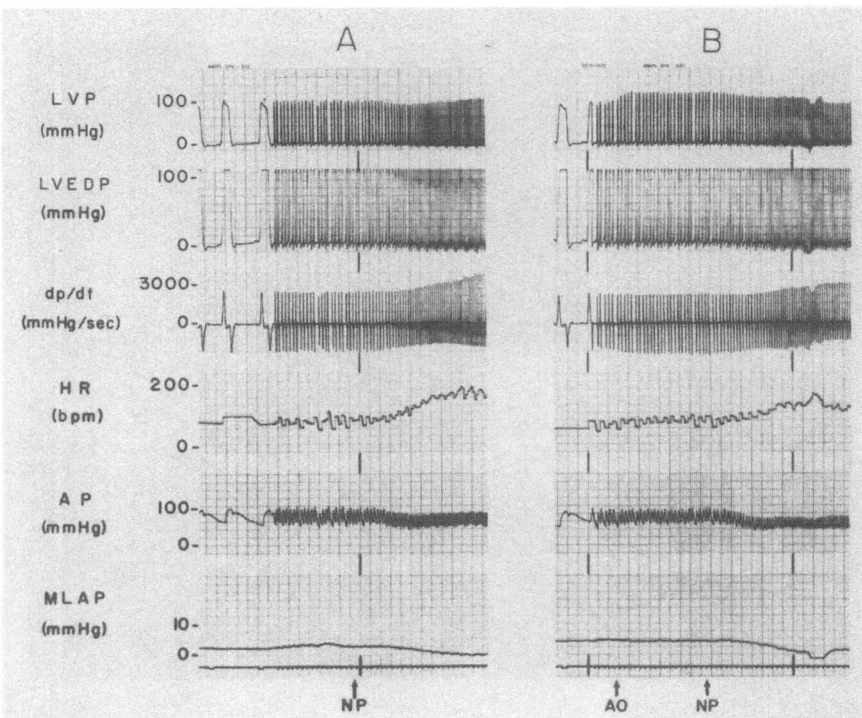

**Figure 5.** Original recording of hemodynamic responses to intravenous nitroprusside infusion before (A) and during (B) increased left ventricular pressure. LVP, left ventricular pressure; LVEDP, left ventricular end diastolic pressure; HR, heart rate; AP, arterial pressure; MLAP, mean left atrial pressure; NP, start of nitroprusside infusion; AO, ascending aortic occlusion. Lowest tracing in each panel is 1 s time markings. (reproduced with permission from ref. 24).

vagal ventricular afferents exist; mechanosensitive and chemosensitive. It has been clearly shown that mechanoreceptors from the left and right ventricles, when stimulated reduce vascular resistance to skeletal muscle and decrease sympathetic nerve activity (35,48). Studies carried out in conscious, sino-aortic denervated dogs in this laboratory showed that obstruction of ventricular outflow resulted in a bradycardia which was abolished by atropine (24,59). Furthermore, stimulation of these mechanoreceptors by simulated aortic stenosis caused an inhibition of the arterial baroreflex control of heart rate (24). Figure 5 shows an original record in which the tachycardia mediated by vena caval occlusion is significantly blunted when carried out in the presence of a partial aortic occlusion. The baroreflex slopes for the relationship of pulse interval and systolic arterial pressure during normal and elevated left ventricular pressure are shown in figure 6.

We hypothesized that ventricular mechanoreceptors would behave much like atrial mechanoreceptors in the setting of chronic heart failure. That is, they would lose their sensitivity and the reflexes that they mediate would also be blunted. In these experiments we used the pacing model of heart failure which was originally described by Coleman et al. (7). In this model ventricular pacing at 250 bpm is carried out for a period of 3-5 weeks. A progressive and severe cardiomyopathy results with characteristic clinical symptoms of low cardiac output congestive heart failure. The hemodynamic responses from a set of these dogs are shown in Table 2 (5). We investigated the effects of left ventricular distension on heart rate before and after the establishment of pacing-induced heart failure. We instrumented dogs

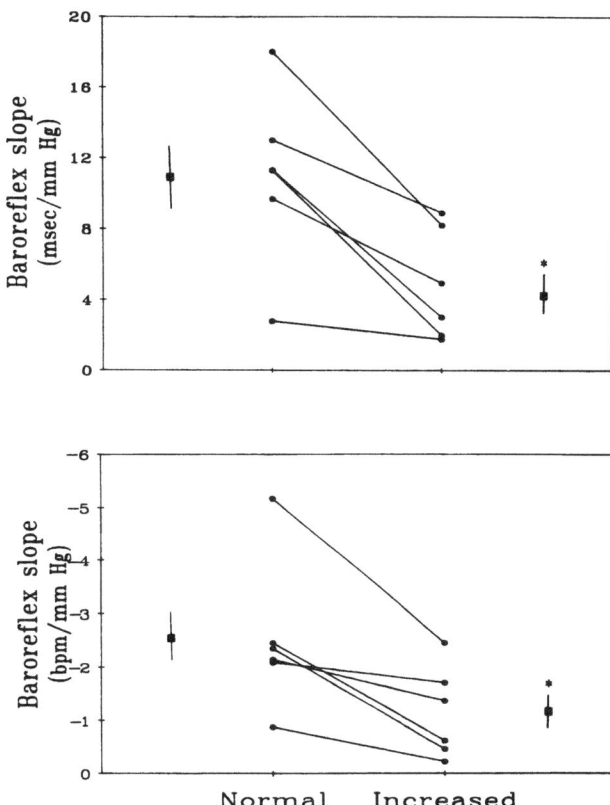

**Figure 6.** Individual (closed circles) and mean (closed squares) data± SE for baroreflex slope of pulse interval (A) and heart rate (B) in dogs infused with nitroprusside during normal and increased left ventricular pressure. * Significantly different from normal left ventricular pressure. (reproduced with permission from ref. 24).

**Table 2.** Hemodynamics before and after 4-6 weeks of left ventricular pacing

|  | Pre Pacing | Post Pacing |
| --- | --- | --- |
| LVSP (mm Hg) | 116.0±18.3 (7) | 105.9±9.5* |
| LVEDP (mm Hg) | 4.6±3.4 (7) | 21.5±6.6+ |
| dp/dtmax (mm Hg/s) | 2250±272 (7) | 1534±447++ |
| SAP (mm Hg) | 121.9±8.3 (13) | 101.4±8.3+ |
| DAP (mm Hg) | 80.9±7.6 (13) | 68.8±7.9+ |
| PP (mm Hg) | 40.9±6.5 (13) | 32.6±6.3+ |
| MAP (mm Hg) | 96.8±7.6 (13) | 80.8±7.6+ |
| LAP (mm Hg) | 2.8±3.5 (12) | 22.2±5.5+ |
| HR (bpm) | 77.1±10.8 (13) | 124.3±13.0+ |

LVSP=left ventricular peak systolic pressure; LVEDP=left ventricular end diastolic pressure; dp/dtmax=maximum of the first derivative of left ventricular pressure; SAP= peak systolic arterial pressure; DAP= diastolic arterial pressure; MAP= mean arterial pressure; PP= pulse pressure; HR= heart rate.
Data are given as mean±SD (n dogs) * $p<0.05$; +$p<0.005$; ++$p<.01$ compared with pre pacing
From Ref. 5

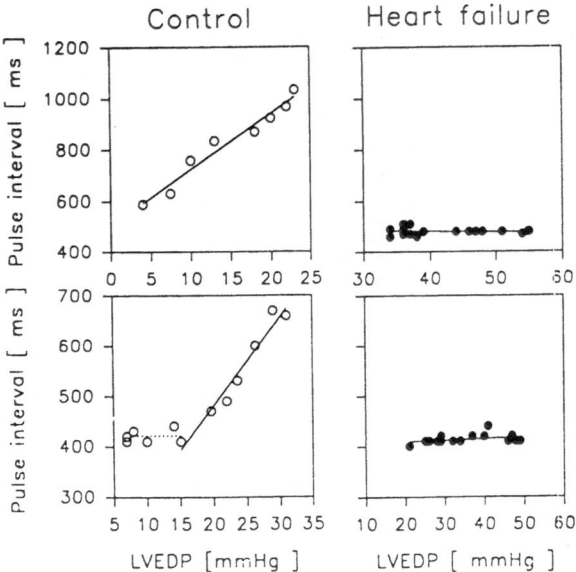

**Figure 7.** Graphs showing the heart rate (pulse interval) response to occlusion of the ascending aorta in two dogs with sino-aortic denervation before (O) and after (●) heart failure was induced by pacing. Pulse interval is plotted against left ventricular end diastolic pressure (LVEDP) on a peat to beat basis. Note that two types of response patterns were observed. The dog represented in the top panels shows a linear decrease in heart rate as LVEDP is increased. This dog also shows a low threshold for this response. The dog represented in the bottom panels shows a high threshold for the bradycardia in response to aortic occlusion. In this dog, heart rate did not change until LVEDP was about 20 mm Hg. In both dogs, there was no response to aortic occlusion after heart failure was induced. (reproduced with permission from ref. 3).

so that left ventricular pressure could be manipulated by occlusion of the ascending aorta (3). Studies were carried out in the intact and sino-aortic denervated state. Figure 7 shows the two types of response patterns that were observed in the control and heart failure states. One out of three dogs responded with a linear increase in pulse interval when left ventricular pressure was raised in the control state (top of figure 7). The others responded with the pattern shown in the bottom of figure 7. There was no change in heart rate until a threshold level of left ventricular pressure was reached. From that point, there was a linear increase in pulse

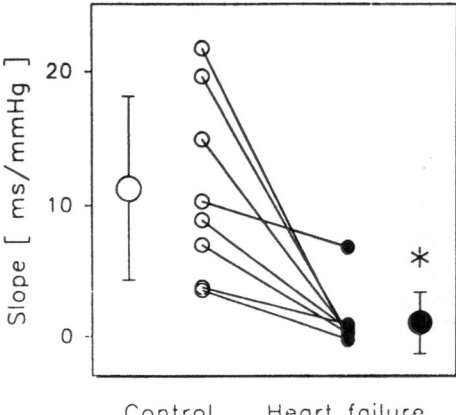

**Figure 8.** Mean and individual slopes of the linear portion of the relationship between left ventricular end-diastolic pressure and pulse interval before (O) and after (●) heart failure. In all but one dog, the slope was reduced to zero after heart failure was induced. The error bars denote ± 1 SD. * P<.05 vs control. (reproduced with permission from ref. 3).

interval. Interestingly, after heart failure was induced, the bradycardia in response to aortic occlusion was abolished irrespective of the initial pattern of response (right side of figure 7). The slope of the linear portion of this relationship was determined for each dog and the mean data are shown in figure 8. A reduction in the slope was consistently observed. In all but one dog, the slope was reduced to near zero. Figure 9 shows the hemodynamic responses in dogs before and after establishment of heart failure and after recovery from heart failure for approximately 4 weeks (i.e. after reprogramming the pacemaker to a rate of 30 bpm). Before sino-aortic denervation, occlusion of the ascending aorta evoked a biphasic response. Initially a tachycardia was evoked as arterial pressure fell. The tachycardia then decreased even though arterial pressure continued to fall. After sino-aortic denervation only a brady-cardia was evoked by aortic occlusion. Following chronic ventricular pacing, the response to ascending aortic occlusion was abolished at 1 week (early) and at approximately 4 weeks (late) of pacing. In three dogs the response was restored after 1 and 4 weeks of recovery from the heart failure state.

From the response of mechanosensitive reflexes and endings we had expected a similar blunted response when chemically sensitive vagal afferents were stimulated. To our surprise, we found that vagal afferent reflexes evoked by chemical stimuli were not attenuated in heart failure, in fact there was some evidence that these reflexes are augmented.

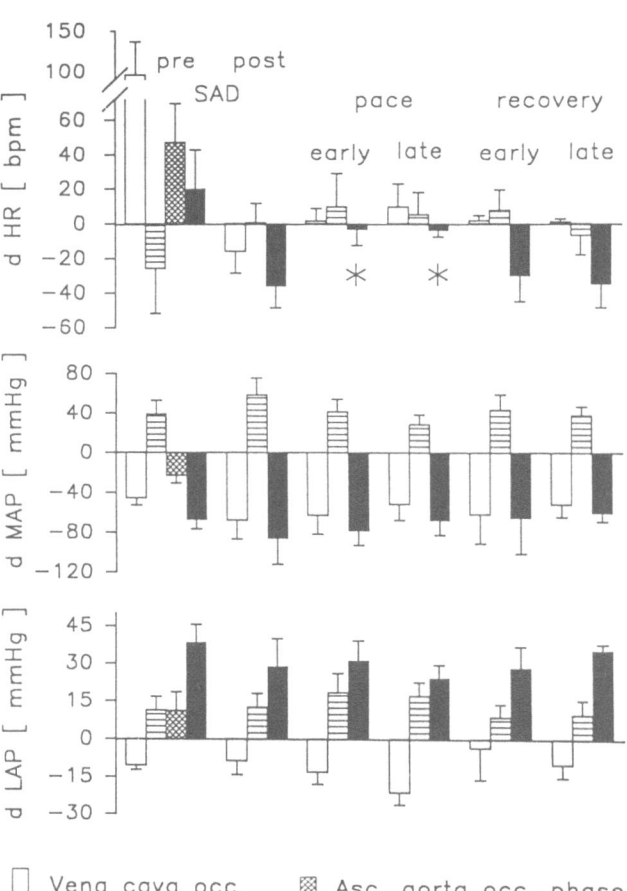

**Figure 9.** Maximum change in heart rate (d HR), mean arterial pressure (d MAP), and mean left atrial pressure (d LAP) in response to various vascular occlusions before and after si-noaortic denervation (SAD) and after pacing was begun. In addition, responses to vena caval occlusions (Vena cava occ.), descending aortic occlusions (Desc. aorta occ.), and ascending aortic occlusions (Asc. aorta occ.) in three dogs after recovery from pacing are shown. Values are mean ± 1 SD. (reproduced with permission from ref. 3).

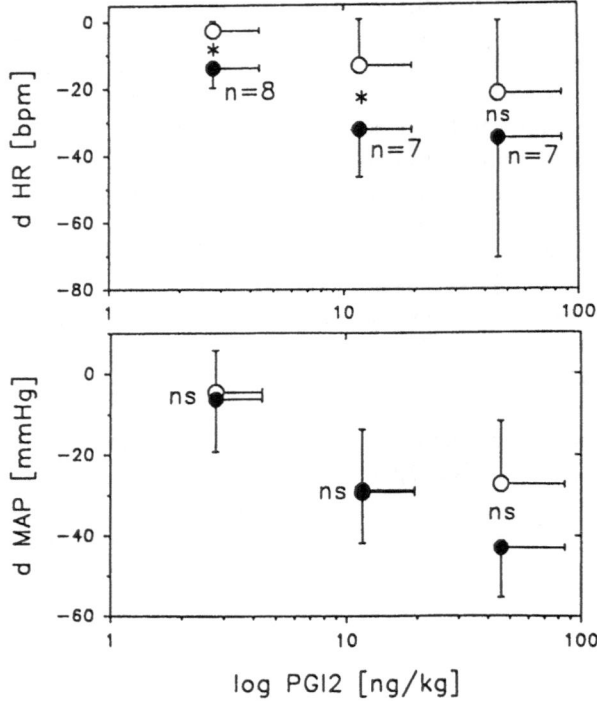

**Figure 10.** Intracoronary prostacyclin (PGI$_2$) dose-response data for the change in heart rate (d HR) and mean arterial pressure (d MAP) before (O) and after (●) induction of heart failure. n indicates the number of dogs for each mean dose-response value. Because each dog had a different sensitivity to PGI$_2$, the data are plotted as the mean log dose against the mean change in response. Error bars indicate ± 1SD; NS, not significant; *= P<.05 vs the pre-paced state. (reproduced with permission from ref. 3).

Figure 10 shows the heart rate and arterial pressure responses to intracoronary administration of various doses of prostacyclin (PGI$_2$). We and others have shown that PGI$_2$ evokes a vagally mediated, Bezold-Jarisch like response in the dog (21,22,40,42,43,60,61). As can be seen, the responses to intracoronary PGI$_2$ were the same or slightly greater in the dogs in the heart failure state, certainly no smaller. This confirms an earlier study from our laboratory in which intracoronary veratridine was used as a stimulus and was found not to evoke a depressed response in dogs with heart failure (6).

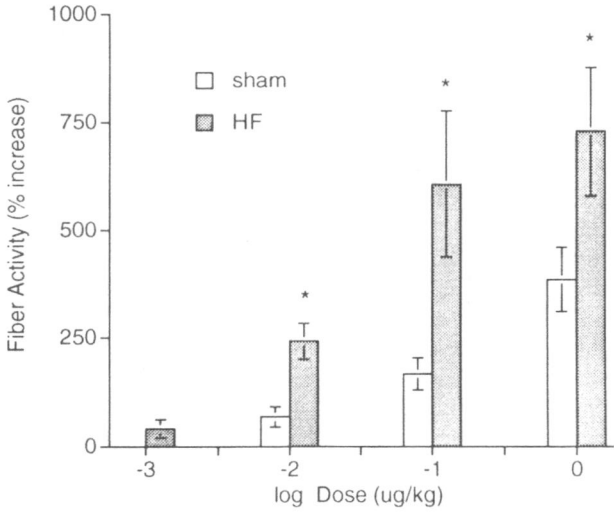

**Figure 11.** A dose-response relationship for vagal c-fiber afferent discharge in response to left ventricular injection of bradykinin in normal sham operated dogs and in pacing-induced heart failure dogs. *= significantly different from sham dogs.

These data indicate that the efferent vagus is capable of reducing heart rate normally during heart failure in these animals. Because both the ventricular mechano and chemoreflex evoke bradycardia by activation of the efferent vagus, the abnormality in the mechanoreflex must be due to either an afferent or central abnormality. This is in contrast to what has been described for the arterial baroreflex control of heart rate in heart failure (10). In fact, muscarinic receptor density is reduced in heart failure (52). The difference between the data obtained in our experiments and those evaluating the arterial baroreflex is not clear. It may be that the two reflexes are organized differently in the CNS. Perhaps different pools of vagal motor neurons are activated when chemically sensitive afferent are stimulated compared to mechanoreceptors such as may reside in the aortic arch, carotid sinus, atria or ventricles. However, one thing seems clear; animals with pacing-induced heart failure do not have an abnormal SA nodal response to activation of left ventricular chemoreflexes.

In order to provide additional evidence for enhanced sensitivity of vagal chemically sensitive afferents from the left ventricle in heart failure, we recorded from vagal c-fiber afferents which responded to both capsaicin and bradykinin (44). All receptors were located in the left ventricle. Figure 11 shows a dose-response relationship for receptor discharge in response to left ventricular injection of bradykinin. As can be seen, receptors from dogs with pacing-induced heart failure had a significantly greater response to bradykinin. This sensitization appears to be receptor specific since there was no enhancement of the response to capsaicin. This sensitization may be mediated by increased levels of prostaglandins since it was reduced by indomethacin.

## VENTRICULAR RECEPTORS: SYMPATHETIC

It is well known that afferent endings from the ventricle also traverse sympathetic pathways. These so called "sympathetic afferents" evoke sympathoexcitatory reflexes which result in increases in blood pressure, heart rate, left ventricular contractility and renal sympathetic nerve activity (16,33,37,45,46). They are also thought to mediate cardiac pain sensation during coronary occlusion (23). In the 1970's, Uchida and coworkers published a series of studies showing that various substances which would be augmented in the myocardium during ischemia were capable of stimulating cardiac sympathetic afferents (49-51). These studies clearly demonstrated that hydrogen ion, potassium and bradykinin could stimulate sympathetic afferents. In addition, others have shown that prostaglandins stimulate sympathetic afferents (39,45). If sympathetic afferents mediate sympathoexcitation during coronary ischemia (38,47) they may also contribute to the chronic elevation in heart failure. Because the heart is dilated and resting heart rate is increased in most models of heart failure, these hearts may be viewed as being in a state of relative ischemia. The reflex effects of cardiac sympathetic afferent stimulation was evaluated in dogs with pacing-induced heart failure in our laboratory (54). The renal sympathetic nerve, arterial pressure and heart rate responses to epicardial administration of capsaicin and bradykinin were tested. Figure 12 shows the responses to two doses of capsaicin applied to the anterior epicardial surface of the left ventricle with small pieces of filter paper. As can be seen, the response is augmented in the dogs with heart failure. A similar response was seen for bradykinin. Pretreatment of the heart with epicardial lidocaine (applied with a patch of filter paper) abolished the response to both bradykinin and capsaicin. Furthermore, lidocaine reduced renal sympathetic nerve activity in the heart failure dogs but not in the normals (Figure 13). These data strongly suggest that cardiac sympathetic afferents provide a tonic level of input to the CNS and can act as an excitatory stimulus to sympathetic outflow in heart failure.

The question of whether or not cardiac receptors are necessary for the sympathoexcitation in heart failure remains open. While the changes described above suggest that these

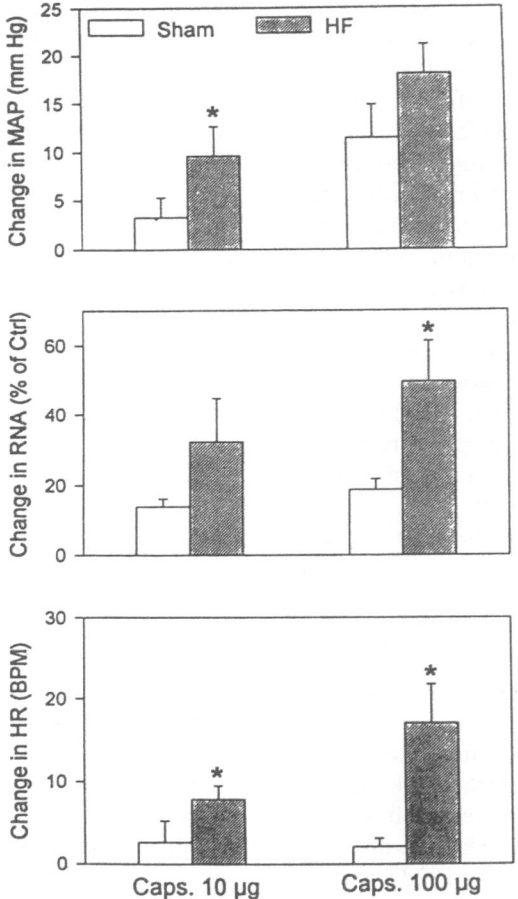

**Figure 12.** The reflex response to two doses of capsaicin applied to the epicardial surface of the left ventricle in sham-operated dogs and in dogs with pacing-induced heart failure. *= significantly different from sham dogs.

receptors play a role, there has not been a definitive cause and effect relationship established. Recently, Levett et al. (30) showed that plasma norepinephrine increases in dogs with cardiac denervation paced into heart failure. One could argue that since the arterial baroreceptors were intact in these animals, cardiac denervation alone would not be sufficient to alter the sympathetic response which occurs in chronic heart failure. However, in a recent study carried out in this laboratory we evaluated the plasma norepinephrine response in paced dogs which were chronically sino-aortic denervated (63). Plasma norepinephrine increased to the same level in intact and sino-aortic denervated dogs even though the severity of the failure was essential identical. What has not been done, of course, is a study of combined cardiac and arterial baroreceptor denervation. This will be a critical experiment to determine if altered reflexes are responsible for the sympathoexitation of heart failure. If reflex mechanisms are not responsible then one must postulate a humoral link between the peripheral circulation and the vasomotor centers in the medulla or perhaps in the hypothalamus. Possible humoral mediators are angiotensin II, nitric oxide, atrial natriuretic peptide and perhaps other as yet to be identified substances. Additional investigation into this intriguing aspect of the reflex control of the circulation will be needed to answer this basic and clinically relevant problem.

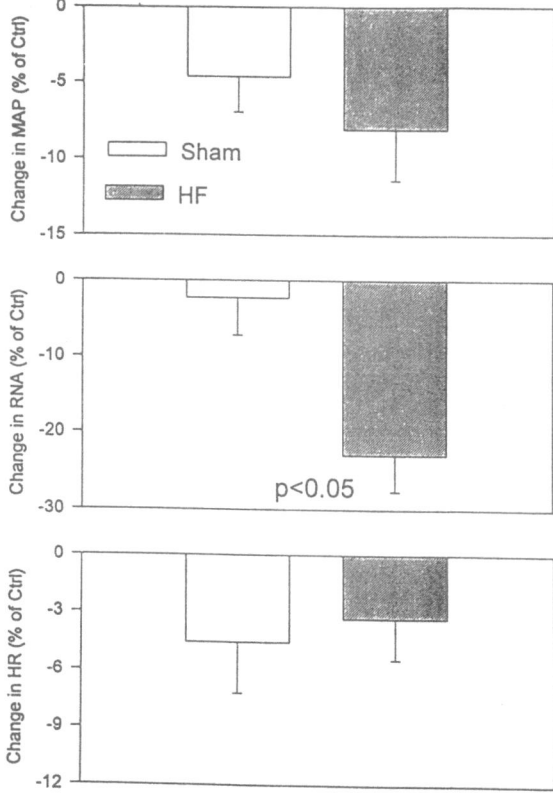

**Figure 13.** The effects of epicardial application of lidocaine on mean arterial pressure (MAP), renal nerve activity (RNA) and heart rate (HR) in sham operated dogs and in dogs with pacing-induced heart failure. These dogs were sino-aortic denervated and vagotomized. A significantly greater decrease in resting RNA was seen in the dogs with heart failure after application of lidocaine.

## ACKNOWLEDGEMENTS

Many of the experiments described in this paper were carried out with the expert technical assistance of Ms. Johnnie F. Hackley and Mrs. Pamela Curry. The authors express their gratitude for their help.

## REFERENCES

1. ANJUCHOVSKIJ, E. P., G. G. BELOSHEPKO, AND F. P. YASINOVSKUYA. Characteristics of atrial mechanoreceptor and elastic features following heart failure. *Sechenov Physiol. J. U. S. S. R.* 62: 1210-1215, 1976.
2. BAINBRIDGE, F. A. The influence of venous filling upon the rate of the heart. *J. Physiol. (Lond.)* 50: 65-84, 1915.
3. BRÄNDLE, M., W. WANG, AND I. H. ZUCKER. Ventricular mechanoreflex and chemoreflex alterations in chronic heart failure. *Circ. Res.* 74: 262-270, 1994.
4. CARR, D. H., D. B. JENNINGS, T. N. THRASHER, L. C. KEIL, AND D. J. RAMSAY. Role of right heart receptors in the control of renin, vasopressin, and cortisol secretion in dogs. *Am. J. Physiol. Regul. Integr. Comp. Physiol.* 263: R1071-R1077, 1992.
5. CHEN, J-S., W. WANG, T. BARTHOLET, AND I. H. ZUCKER. Analysis of baroreflex control of heart rate in conscious dogs with pacing-induced heart failure. *Circulation* 83: 260-267, 1991.
6. CHEN, J-S., W. WANG, K. G. CORNISH, AND I. H. ZUCKER. Baro and ventricular reflexes in conscious dogs subjected to chronic tachycardia. *Am. J. Physiol.* 263: H1084-H1089, 1992.

7. COLEMAN, H. N., R. R. TAYLOR, P. E. POOL, G. H. WHIPPLE, J. W. COVELL, J. JR. ROSS, AND E. BRAUNWALD. Congestive heart failure following chronic tachycardia. *Am. Heart J.* 81: 790-798, 1971.

8. COLERIDGE, H. M., J. C. G. COLERIDGE, AND C. KIDD. Cardiac receptors in the dog, with particular reference to two types of afferent ending in the ventricular wall. *J. Physiol. (Lond.)* 174: 323-339, 1964.

9. COLERIDGE, H. M., J. C. G. COLERIDGE, E. R. POORE, A. M. ROBERTS, AND H. D. SCHULTZ. Aortic wall properties and baroreceptor behaviour at normal arterial pressure and in acute hypertensive resetting in dogs. *J. Physiol. (Lond.)* 350: 309-326, 1984.

10. ECKBERG, D. L., M. DRABINSKY, AND E. BRAUNWALD. Defective cardiac parasympathetic control in patients with heart disease. *New Engl. J. Med.* 285: 877-883, 1971.

11. GILMORE, J. P. Neural control of extracellular volume in the human and nonhuman primate. In: *Handbook of Physiology- The Cardiovascular System III*, Bethesda, Md.: American Physiological Society, 1983, p. 885-915.

12. GILMORE, J. P. AND I. H. ZUCKER. Discharge of type B atrial receptors during changes in vascular volume and depression of atrial contractility. *J. Physiol. (Lond.)* 239: 207-223, 1974.

13. GOETZ, K. L., G. C. BOND, AND D. D. BLOXHAM. Atrial receptors and renal function. *Physiol. Rev.* 55: 157-205, 1975.

14. GOETZ, K. L., B. C. WANG, P. BIE, R. J. LEADLEY,JR., AND P. G. GEER. Natriuresis during atrial distension and a concurrent decline in plasma atriopeptin. *Am. J. Physiol.* 255(24): R259-R267, 1988.

15. GOETZ, K. L., B. C. WANG, P. G. GEER, R. J. JR. LEADLEY, AND H. W. REINHARDT. Atrial stretch increases sodium excretion independently of release of atrial peptides. *Am. J. Physiol.* 250: R946-R950, 1986.

16. GORMAN, A. J., I. H. ZUCKER, AND J. P. GILMORE. Renal nerve responses to cardiac receptor stimulation with bradykinin in monkeys. *Am. J. Physiol.* 244: F659-F665, 1983.

17. GREENBERG, T. T., W. H. RICHMOND, R. A. STOCKING, P. D. GUPTA, J. P. MEEHAN, AND J. P. HENRY. Impaired atrial receptor responses in dogs with heart failure due to tricuspid insufficiency and pulmonary artery stenosis. *Circ. Res.* 22: 424-433, 1973.

18. HAINSWORTH, R. Atrial Receptors. In: *Reflex control of the circulation*, edited by I. H. Zucker and J. P. Gilmore. Boca Raton: CRC Press, 1991, p. 273-289.

19. HENRY, J. P., O. H. GAUER, AND J. L. REEVES. Evidence of the atrial location of receptors influencing urine flow. *Circ. Res.* 4: 85-90, 1956.

20. HENRY, J. P., O. H. GAUER, AND H. O. SIEGER. The effect of moderate changes in blood volume on left and right atrial pressures. *Circ. Res.* 4: 91-94, 1956.

21. HINTZE, T. H. AND G. KALEY. Ventricular receptors activated following myocardial prostaglandin synthesis initiate reflex hypotension, reduction in heart rate, and redistribution of cardiac output in the dog. *Circ. Res.* 54: 239-247, 1984.

22. HINTZE, T. H., X. XU, AND J. WANG. Enhanced ventricular reflex prostaglandin mediated bradycardia during pacing induced heart failure. *Circulation* 84:: II-0389, 1991.(Abstract)

23. HOKA, S., Z. J. BOSNJAK, J. L. SEAGARD, D. SIKER, AND ET AL.. Left ventricular reflex control of venous return and systemic vascular capacitance in dogs. *Can. J. Physiol. Pharmacol.* 66: 112-118, 1988.

24. HOLMBERG, M. J. AND I. H. ZUCKER. Increased left ventricular pressure attenuates the baroreflex in unanesthetized dogs. *Am. J. Physiol.* 251: R23-R31, 1986.

25. JARISCH, A. AND A. RICHTER. Die afferenten Bahnen des Veratrineffektes in den Herznerven. *Archiv. Exp. Path. Pharmak.* 193: 335, 1939.

26. KARIM, F., C. KIDD, C. M. MALPUS, AND P. E. PENNA. The effects of stimulation of the left atrial receptors on sympathetic efferent nerve activity. *J. Physiol. (Lond.)* 227: 243-260, 1972.

27. KAUFMAN, S. AND J. STELFOX. Atrial stretch-induced diuresis in brattleboro rats. *Am. J. Physiol.* 252(21): R503-R506, 1987.

28. KRIEGER, E. M. Time course of baroreceptor resetting in acute hypertension. *Am. J. Physiol.* 218: 486-490, 1970.

29. LEDSOME, J. R. Atrial receptors, vasopressin and blood volume in the dog. *Life Sciences* 36: 1315-1330, 1985.

30. LEVETT, J. M., C. C. MARINELLI, D. D. LUND, B. J. PARDINI, S. NADER, B. D. SCOTT, N. V. AUGELLI, R. E. KERBER, AND P. G. SCHMID, JR. Effects of β-blockade on neurohumoral responses and neurochemical markers in pacing-induced heart failure. *Am. J. Physiol. Heart Circ. Physiol.* 266: H468-H475, 1994.

31. LINDEN, R. J. AND C. T. KAPPAGODA. *Atrial Receptors*. Cambridge: Cambridge University Press, 1982, p. 1-363.

32. LINDEN, R. J., D. A. S. G. MARY, AND D. WEATHERILL. The nature of the atrial receptors responsible for a reflex decrease in activity in renal nerves in the dog. *J. Physiol. (Lond.)* 300: 31-40, 1980.

33. MALLIANI, A. Cardiovascular sympathetic afferent fibers. *Rev. Physiol. Biochem. Pharmacol.* 94: 11-74, 1982.

34. MARK, A. L. The Bezold-Jarisch reflex revisited: Clinical implications of inhibitory reflexes originating in the heart. *J. Am. Coll. Cardiol.* 1: 90-102, 1983.

35. MARK, A. L., J. M. KIOSCHOS, F. M. ABBOUD, D. D. HEISTAD, AND P. SCHMID. Abnormal vascular responses to exercise in patients with aortic stenosis. *J. Clin. Invest.* 52: 1138-1146, 1973.

36. MIKI, K., Y. HAYASHIDA, AND K. SHIRAKI. Cardiac-renal-neural reflex plays a major role in natriuresis induced by left atrial distension. *Am. J. Physiol.* 264: R369-R375, 1993.

37. MINISI, A. J. AND M. D. THAMES. Activation of cardiac sympathetic afferents during coronary occlusion: Evidence for reflex activation of sympathetic nervous system during transmural myocardial ischemia in the dog. *Circulation* 84: 357-367, 1991.

38. MINISI, A. J. AND M. D. THAMES. Distribution of left ventricular sympathetic afferents demonstrated by reflex responses to transmural myocardial ischemia and to intracoronary and epicardial bradykinin. *Circulation* 87: 240-246, 1993.

39. NERDRUM, T., D. G. BAKER, H. M. COLERIDGE, AND J. C. G. COLERIDGE. Interaction of bradykinin and prostaglandin E1 on cardiac pressor reflex and sympathetic afferents. *Am. J. Physiol.* 250: R815-R822, 1986.

40. NGANELE, D. E. AND T. H. HINTZE. Prostacyclin reduces preload in conscious dogs via a vagal reflex mechanism. *Am. J. Physiol.* 253(22): H1477-H1483, 1987.

41. NONIDEZ, J. F. Identification of the receptor areas in the venae cavae and the pulmonary veins which initiate reflex cardiac acceleration (Bainbridge's reflex). *Am. J. Anat.* 61: 203-231, 1937.

42. PANZENBECK, M. J., W. TAN, M. A. HAJDU, AND I. H. ZUCKER. Prostaglandins mediate the increased sensitivity of left ventricular reflexes after captopril treatment in conscious dogs. *J. Pharmacol. Exptl. Therap.* 244(1): 384-390, 1988.

43. PANZENBECK, M. J., W. TAN, M. A. HAJDU, AND I. H. ZUCKER. Intracoronary infusion of prostaglandin I2 attenuates arterial baroreflex control of heart rate in conscious dogs. *Circ. Res.* 63(5): 860-868, 1988.

44. SCHULTZ, H. D., W. WANG, E. USTINOVA, AND I. H. ZUCKER. Enhanced responsiveness of cardiac vagal chemosensitive endings to bradykinin in heart failure. *The FASEB J.* 8: A332, 1994.(Abstract)

45. STASZEWSKA-BARCZAK, J. Prostanoids and cardiac reflexes of sympathetic and vagal origin. *Am. J. Cardiol.* 52: 36A-45A, 1983.

46. STASZEWSKA-WOOLLEY, J., G. WOOLLEY, AND D. REGOLI. Specific receptors for bradykinin-induced cardiac sympathetic chemoreflex in the dog. *Euro. J. Pharm.* 156: 309-314, 1988.

47. THAMES, M. D., T. KINUGAWA, AND M. E. DIBNER-DUNLAP. Reflex sympathoexcitation by cardiac sympathetic afferents during myocardial ischemia: Role of adenosine. *Circulation* 87: 1698-1704, 1993.

48. THORÉN, P. Role of cardiac vagal c-fibers in cardiovascular control. *Rev. Physiol. Biochem. Pharmacol.* 86: 3-94, 1979.

49. UCHIDA, Y. AND S. MURAO. Bradykinin-induced excitation of afferent cardiac sympathetic nerve fibers. *Jap. Heart J.* 14: 84-91, 1974.

50. UCHIDA, Y. AND S. MURAO. Potassium-induced excitation of afferent cardiac sympathetic nerve fibers. *Am. J. Physiol.* 226: 603-607, 1974.

51. UCHIDA, Y. AND S. MURAO. Acid-induced excitation of afferent cardiac sympathetic nerve fibers. *Am. J. Physiol.* 228: 27-33, 1975.

52. VATNER, D. E., D. L. LEE, K. SCHWARZ, J. LONGABAUGH, A. FUJII, AND ET AL.. Impaired cardiac muscarinic receptor function in dogs with heart failure. *J. Clin. Invest.* 81: 1836-1842, 1988.

53. VON BEZOLD, A. AND L. HIRT. Über die physiologischen Wirkungen des essigsauren Veratrin. *Unters Physiol. Lab. Wurzberg* 1: 75-156, 1867.

54. WANG, W. AND I. H. ZUCKER. Enhanced cardiac sympathetic afferent reflex in dogs with heart failure. *The FASEB J.* 8: A603, 1994.(Abstract)

55. ZUCKER, I. H. Left ventricular receptors: Physiological controllers or pathological curiosities? *Basic Res. Cardiol.* 81: 539-557, 1986.

56. ZUCKER, I. H., A. M. EARLE, AND J. P. GILMORE. The mechanism of adaptation of left atrial stretch receptors in dogs with chronic congestive heart failure. *J. Clin. Invest.* 60: 323-331, 1977.

57. ZUCKER, I. H., A. M. EARLE, AND J. P. GILMORE. Changes in the sensitivity of left atrial receptors following reversal of heart failure. *Am. J. Physiol.* 237: H555-H559, 1979.

58. ZUCKER, I. H., A. J. GORMAN, K. G. CORNISH, AND M. LANG. Impaired atrial receptor modulation of renal nerve activity in dogs with chronic volume overload. *Cardiovas. Res.* 19: 411-418, 1985.

59. ZUCKER, I. H., M. J. NIEBAUER, AND K. G. CORNISH. Acute aortic stenosis in the conscious dog: effects of inotropic state on heart rate. *Am. J. Physiol.* 250: H159-H166, 1986.

60. ZUCKER, I. H., M. J. PANZENBECK, S. BARKER, W. TAN, AND M. A. HAJDU. PGI2 attenuates baroreflex control of renal nerve activity by a vagal mechanism. *Am. J. Physiol.* 254(23): R424-R430, 1988.

61. ZUCKER, I. H., M. J. PANZENBECK, J. F. HACKLEY, AND K. HAIDERZAD. Baroreflex inhibition during coronary occlusion is mediated by prostaglandins. *Am. J. Physiol.* 257(R26): R216-R223, 1989.

62. ZUCKER, I. H., L. SHARE, AND J. P. GILMORE. Renal effects of left atrial distension in dogs with chronic congestive heart failure. *Am. J. Physiol.* 236: H554-H560, 1979.

63. ZUCKER, I. H., W. WANG, M. BRÄNDLE, AND K. P. PATEL. Hemodynamic and norepinephrine responses to pacing - induced heart failure in conscious intact and sino-aortic denervated dogs. *The FASEB J.* 7: A532, 1993.(Abstract)

# 12

# ATRIAL RECEPTORS AND HEART VOLUMES

R. J. Linden[*]

Biomedical Sciences Division
King's College London
University of London, United Kingdom

## THE FUNCTION OF ATRIAL RECEPTORS

The atrial receptors have reflex connections to the heart, such that on stimulation of the receptors there is an increase only in the heart rate; reflex connections to the kidney, causing an increase in water and sodium excretion and to the brain causing an inhibition of water intake.

Microscopically these receptors are unencapsulated nerve endings found in the subendocardial tissue at the junctions of the superior and inferior vena cavae and the right atrium, and the pulmonary veins and the left atrium. Methods of stimulating these receptors without interfering with the remainder of the circulation have been devised and result in an increase in heart rate (1,2) ranging from 10 to 90 beats/min. The afferent pathway is solely in the vagal nerves; a technique of cooling the vagal nerves showed that the increased activity in afferent vagal myelinated fibres, and the increase in heart rate, from the stimulation of atrial receptors could both be gradually blocked during stepwise cooling below 18 degrees celsius and completely blocked at 9 degrees celsius (3). The efferent limb of the reflex was solely in the sympathetic nerves to the heart and there was, surprisingly, no positive inotropic effect (1). No effect on efferent vagal nerves has been observed (2). The efferent sympathetic pathway of the reflex to the heart constituted a single bundle of nerves, and these afferent nerves were not involved in any other reflexes (4).

Stimulation of the atrial receptors was shown to inhibit the activity in efferent sympathetic nerves to the kidney, increase the activity in efferent sympathetic nerves to the heart but to have no effect on any other efferent sympathetic nerves (5). However distension of balloons in the left atrium resulted in an increase in urine flow. Using the stepwise cooling technique this reflex response was also shown to be a result of increase activity in atrial receptors and vagal afferent myelinated fibres (1). Right atrial receptors were shown to have the same effect.

[*] Address for correspondence: Prof. R.J. Linden, Physiology Group, Biomedical Sciences Division, King's College London, Strand, London, WC2R 2LS.; Tel. 44-71-873-2541,Fax. 44-71-873-2541.

*Control of the Cardiovascular and Respiratory Systems in Health and Disease*
Edited by C. T. Kappagoda and M. P. Kaufman, Plenum Press, New York, 1995

125

The efferent limb to the kidney of the atrial receptor reflex consists of three components. The nervous component, a depression of efferent sympathetic activity, causes an increase in urine flow and a small increase in sodium excretion; a haemodynamic component, resulting from changes in heart rate, causes a small increase in urine flow and sodium excretion; a hormonal component which causes an increase in excretion only of water and consists of a depression of the secretion of ADH and the increase in the concentration of a diuretic substance, as yet of unknown identity (2,6). There is also a decrease in renin concentration (7).

Some support for this concept that these receptors are volume receptors was supplied by the experiments of Gupta et al (8) who prepared two group of dogs with different blood volumes; one group had a mean blood volume of 95 ml.kg$^{-1}$ and the other 57 ml.kg$^{-1}$. The systemic blood pressure, atrial pressure, heart rate, plasma concentrations of sodium and plasma protein were not different between the two groups. Stimulation of atrial receptors resulted in greater diuretic and natriuretic responses in the dogs with high blood volume than in those with lower blood volume; results which support the hypothesis.

Stimulation of atrial receptors also inhibits drinking in the dog. Fitzsimons and Moore-Gillon (9) distended a small balloon in a pulmonary vein-left atrial junction of the conscious dog which caused a reduced spontaneous water intake without affecting food intake. This inhibition of thirst was not observed when the left vagosympathetic trunk was blocked. The most likely receptors to be involved are the left atrial receptors. Similar effects on drinking have been observed by distending the junction of the right superior vena cava and the right atrium in the rat (10).

## Summary

Stimulation of atrial receptors, discharging into vagal myelinated afferent nerves, has resulted in:

1. An increase in heart rate by efferent sympathetic nerves with no changes in inotropic state.

2. An increase in urine flow by hormonal mechanisms with no change in sodium excretion.

3. An increase in urine flow and sodium excretion by haemodynamic changes and by changes in renal efferent nerves.

4. An inhibition of water intake.

## PROPOSED CONTROL SYSTEM

As part of a control system the above responses can be included in only one hypothesis; that atrial receptors, attached to myelinated vagal afferent fibres, are the first link in a negative feedback mechanism controlling heart volumes. Such a mechanism has already been postulated (1,2,11,12,13). It is suggested that, without such a reflex control system, an increased inflow of blood to the heart, caused by an increase in venous return and/or blood volume, would cause the heart to increase its volumes and work over a range of larger volumes. This proposed control system would cause changes such that the heart would function over a range of small volumes. The heart would be smaller and this would be advantageous.

# ADVANTAGES OF A SMALLER HEART

There appear to be good physical reasons why a small heart is advantageous. As long ago as 1892 it was argued, assuming a spherical ventricle, that in a heart dilated to twice its normal size, the muscle would have to exert itself eight times as much in order to exert the same pressure on the contained blood (14). Although the supposition that the number of muscle fibres per unit are would vary inversely as the square of the radius is probably an exaggerated claim quantitatively, the argument does emphasize that there is a great disadvantage in a dilated heart.

Katz (15) has also argued that there are physical reasons why it is advantageous for the heart to function over a small range of volumes. He pointed out that in frog sartorius muscle (where muscle fibres are arranged in parallel fashion) the amount of heat liberated by the contracting muscle is totally independent of load; but this is not so in muscle where the fibres are not so arranged and where on contraction there are significant shape changes within the muscle, such as frog gastrocnemius and heart muscle. Here the liberation of heat increases with the load, there being extra liberation of heat when the loads are high. Katz (15) goes on to point out that there is an energy wastage when the ventricle increases its size, when it becomes dilated. As the ventricular cavity becomes enlarged the radius of curvature of the muscular walls increases; according to the law of Laplace even if the intraventricular pressure is constant, dilatation decreases mechanical efficiency by increasing the load on the individual muscle fibres.

In addition there is another fact. In the normal heart there is an "unloading" of the myocardium during ejection; when the ventricle ejects the stroke volume the radius decreases significantly. According to the law of Laplace this reduction in ventricular volume leads to a reduction in wall tension even though the intraventricular pressure may remain essentially unchanged. Ejection of the same stroke volume in a dilated heart causes much less change in ventricular dimensions, so that the normal decrease is much reduced. He points out that in the normal heart the reduction in circumference would be about 40% whereas the dilated heart ejects the same amount of blood with a reduction of 5% but at much higher wall tension (in his example, almost double). There are many assumption in his calculations but what they illustrate is that there are good physical reasons why it is advantageous for the heart to work over its normal small range.

## Physiological Mechanisms Affecting Heart Volumes

To respond to an increase in size caused by an increased amount of blood returning to it, or an increase in blood volume, the heart has three main mechanisms, the Frank-Starling hypothesis, an increased activity in sympathetic nerves and an increase in the heart rate. The first two mechanisms are well known. Briefly the Frank-Starling hypothesis allows the simple statement that, an increased inflow to the heart increases the end-diastolic pressure and volume of the ventricle which stretches the muscle fibres and results in an increased force of contraction and stroke volume. Stimulation of sympathetic nerves to the heart, at constant heart rate, causes an increase in its force of contraction from the same end-diastolic pressure (and volume) and an increase stroke volume by encroaching on the systolic reserve volume. The difficulty in demonstrating exactly these relationships resides in the fact that it is not possible to measure the length of, or the force of contraction of, the muscle fibres; therefore we rely on indices of each parameter, which are numerous (16).

One way of illustrating two of the three mechanisms is to consider in the isolated blood perfused heart the relationship between the stroke work (as an index of the force of contraction ) and the end-diastolic pressure (as an index of muscle fibre length) in the left ventricle (see Linden (16), Fig. 6, p345). Clearly both during stimulation of the sympathetic nerves and in its absence, left ventricular output in terms of work/beat and end-diastolic pressure are related in accordance with the Frank-Starling hypothesis. In addition increase activity in sympathetic nerves causes the heart to do more work at the same or lower end-diastolic pressure in the ventricle i.e. the contractility of the muscle has increased; at any even level of work output the end-diastolic pressure is less and thus the heart is contracting from a smaller end-diastolic volume. Thus one of the advantages of this mechanism seems to be that it maintains the heart volumes working over a small range in the face of the increased input and output. It can be concluded that such a mechanism enabling the heart to reduce to its advantageous operating volume is particularly applicable over the normal ranges of volumes and end-diastolic pressures.

However probably the most important mechanism controlling the size of the heart is the third mechanism, the change in heart rate, which is considered below.

## HEART RATE AND HEART VOLUMES

Measurement of the various volumes has been attempted, under various volumes has been attempted, under various conditions of exercise and during simple increases in heart rate, using many different techniques with varying results. Stroke volume is fairly easy to measure during steady states as it may be calculated by measuring the cardiac output and dividing by the heart rate. Other volumes are much more difficult to estimate. Rushmer has recorded incidences of volume in the unanesthetized dog by means of various gauges attached so as to give changes in circumference or diameter (see (16) for refs.).

However changes in volume in dog and man can not be measured accurately either at rest or during exercise, but reasonable estimates can be made. Various X-ray techniques, with and without the injection of contrast media, have been used (for refs. see e.g. 16,17,18,19,20 & 21). It seems that, over a large range of activity, roughly 50 to 60 % of the end-diastolic volume is ejected as stroke volume.

As stated above, the stroke volume (SV) can be calculated from the equation, CO = HR x SV, knowing the cardiac output (co) and heart rate (hr). It may be inferred that if the heart rate increases, the cardiac output must increase, which assumes that the HR and SV are independent variables - which they are not. Roughly, if the heart rate doubles, the stroke volume halves and the cardiac output stays the same or decreases. The effects of an increase in heart rate are considered in more detail.

It has been pointed out (22) that all the evidence suggests that a simple increase in heart rate does not result in an increase in cardiac output; it may be shown that the increase in heart rate from 40 beats/min up to 70 to 90 beat/min produces a small increase in output, but above this rate the cardiac output decreases. An example of an investigation illustrating this type of response is that of Miller et al (23) who stimulated the ventricles of anaesthetized and unanesthetized dogs with heart block at different frequencies and showed that an increase in heart rate from 60 to 90 beats/min resulted in a small increase in cardiac output. At rates between 90 and 150 beats/min the output fell a little and above 150 beats/min fell even more. Similar results have been reported in man by Beevegarde et al (24). Another investigation, that of Rushmer (25) again shows the effect of an almost pure increase in heart rate. Rushmer stimulated the right atrium at different frequencies in unanesthetized dogs and at the same time recorded changes in the diameter of the left ventricle. He observed changes in end-diastolic diameter and stroke diameter and found that with an increase in heart rate from 90 to 180 beats/min there was a progressive decrease in end-diastolic diameter to about half that observed in the control period; this result must mean an even greater decrease in the respective volumes because the diameter is related to the radius whereas the volume is related to a power function of the radius ($r^2$ or $r^3$). Most of the decrease in stroke diameter is brought about by a decrease in diastolic size. This relationship between diastolic volume of the left ventricle and heart rate seem to extend from dog to dog(26). Some support for the inverse relationship between the heart size and heart rate is given by Hamilton et al (27) in the intact dog, who showed that over the range of normal diastolic times of 100-160 msec(heart rate, 180/min) to 700-750 msec(heart rate, 60/min) there was a reasonably linear relationship. There is a little evidence, based on the x-ray technique for measuring heart volumes and the administration of atropine for altering the heart rate (28), to show that in man in the upright posture the heart volume decreases as the heart rate increases above the rate of 75 beats/min.

An important investigation was that of Brutsaert (29) who obtained similar results in anaesthetized and unanesthetized dogs with heart block; in addition he exercised his dogs with heart block at a constant moderate degree of work, so that the oxygen uptake during the exercise was 3 to 3.5 times that at rest and the cardiac output was doubled. Dogs were exercised so that the work load remained constant and the heart rate was varied by pacing the ventricle at 50 beats/min, 120 beats/min and 180 beats/min. During this degree of exercise there was no difference in the cardiac output whether the heart rate was 120 beats/min or 180 beats/min. It is probable though that an increase in the heart rate although decreasing the heart size, would increase the maximum output attainable and, conversely, that a relatively low heart rate for any degree of exercise would limit the maximum output attainable. However, there is little evidence at the moment to allow these conclusions to be drawn. In the studies of Brutsaert (29) the cardiac output appeared to be less at the heart rate of 50 beats/min thus suggesting that this hypothesis might be correct; but the evidence is inadequate to allow a definite conclusion.

An explanation of this relationship between heart rate and heart volumes lies in the rate of filling of the ventricle at different heart rates. When the heart rate is slow, 40-60 beats/min, following the rapid inflow phase there is a period of slow filling (diastasis) during which the volume of the ventricle changes little. When the heart rate increases the time of systole remains relatively constant and it is the period of slow filling which is encroached upon as diastole is shortened; thus the ventricular end-diastolic volume changes little and the output increases. However as the heart rate increases above 70 beats/min the time of rapid filling is encroached upon and becomes less, end-diastolic volume and stroke volume decrease and the output stays the same, or decreases only a little, until the heart rate reaches 150 beats/min. Above 150 beats/min there is further encroachment of the rapid inflow phase (rapid filling time), the inflow is decreased by a relatively larger amount and the cardiac output decreases.

Having considered the three main mechanisms available to the heart in dealing with an increase in venous return we can now examine the changes in heart rate and heart volumes during exercise.

# THE HEART IN EXERCISE

It is important to remember that posture has a great effect on heart size and volumes; most of human physical activity occurs in the upright posture. Therefore the effect of posture on cardiac output and heart volumes emphasized that it is necessary, when examining changes in heart volumes during exercise, to consider only those bouts of exercise in which the subjects were in the erect posture both during the pre-exercise (control) and the exercise periods (16); selection of only such bouts of exercise for consideration removes many of the inconsistences in the previous explanations of the effect of exercise (and heart rate) on the heart volumes. Such an attitude was also emphasized by Gauer and Thron (30) who, having explained that "Clinicians and physiologists for generations worked predominantly on recumbent subjects and generally considered that this posture represents the physiological baseline", state that it would be "more reasonable to consider orthostatic (upright standing) as the resting 'physiologic' posture". Gauer and Thron (30) also comment that this was the opinion of Groedel as early as 1908.

The main reasons for this attitude concern the dramatic changes in the circulatory system of rising from the supine to the upright posture in man. The many reports on these effects have been summarized (30,31,32,33): there is a reduction in the end-diastolic volume leading to a decrease of 40-50% in stroke volume, a decrease in cardiac output of about 30% but an increase in heart rate of 10 to 20 beats/min. The cause of these changes is a shift of blood to the dependent parts, the lower limbs up to the groin receiving about 500 ml., 78% of which comes from the intrathoracic vascular compartments, reducing the quantity of blood in the heart and pulmonary circuit by about 25%.

The heart volumes, calculated from the change in densities in the heart and lungs measured by X-ray techniques, were said to be reduced during the change from the supine to the upright posture by about 130 ml. in men and 90 ml. in women. Given that there is approximately a 500 ml. shift from the intrathoracic vascular bed to the lower limbs, 78 % (400ml.) of which comes from the heart and pulmonary circulation allowing less that 1/4 of this to derive from the left ventricle, a calculation of a reduction in volume of about 70 ml. from the left ventricle seems to be a reasonable value. It also seems that this reduction in volume would occur from a maximal value for the end-diastolic volume. Sjostrand (for ref see (16)) has suggested that the size of the cardiac silhouette in man is maximal in the recumbent posture; he was unable to cause the heart to expand beyond the size observed in the supine position. Rushmer (34) in the resting inattentive,recumbent, unanesthetized dog infused blood to raise the effective atrial pressure by 15 mmHg, but failed to increase the left ventricular diameter by as much as 1 mm. It is tempting to assume that in the lying position the heart, in dog and man, is at its maximal volume at the end of diastole because of the limitation imposed by the pericardium. It also suggest that exercise in the upright posture allows end-diastolic volume to increase whereas in the supine position it can only stay the same or become smaller.

## Changes in Heart Rate and Volumes during Exercise in the Erect Posture

In the erect posture the heart is poised to deal with the amount of blood returning to it. Some justification for this view is provided by accepting the importance of the muscle pumps in returning the blood to the heart in the upright posture. Most arguments, and in most textbooks, for the role of the heart as a pump primarily responsible for the circulation of the blood have been based on experiments in anaesthetized animals lying on their backs with their chest and abdomen open - a condition where *vis a tergo*, the force from behind, attributed to the contraction of the heart muscle, can easily be demonstrated to be the only important force. But once the upright posture is attained the muscle pumps must be more involved.

Indeed there is evidence from experiments completed in the dependent leg of man (35) as well as the dependent limb of the cat (36) that the muscles in heavy rhythmic exercise can allow an input and sustain an output from the muscles of the lower limb equivalent to 20-25 L/min in man at about 60 beats/min. The evidence also suggest that this effect is purely a function of the muscle pump because the arterial inflow to the muscular vascular bed occurred only between contractions, whereas nearly all the venous outflow (pump outflow) from the muscles occurred during the contractions. Support for this argument comes from patients with incompetent venous valves in the deep veins of their legs who have much reduced aerobic capacity (37,38).

## Evidence of Change in Heart Volumes in Exercise

First, what would we expect the maximum increase in stroke volume to be? A subject with a heart rate of 70 beats/min might have a cardiac output of 5 L/min at rest thus giving a stroke volume of about 70 ml. When exercising he may have maximal output of 25 L/min at a heart rate of 180 beats/min, thus giving a stroke volume of 140 ml. - double that at rest. Such results have been observed experimentally in man (e.g. 39,40). From the work of Lilijestrand (39) who estimated the change in volumes of the heart in the upright posture by X-ray in two dimensions it is also possible to conclude that there was an increase in the ventricular size during exercise; but the increase in end-diastolic volume was only equal to about half the increase in stroke volume, so that about half the increase in stroke volume had been provided by an encroachment on the end-systolic reserve volume (16). In the same review after examining the publications of Rushmer in conscious dogs it was possible draw similar conclusions. These conclusions are supported by considering evidence from the use of B-blocking drugs in man (e.g. 41) and dog (e.g. 42).

Thus in man, and conscious dog, from resting standing, to a steady state of maximal exercise standing, the heart rate increases to about 2 1/2 times the resting value, the cardiac output increases up to 5 times the resting value and the stroke volume doubles; about half the increment in stroke volume arises from an increase in end-diastolic an half from an encroachment on end-systolic reserve volume.

## SUMMARY

So now we have a view of the circulation in exercise in the upright posture. The output of the heart depends on the input which is driven by the muscle pumps mainly from the lower limbs but augmented by the upper limbs, and the thoracic and abdominal muscles. The heart thus receives an increasing inflow of blood but the heart volumes change comparatively little - only over a very small range of values. Physiologists, in animals and man, have measured changes in cardiac output and heart volumes - it is difficult to be precise about the results but we can agree on order of magnitude. Cardiac output can increase 5,6 or 7 times in trained athletes but the stroke volume never more than doubles, the end-diastolic volume increases only 50% and the end-systolic volume decreases by about the same amount (16,43). The heart rate increases about 2 1/2 times in the untrained to 5 times in the trained physically fit athlete.

It would also seem that the main influence causing an increase in heart size is an increase in venous return mainly caused by the muscle pumps, An increase in activity in sympathetic nerves to the heart muscle forms a mechanism which assists in the control of heart volumes, decreasing end-systolic volume and increasing stroke volume. It is not known what causes this increase in sympathetic activity during exercise but it may just be concomitant with the increase in activity in nerves to the skeletal muscles and originate in a similar area of the brain.

There also seems no doubt that as the heart rate increases, with no other concomitant changes, the heart volumes decrease. Thus is may concluded that any increase in heart rate during exercise would limit the increase in the size of the heart which results from the increase in venous return. Next we must consider mechanisms which might affect the two efferent nerves to the heart, the sympathetic and the vagal nerves, these being the final common path of changes in heart rate.

# REFLEXES AFFECTING HEART RATE

What we have to consider is which, of all the reflexes affecting heart rate, could act as a negative feed-back on heart volumes in the face of an increase in venous return (cardiac output), this being the commonest cause of an increase in heart volumes. We shall only consider arterial baroreceptors and atrial receptors. It has been claimed that the baroreceptors could be sensitive to changes in aortic blood flow and therefore could signal changes in stroke volume (and therefore venous return and cardiac output) (44) but the baroreceptors are probably more importantly related to the peripheral vasculature (45,46), and in addition baroreceptor reflexes appear not to be effective during exercise in man (47) and in the dog (48).

The atrial receptor reflex, described earlier, could be the reflex involved in controlling heart volumes. Atrial receptors increase their discharge with increase stretch (1) and with an increase in volume (49). They also can signal rates of change; atrial receptors are sensitive to the rate of rise of atrial pressure during the 'v' wave of the atrial pressure pulse (49), to the rate of change stretch of the atrial wall (50), to the rate of change of atrial diameter (51), and to the rate of change of atrial volume as well as to atrial volume itself (52). Thus they could signal the rate of inflow to the atria as well as the average volume.

# NEGATIVE FEED-BACK ON HEART VOLUMES

Because the time of ventricular systole is relatively constant, the rate of change of change of pressure up to the 'v' wave of the atrial pressure pulse (at the end of ventricular systole ) is directly related to the inflow into the atrium, i.e. more flow over a constant time causes an increase in the rate of change of pressure. The atrial receptors discharge mainly during ventricular systole. Thus an increase in blood flow into the atria, particularly during ventricular systole, with the atrioventricular valves closed, causes an increase rate of change of pressure in the atria, which in turn causes an increase discharge from the atrial receptors and thus an increase heart rate by this reflex: and this increase in heart rate would then reduce the time of filling and maintain the end-diastolic volume (and thus atrial pressure and volume) at a relatively constant value despite the increase in venous return. It has thus been argued that the main function of changes in heart rate *is* the control of heart volumes, particularly in exercise (2).

Atrial receptors, as well as responding with an increased discharge to an increased rate of change of pressure (volume), also respond in the same way to an increase in mean pressure (volume) of the atria. Thus it is possible also to speculate that this mechanism maintains the heart sizes within narrow limits not only in face of increase venous return but also during an increase in extracellular volume and blood volume.

This increase in discharge of atrial receptors will, in the short term, cause an increase in heart rate and a decrease to smaller heart volumes; in the long term the reflex mechanism involving the kidney (water and sodium) may adjust the extracellular fluid volume so that the blood volume becomes less and thus the heart volumes less. In addition, because of the inhibition of thirst there would be less intake of water, greater concentration of sodium, more output of sodium, and a relatively contracted extracellular volume and blood volume by this route.

It is therefore postulated that together, the atrial receptors and the two responses which can be elicited during their stimulation (the effect on blood volume and the effect on heart rate), form a remarkable control system which controls the size of the heart - and keeps the size of the heart within narrow limits even during severe exercise when the heart pump may have to increase its output from 5 liters/min at rest to 35 liters/min in severe exercise - an enormous increase.

## ACKNOWLEDGEMENT

I would like to thank the British Heart Foundation for support for some of this work.

## REFERENCES

1. Linden, R.J. and Kappagoda, C.T. (1982). Atrial receptors. Cambridge Univ. Press, Cambridge.
2. Linden, R.J. (1987). The function of atrial receptors. In Cardiogenic Reflexes.ed. Hainsworth, R., McWilliam, P.N. and Mary, D.A.S. pp. 18-39. Oxford Univ. Press, Oxford.
3. Kappagoda, C.T., Linden, R.J. and Sivananthan, N. (1979). The nature of the atrial receptors responsible for a reflex increase in heart rate in the dog. J. Physiol *291*, 393-412.
4. Linden, R.J., Mary, D.A.S. and Walters, G.E. (1982). The response in efferent cardiac sympathetic nerves to stimulation of atrial receptors, carotid sinus baroreceptors and carotid chemoreceptors. Quart.J. exp. Physiol. *67*, 151-163.
5. Karim, F., Kidd, C., Malpus, C.M. and Penna, P.E. (1972). The effect of stimulation of the left atrial receptors on sympathetic efferent nerve activity. J. Physiol. *227*, 243-260.
6. Bennett, K.L. and Linden, R.J. (1988). Diuresis from atrial receptors after hypophysectomy and local ablation with no changes in plasma vasopressin. Quart,J.exp. Physiol.*73*, 755-765.
7. Drinkhill, M.J., Hicks, M.N., Mary, D.A.S. and Pearson, M.J. (1988). The effect of stimulation of the atrial receptors on plasma renin activity in the dog. J. Physiol. *398*, 411-421.
8. Gupta, B.N., Linden, R.J., Mary,D.A.S. and Weatherill, D. (1982). The diuretic and natriuretic responses to stimulation of left atrial receptors in dogs with different blood volumes. Quart,J. exp. Physiol. *67*,235-258.
9. Fitzsimons, J.T. and Moore-Gillon, M.J. (1982). Pulmonary vein-atrial junction stretch receptors and the inhibition of drinking. Amer.J.Physiol.*242*, R453-R457.
10. Kaufman, S. (1984). Role of right atrial receptors in the control of drinking in the rat. J.Physiol. *349*, 389-396.
11. Furnival, C.M., Linden, R.J. and Snow, H.M. (1971). Reflex effects on the heart of stimulating left atrial receptors. J.Physiol. *218*, 447-463.
12. Linden, R.J. (1973). Function of cardiac receptors. Circulation. *48*, 463-480.
13. Linden, R.J. (1975). Reflexes from the heart. Progress in Cardiovascular Research. *18*,201-221.
14. Woods, R.H. (1892). A few applications of a physical theorem to membranes in the human body in a state of tension. J. Anat. & Physiol.*26*,362-370.
15. Katz, A.M. (1977). Physiology of the Heart. New York, Raven Press.
16. Linden, R.J. (1963). The control of the output of the heart. In recent advances in Physiology.ed. Creese.R.pp.330-381. Cambridge Univ. Press, Cambridge.
17. Brecher, G.A. and Galletti, P.M. (1963). Functional anatomy of cardiac pumping. In Handbook of Physiology, Section 2: Circulation, Vol. II. eds. Hamilton, W.F. and Dow, P. Amer. Physiol.Soc., Washington, D.C.
18. Gauer, O.H. (1955). Volume changes in the left ventricle during blood pooling and exercise in the intact animal. Their effects on left ventricular performance. Physiol.Rev. *35*, 143-155.
19. Gribbe, P.L., Hirvonen, J.L. and Wegilius, C. (1959). Cineangiocardiographic recordings of the cyclic changes in volume of the left ventricle. Cardiologia. *34*, 348-366.
20. Holt, J.P. (1956). Estimation of the residual volume of the ventricle of the dog's heart by two indicator dilution techniques. Circulation Res.*7*, 187-195.
21. Folse, R., Braunwald,E and Aygen, M.M. (1961). Clinical technique for determining the fraction of left ventricular end-diastolic volume ejected per beat. Circulation. *24*, 934.
22. Linden, R.J. (1968). The heart-ventricular function. Anaesthesia. *23*, 566-584.
23. Miller, D.E., Gleason, W.L., Whalen, R.E., Morris, J.J. and McIntosh, H.D. (1962). Effect of ventricular rate on the cardiac output in the dog with chronic heart block. Circulation Res. *10*, 658-663.
24. Bevegarde, S., Johnsson, B., Karlof, I., Lagergren, H, and Sowton, E. (1967). Effect of changes in ventricular rate on cardiac output and central pressures at rest and during exercise in patients with artificial pacemakers. Cardiovascular res. *1*,21-33.
25. Rushmer, R.F. (1959). Consistency of stroke volume in ventricular responses to exertion. Ameri. J. Physiol.*196*,745-750.
26. Chapman, C.B., Baker, O. and Mitchell, J.H. (1959). Left ventricular functin at rest and during exercise.J. Clin. Invest.*38*,1202-1213.

27. Hamilton, W.F., Remington, J.W. and Hamilton W.F., Jr. (1950). Factors relating to heart size in the intact animal. Amer. J.Physiol.163, 260-267.

28. Larsson, H. and Kjellberg, S.R. (1948). Roentgenological heart volume determination with special regard to pulse rate and the position of the body. Acta.Radiol.Scand.29,159-177.

29. Brutsaert, D. (1965). Influence of different stimulation frequencies on the cardiac output at rest and during moderate exercise in dogs with chronic atrioventricular heart block. Acta Cardiologica.20,469-498.

30. Gauer, O.H. and Thron, H.L. (1965). Postural changes in the circulation. In Handbook of Physiology, Section 2: Circulation Vol. III. eds. Hamilton, W.F. and Dow, P. Amer.Physiol.Soc., Washington, D.C.

31. Rowell, L.B. (1986). Human Circulation Regulation During Stress. New York, Oxford University Press.

32. Rowell, L.B. (1993). Human Cardiovascular Control. New York, Oxford University Press.

33. Wieling, W., and Wesseling, K.H. (1993). Importance of reflexes in the circulatory adjustments to postural changes. In Cardiovascular Reflex Control in Health and Disease. eds. Hainsworth, R. and Mark, A.L., London, Saunders.

34. Rushmer, R.F., Smith, O. and Franklin, D. (1959). Mechanisms of cardiac control in exercise. Circulation Res. 7,602-627.

35. Folkow, B., Hagland, U., Jodal, M. and Lundgren, O. (1971). Blood flow through the calf of man during heavy rhythmic exercise. Acta.Physiol.Scand.81,157-163.

36. Folkow, B., Gaskell, P. and Waaler, B.A. (1970). Blood flow through limb muscles during heavy rhythmic exercise. Acta.physiol.scand. 80,61-72.

37. Arenander, E. (1960). Hemodynamic effects of varicose veins and results of radical surgery. Act.chir.scand. Supply.260.

38. Bevegarde, S. and Lodin, A. (1962). Postural circulatory changes at rest and during exercise in five patients with congenital absence of valves in deep veins of the legs. Acta.med.scand.172, 21-29.

39. Liljestrand, G., Lysholm, E and Nylin, G. (1938). The immediate effects of muscular work on the stroke and heart volumes in man. Skand.Arch.Physiol. 80,265-282.

40. Mitchell, J.H., Sproule, B.J. and Chapman, C.B. (1958). The physiological meaning of the maximal oxygen intake test.J.clin. Invest.37, 537-547.

41. Epstein, S.E., Robinson, B.F., Kahler, R.L. and Braunwald, E. (1965). Effects of beta-blockade on the cardiac response to maximal and submaximal exercise in man.J.Clin.Invest. 44, 1745-1753.

42. Donald, D.E., Fergusion, D.A. and Milburn, S.E. (1968). Effect of beta-adrenergic receptor blockade on racing performance of greyhounds with normal and denervated hearts. Circulation Res. 22, 127- 134.

43. Linden, R.J. and Snow, H.M. (1974). The inotropic state of the heart. In Recent Advances in Physiology. ed. Linden, R.J. pp.148-190. Cambridge Univ. Press, Cambridge.

44. Baertschi, A.J. and Charlton, J.D. (1982). Cardiac output: Is it signalled to the brain? In Cardiovascular System Dynamics. Eds. Kenner, T., Basse, R. and Hinghafer-szalkay, H. pp.471- 478. Plenum Press, London.

45. Heymans, C. and Neil, E. (1958). Reflexogenic areas of the cardiovascular system. Churchill, London.

46. Hainsworth, R., Ledsom, J.R. and Carswell, F. (1970). Reflex responses from aortic baroreceptors. Amer.J. Physiol.218,423-429.

47. Bristow, J.D., Brown, E.B., Cunningham, D.J.C., Howson, M.G., Peterson, E.S., Pickering, T.G. and Sleight, P. (1971). Effect of bicycling on the baroflex regulation of pulse interval. Circulation Res.28,582-592.

48. McRitchie, R.J., Vatner, S.F., Boettcher, D., Heydrickx, G.R., Patrick, T.A. and Braunwald, E. (1976). Rose of arterial baroreceptors in mediating cardiovascular response to exercise. Amer.J.Physiol. 230,85-89.

49. Paintal, A.S. (1973). Vagal sensory mechanisms and their reflex effects. Physiol.Rev.53,159-227.

50. Arndt, J.O., Bamberg, P., Hindorf, K., and Roehnelt, M. (1974). The afferent dischange pattern of atrial mechanoreceptor in the cat during sinusoidal stretch of atrial strips in situ. J.Physiol.240,33-52.

51. Recordati, G., Lombardi, F., Bishop, V.S. and Malliani, A. (1975). Response of type B atrial vagal receptors to changes in wall tension. Circulation Res. 36, 682-691.

52. Baertschi, A.J. and Gann, D.S. (1977). Responses of atrial mechanoreceptor to pulsation of atrial volume. J.Physiol.273, 1-21.

# 13

# ASPECTS OF CORONARY VASOMOTOR REGULATION[*]

Julien I. E. Hoffman[†], Giovanni Piedimonte, Andrew J. Maxwell, Jay A. Nadel, Shiro Iwanaga, and Waleed K. Husseini

Cardiovascular Research Institute and
Departments of Pediatrics and Internal Medicine
University of California, San Francisco, California 94143

## INTRODUCTION

Matching the supply of blood and oxygen to the myocardial demand for oxygen is an essential function of the coronary regulatory system. The heart cannot tolerate a deficit of oxygen for more than a few seconds without harmful consequences, so that the regulatory system must respond rapidly to changes in oxygen demand or perfusion pressure. The matching of oxygen supply to myocardial oxygen demand can be examined in three main ways. One of them is to test autoregulation which may be defined as the response to a change in perfusion pressure when myocardial oxygen demand remains unchanged. To do this in the heart, the coronary artery must be cannulated so that changes in perfusion pressure have no or relatively little effect on myocardial oxygen demand. The second method is to examine the response during constant perfusion pressure to changed myocardial oxygen demand which may either be increased by pacing, increases in cardiac pressure or volume work, or infusion of metabolic stimulators like calcium or catecholamines, or may be decreased by decreasing cardiac pressure or volume work or by decreasing contractility. A subset of these experiments is to examine maximal flows induced by severe exercise or very rapid pacing. The third method is to examine the reactive hyperemic response; coronary flow is stopped for several seconds, and the hyperemic response after flow is restored is examined. It is possible that all these three methods are examining the same regulatory systems, except that reactive hyperemia only examines decreases in oxygen supply and definitely produces transient ischemia.

On general principles, contraction of vessels is due to increased calcium entry and activity, and dilatation of vessels is due to removal of calcium and a decrease in its activity.

[*] Supported in part by Program Project Grants HL 25847 and HL-24136 and Training Grant HL-07192 from the National Institutes of Health

[†] Address for correspondence: Julien I.E.Hoffman, M.D., Box 0544, University of California, San Francisco, CA, 94143; Telephone: 415-476-9313; FAX: 415-476-0676

*Control of the Cardiovascular and Respiratory Systems in Health and Disease*
Edited by C. T. Kappagoda and M. P. Kaufman, Plenum Press, New York, 1995

135

All vessels are exposed to forces that tend to widen them and other forces that tend to narrow them, and at any moment the resistance offered to flow through the vessel depends on the balance of forces. Changes in calcium action may be mediated by a wide variety of mechanisms. These include changes in the amount of cyclic AMP due to altered activities of membrane bound receptor-adenylate cyclase-G protein complexes, related to occupancy of $\beta_1$-adrenergic, $\beta_2$-adrenergic, histamine $H_2$, vasoactive intestinal polypeptide, adenosine $A_1$, acetylcholine $M_2$, and somatostatin receptors by their specific agonists. Another mechanism involves $\alpha_1$-adrenergic receptors which activate calcium through changes in phospholipase C and so increase concentrations of diacylglycerol and inositol phosphate. A third mechanism involves increased cyclic GMP due to stimulation of soluble guanylate cyclase by nitric oxide, a reactive radical produced by endothelial stimulation of a nitric oxide synthase by a number of stimuli, including increased shear stresses, substance P, possibly calcitonin gene-related peptide, acetylcholine, bradykinin and many others. A fourth mechanism involves activation of particulate, membrane bound guanylate cyclase that is stimulated by atrionatriopeptide to increase cyclic GMP. Then there are ATP-dependent $K^+$ channels which, when stimulated, open, hyperpolarize the cell, and inhibit calcium entry into it. There are a number of potent vasoconstrictors like endothelin and neuropeptide Y, the latter co-localized with norepinephrine in adrenergic nerve endings. Other less well characterized dilators and constrictors have been described. Finally, intrinsic myogenic mechanisms that include stretch activated calcium channels and may or may not include some of the above mechanisms can also affect vessel tone. The likelihood exists that several of these mechanisms interact.

Many of the known agonists have been evaluated in the coronary vascular bed by a variety of means, including blocking the relevant receptors, preventing release of the agonist, or destroying all or most of the agonist. The results of such experiments have generally shown no or, only a partial effect on coronary vasomotor regulation. Because the matching of flow and oxygen supply to myocardial oxygen needs is so vital to the organisms's preservation, it is likely that several back-up systems will be present so that the heart does not have to rely on a single regulatory system that might become deranged or inhibited. Therefore there is good reason to study the effects of interfering with two or more agonists at one time in an attempt to elucidate what regulatory mechanisms are in play, although this has seldom been done. One other major deficiency in the existing studies of coronary vasomotor regulation is that the vast majority of studies examine steady-state effects rather than transients. It is entirely possible that eliminating an agonist has no effect on the steady-state response but alters the time course of the changes, and that knowing what these transients are might help to evaluate what systems are operating under any given set of circumstances. This issue becomes important when considering the differences between isolated and in vivo hearts. Inasmuch as isolated hearts appear to maintain coronary vasomotor regulation (in fact, these preparations were the basis for much of the early work in the coronary circulation), regulation must be intrinsic to the heart and independent of circulating hormones or nervous influences. That does not mean, however, that these extrinsic influences do not play a role, possible even a dominant role, in the normal heart.

Another major problem in evaluating coronary vasomotor regulation is determining what the signal is that sets the vasomotor changes in motion. Nothing changes instantaneously, and there must be at the onset of the change in pressure or metabolism a discrepancy between oxygen supply and demand that is sensed and produces a response. For sudden step changes in heart rate, flow or pressure, the response of the normalized resistance takes several seconds until a new steady state is reached, with half times typically in the $5-15$ second range (22, 23). The responses are not identical for increases and decreases of these stimuli. These findings suggest that the imbalance between oxygen supply and demand generates an error signal that causes a response, the time course being that required to generate the error signal and produce the response. That this error signal is oxygen related is clear but too diffuse a hypothesis to be useful. What we need is an understanding of what specific steps take place when this imbalance occurs to set the necessary vasomotor changes in motion. Then, once the flow is matched to metabolism, the error signal presumably disappears, so that the mechanism for maintaining the increased flow needs to be determined. Finally, when metabolism decreases and flow is correspondingly reduced, there must be another signal involved. Whether these different phases can all be directed by one mechanism will be discussed in detail below.

One limited exception to the feedback concept is the feed forward concept put forward by Miyashiro and Feigl (65). They considered that sympathetic stimulation, for example, at the onset of exercise, would dilate coronary vessels by β-adrenergic receptor stimulation without requiring any feed back, so that in effect there was anticipatory vasodilatation.

## POSSIBLE REGULATORY MECHANISMS

The signals that couple changes in oxygen supply and consumption may be considered as unitary or multifaceted. Unitary hypotheses feature a single vasoactive agent that has a steady state concentration in the myocardial interstitial space, a concentration that changes in one direction when demand exceeds supply and in the other direction when supply exceeds demand. Once an error signal is generated, the steady state concentration of the agonist changes and the appropriate vasomotor response occurs. By contrast, multifaceted hypotheses regard vasomotion as the concerted activation of several vasoactive mechanisms, some to initiate vasomotion, some to sustain it, and some to restore it to its former state. Such a multiple system requires coordinating mechanisms, but there are precedents for this in the coagulation cascade, the function of neutrophils in defense against bacteria, and many other systems in the body.

One of the best examples of a unitary hypothesis is the adenosine hypothesis (4,7-9) During muscle contraction myokinase forms AMP through the dismutation of ADP generated by ATP consumption. The action of 5'-nucleotidase on AMP produces adenosine which passes into the interstitial space where it can vasodilate. The adenosine is then rapidly removed by several mechanisms, so that there is a steady state concentration of adenosine in the interstitial space. A decreased oxygen supply leads to more ATP breakdown and more adenosine, so that the coronary vessels dilate and restore oxygen supply towards normal. An increased blood flow (and oxygen supply) following suddenly increased coronary perfusing pressure washes out some of the excess adenosine and reduces its formation, thereby reducing the vasodilator stimulus and preventing coronary vessels from dilating further. Similarly, a decrease in myocardial oxygen consumption at a given flow leads to a decrease in the amount of adenosine produced, so that the vessels constrict. This system has all the ingredients necessary for coronary control: a steady state concentration of the prime agonist at rest, a way of increasing its concentration when vasodilatation is needed and a way of decreasing its concentration when vasoconstriction is needed. An important variant of this hypothesis is that the production of adenosine is coupled to the cellular energy state rather than to oxygen directly (24, 32, 45, 72).

Interstitial adenosine levels are difficult to measure. One of the first methods to do this concluded that the concentration of interstitial adenosine was too low to be a vascular regulator at rest (44), but these measurements are very technique dependent, and other studies (25, 40) have found adequate concentrations of interstitial adenosine. There is general agreement that adenosine concentrations are high enough in ischemia to have a significant effect. A second reason to question whether adenosine is the main coronary vascular regulator is that giving low molecular weight adenosine deaminase decreases adenosine concentrations to 10% of their resting values (43) or one-third of their resting values (103) yet does not alter autoregulation (28, 43) nor change resting coronary flow (55, 103). These findings are not, however, definitive. They might merely indicate that some other dilator could take over the role of adenosine, and the studies cited measured only steady state response and not the time courses of the transients. Furthermore, the possibility exists that the concentration of adenosine is not the same at the active sites as in the interstitial fluid because of localization or protein exclusion regions. Finally, there are differences in the concentrations of adenosine required to dilate vessels applied adventially rather than endothelially (46), so that responses based on concentrations achieved by intravenous injection may not be relevant. That adenosine plays some role in ischemia is more certain, but even then it is not dominant. Giving adenosine deaminase (85) or the adenosine receptor blocker theophylline (39) reduces reactive hyperemia by about 40 – 50%, so that there must be other dilators present. It may be that adenosine is a secondary regulator that intervenes when primary mechanisms are inadequate to prevent ischemia, a role to which it is particularly well suited because of its likely protective role against ischemic damage (56).

Modulation of tone by adrenergic nerves could fit the model described above, although the error signal that increases or decreases their activity is not well defined. However, adequate regulation is observed in denervated or transplanted hearts, so that nerve activity is not essential for regulation even if it does play a role in normal hearts.

Recently, a multitude of vasoactive agonists with an effect on coronary vascular tone have been described. Substance P and calcitonin gene-related peptide (CGRP), together with neurokinin A, are present in varicosities in about 10–15% of the small myelinated and unmyelinated sensory nerves (21, 36, 48, 66, 95-97) The discrete packets of peptides are particularly concentrated in the atria, conducting system of the heart, around the coronary arteries, and beneath the endocardium and epicardium (94). They also surround the cardiac ganglia (91, 97) which, however, do not make these peptides. These peptides are made in the cell bodies of the thoracic and nodose ganglia (58) and transported down the axons to the peripheral nerves (100, 101). They can be released, together with neurokinin A, (27, 58) by capsaicin (36-38, 48, 63), and also by nicotine, bradykinin and ischemia (38). In fact, these peptides are among the candidates for the nociceptive system in the heart (3). Substance P and CGRP are present in the circulation at concentrations of about 5 nmol/l and 25 pmol/l respectively (63). They probably enter the circulation by spill over from the nerves when they are released.

Substance P (12, 20, 68) and CGRP (11, 34, 35, 37, 41) are powerful vasodilators. Substance P acts through release of nitric oxide, probably via stimulation of specific $NK_1$ receptors (see below). CGRP has both $CGRP_1$ and $CGRP_2$ receptors, the former of which are mainly on blood vessels; they are in high concentration in the smaller intramyocardial arteries. CGRP is thought to exert its effect predominantly by opening ATP-dependent $K^+$ channels (52, 69); however, some investigators find an endothelial dependent component to the action of CGRP (1, 80), even though the effect is mediated through an increase in cyclic AMP and not cyclic GMP as would occur if nitric oxide were the messenger.

ATP-dependent $K^+$ channels are involved in controlling smooth muscle tone (26, 83, 89). Agonists like cromakalim (and its active enantiomer, lemakalim), pinacidil, and nicorandil (10, 13, 16-19, 47, 57, 64, 74, 81) dilate vessels in many organs and species, and specific antagonists like the sulphonylurea glibenclamide can be used to explore their actions. These channels have a role in maintaining basal coronary tone (50, 86) and in reactive hyperemia (49, 53), and have been invoked in explaining part or all of the vasodilator effects of CGRP(69).

Endothelial derived relaxing factor (EDRF) is another potent vasodilator (probably nitric oxide) produced in many cells, including endothelial cells, by a nitric oxide synthase (a cytochrome P450-like enzyme) from l-arginine. Substance P is thought to exert its effect by stimulating release of nitric oxide. Inhibition of nitric oxide release has been shown to reduce reactive hyperemia (51, 93). Nitric oxide is involved in maintenance of normal coronary vascular tone (5, 6), and is a major mediator in the responses of other agonists (60, 67). It improves endothelial function in the coronary microcirculation of patients with hypercholesterolemia (29).

There are also many other vasoactive agents. These include neuropeptide Y which is in separate storage granules in the adrenergic efferent nerve endings (42, 59) and which may affect $\alpha_1$ adrenergic responsiveness (90); vasoactive intestinal polypeptide in vagal nerve endings where it is co-localized with acetylcholine (95); angiotensin II present both in plasma and produced locally in tissues and blood vessels (31); endothelin released from the endothelial cells; endothelial-derived constricting factor; atrionatriopeptides A, B and C, the first present in atrial and ventricular myocytes (98); bradykinin present in most tissues as the inactive precursor kininogen (15, 82); and the ubiquitous eicosanoids produced from arachidonic acid in the cell membranes of almost all tissues (104). There are receptors for all of these agonists on coronary vessels, and a fairly large amount of pharmacologic testing has been done of these agonists in animals and humans. Marked vasoactivity has been demonstrated in most species, but the physiologic role of these agents is much less well understood because the concentration of these agonists at the receptor sites after local release is usually unknown. To date, no one of these agonists has been shown to be essential for coronary reactivity. Several of them decrease resting coronary flow when blocked, but do not affect maximal flow with pacing, (14), exercise (30), or reactive hyperemia (102). Others, like adenosine or nitric oxide, can be blocked or eliminated, yet reactive hyperemia attains at least 50% of its previous amount. Clearly, replacement agonists abound in this system.

One of the ways of studying the response to locally delivered agonists, as opposed to intravenously injected agents, is to cause release of the agonists from their sites of formation (27, 58). This can be done for substance P and CGRP by injection of capsaicin (37, 38, 80), a neurotoxin that stimulates A$\delta$ and C nerve fibers with release of their contained peptides (48, 63) In fact, capsaicin is the agent that causes the face to flush after eating hot peppers, the mechanism being the local release of substance P. Furthermore, pretreatment of rats with larger doses of capsaicin depletes the sensory nerves of most of their peptides (substance P, CGRP, and neurokinin A) (36, 38, 92, 101) so that the absence of these agonists can be studied. Another useful way of eliminating these peptides is to give colchicine which blocks the axonal transport of the peptides from cell bodies to peripheral nerves (100).

After release, these peptides are rapidly broken down by enzymes, the most important of which are neutral endopeptidase (enkephalinase) (33, 54) and kininase II (angiotensin converting enzyme) (84), both metallo-ectopeptidases. Other peptidases like mast cell tryptase also degrade peptides. A strategy that has been used successfully is to reduce peptide degradation with an inhibitor of one of these enzymes: phosphoramidon, thiorphan, and captopril (84, 87, 99). If a vasodilator response is noted after peptidase inhibition, then it suggests that the agonist is released but not broken down. Of necessity, this approach is indirect. Recently, specific receptor blockers have been produced. It is known that substance P acts on $NK_1$ receptors, neurokinin A on $NK_2$ receptors, and neurokinin B on $NK_3$ receptors (73), and we have shown in airways and nasal mucosa that peptides released from sensory nerves cause vasodilatation (76) and that blockade of the $NK_1$ receptors with the specific blocker CP-99,994 (88) prevents the vasodilatation induced by capsaicin (75, 77). There are also specific receptor blockers for the $CGRP_1$ receptor, namely $CGRP_{8-37}$, as well as monoclonal antibodies that can be used. Neuropeptide Y receptor blockers are also available, although they do have partial agonist action (42). Atrionatriuretic peptide (ANP) is a powerful vasodilator that is one of the major substrates acted upon by neutral endopeptidase. Although it is present in greater concentration in the atria than the ventricles, and although its main method of release is following stretch, there are data to suggest that it can be released by neural stimulation. Recently, selective ANP receptor blockers have been developed (70, 71) so that the role of ANP in coronary vasodilatation can be explored.

## RECENT EXPERIMENTAL STUDIES

In order to study these peptides, we first investigated the application of the microsphere technique to the rat. The rats are anesthetized with sodium pentobarbital 65 mg/kg i.p. Each femoral artery is cannulated with a 0.8 mm id Angiocath to measure pressures and withdraw reference samples. A similar catheter is introduced through the right carotid artery and manipulated into the left ventricle with pressure monitoring. The catheters are filled with heparin; the carotid and one femoral arterial catheters are connected to solid state transducers, and the other femoral arterial catheter to a Harvard reciprocal withdrawal pump. Flow measurements are made with radioactive polystyrene microspheres suspended in 70% dextrose; each injection is of 100,000 microspheres suspended in 0.25 ml of fluid. For the injections, blood is withdrawn from the reference sample catheter at 1.03 ml/min for 60 seconds, and a similar amount of saline is flushed into the injecting catheter by the same pump, so that the rats remain isovolemic. At the end of the experiment the rats are exsanguinated and the positions of the catheters are verified. Selected organs are removed, cut into appropriate pieces, and the count rates from the microspheres are measured. From these rates, and those from the reference samples, flows can be determined (2).

The first set of studies was done to determine how many injections of 200,000 microspheres could be given in one rat after deciding that we needed this number of microspheres to measure accurately flow to the airways. We found that with careful attention to technique, including suspending the microspheres in 70% dextrose and infusing and withdrawing fluids simultaneously, that three successive injections could be given without perturbing any regional flows except for a small decrease in renal flow after the third injection.(75-77). Since then, we have decided that for measuring coronary blood flow, 100,000 microspheres will suffice. When these are injected into the left ventricle, three successive injections can be given without perturbing flow to the heart, and the fourth and fifth injections increase left ventricular myocardial vascular resistance slightly. A sixth, injection, however, almost always reduces coronary flow by over 30% (61, 62).

We have done three specific studies on the interaction of peptides and coronary flow. In one, we first did a dose response curve to capsaicin, which releases these peptides from sensory nerves. In other rats we gave a subthreshold dose of capsaicin which caused no change in coronary flow. Then we pretreated a group of rats with phosphoramidon, an inhibitor of neutral endopeptidase, and gave the same subthreshold dose of capsaicin. Phosphoramidon pretreatment by itself caused a slight increase in coronary flow, but capsaicin caused coronary flow to double, an effect that persisted for at least 5 minutes when the measurements ended. We concluded that one of the peptides released by capsaicin, is normally broken down so rapidly that it did not exert an effect, but when the degrading enzyme was inhibited, the same amount of CGRP had a marked coronary vasodilator effect.

The second and third studies were done while investigating a rat model in which 50 mg/kg of isoproterenol is given intraperitoneally to cause marked myocardial hypoperfusion and ischemia. The first of these two studies examined the possibility that the ischemic stimulus caused release of vasoactive peptides and determined the respective contributions of neutral endopeptidase (NEP) and angiotensin converting enzyme (kininase II, ACE) in degrading these peptides (61, 62). The huge dose of isoproterenol caused blood pressure to fall markedly, after which it recovered to 70% of its former level. Heart rate increased dramatically. Cardiac index increased modestly, and consequently total peripheral vascular resistance decreased markedly and remained low for the 25 minute duration of the experiment. Left ventricular myocardial blood flow decreased to 50% of control immediately after the isoproterenol was given, but then rapidly recovered, almost doubled by 5 minutes after the injection, and remained raised for the rest of the study. Left ventricular myocardial vascular resistance decreased immediately and then remained very low. When the rats were pretreated with phosphoramidon, there was a slight rise in aortic blood pressure and heart rate before the isoproterenol was given. After it was given, the changes in pressure and heart rate paralleled those seen in the control rats. There were only minor differences in the responses of cardiac index and total peripheral vascular resistance. In fact, phosphoramidon had little effect on systemic hemodynamics. By contrast, phosphoramidon caused myocardial blood flow to increase before the isoproterenol was given and to maintain this high level after it had been given, so that left ventricular myocardial vascular resistance decreased more rapidly after than before the phosphoramidon had been given, to reach the very low value seen in the control rats 10 minutes after the isoproterenol had been given.

We interpreted these data to mean that before the isoproterenol was given, there might have been some tonic release of vasodilator peptides (either locally in the heart, or else remotely and traveling to the heart in the blood), but that these were broken down by NEP or other enzymes so that they did not show any effect. After the potent stimulus of the isoproterenol, however, inhibiting the degrading enzymes produced enough myocardial vasodilatation that the myocardial blood flow did not decrease. An interesting side issue, for which we have no explanation, is that the phosphoramidon inhibited the late rise in myocardial blood flow that occurred in the control rats.

Inhibiting ACE decreased peripheral vascular resistance and arterial blood pressure, caused a greater decrease in left ventricular myocardial vascular resistance, and about the same increase in myocardial blood flow as did phosphoramidon. When both enzyme inhibitors were given together, however, all these changes were less than seen for ACE inhibition alone. The two enzymes are thus not synergistic, but presumably preferentially degrade different combinations of dilator and constrictor peptides.

The second study done with this model (78, 79) attempted to assess which peptides were responsible for the vasodilatation that occurred when NEP was inhibited. Our initial belief was that it would not be CGRP which has too long a duration of action and is not a good substrate for NEP, but that it might well be substance P which has a short duration of action and is a good substrate for NEP. When we gave the neurokinin$_1$ receptor blocker, which is a specific receptor blocker for substance P (77, 88), there were no differences in the results of inhibiting NEP and giving isoproterenol. On the other hand, when we gave Hoe-140, a specific blocker of bradykinin$_2$ receptors, the effects of inhibiting NEP were completely reversed.

These results raise the possibility that bradykinin is an important mediator of coronary regulation. The Coleridges had raised the possibility that bradykinin might be part of the nociceptive system in the heart (3). Bradykinin is known to be liberated during ischemia, or rather ischemia is thought to induce bradykinin receptors. It is conceivable that when myocardial oxygen consumption is suddenly increased, there is a transient oxygen deficit which causes the vessels to be stimulated by bradykinin. This dilates the vessels, flow increases, and oxygen supply and demand are brought back into balance. Such a hypothesis is certainly too simple. It does not allow for the fact that degrading enzymes are normally active enough to break down these vasoactive peptides very rapidly, and does not invoke the many other peptides that may well be liberated at the same time. Nevertheless, there is now hard evidence that bradykinin is involved in certain coronary regulatory mechanisms, and opens the way to exploring its role in more physiological states.

# REFERENCES

1. Abdelrahman, A., Y.-X. Wang, S. D. Chang, and C. C. Y. Pang. Mechanism of the vasodilator action of calcitonin gene-related peptide in conscious rats. *Br J Pharmacol* 106: 45 – 48, 1992.

2. Baer, R. W., B. D. Payne, E. D. Verrier, G. J. Vlahakes, D. Molodowitch, P. N. Uhlig, and J. I. E. Hoffman. Increased number of myocardial blood flow measurements with radionuclide-labeled microspheres. *Am J Physiol* 246 (Heart Circ. Physiol. 15): H418 – H434, 1984.

3. Baker, D. G., H. M. Coleridge, J. C. G. Coleridge, and T. Nerdrum. Search for a cardiac nociceptor: stimulation by bradykinin of sympathetic afferent nerve endings in the heart. *J Physiol* 306: 519 – 536, 1980.

4. Bardenheuer, H., and J. Schrader. Supply-to-demand ratio for oxygen determines formation of adenosine by the heart. *Am J Physiol* 250 (Heart Circ Physiol 19): H 173 – H180, 1986.

5. Bassenge, E. Endothelium-dependent control of coronary tone and myocardial perfusion in conscious dogs. *Circulation* 84 (Suppl II): 655, 1991.

6. Benyó, Z., G. Kiss, C. Szabó, C. Csaki, and A. Kovach. Importance of basal nitric oxide synthesis in regulation of myocardial blood flow. *Cardiovasc Res* 25: 700 – 703, 1991.

7. Berne, R. M. Cardiodynamics and the coronary circulation in hypothermia. *Ann N Y Acad Sci* 80: 365-383, 1959.

8. Berne, R. M. The role of adenosine in the regulation of coronary blood flow. *Circ Res* 47: 807-813, 1980.

9. Berne, R. M., and R. Rubio. Coronary circulation., In Berne, R. M., and N. Sperelakis (ed.), *Handbook of Physiology, Section 2: The Cardiovascular System.* Volume 1: The Heart,. Bethesda, Maryland: The American Physiological Society, 1979, pp 873-952.

10. BoSmith, R. E., I. Briggs, T. L. Grant, S. Grimwood, N. J. W. Russell, M. A. Stone, and A. D. Wickenden. A comparison of the effects and interactions of cromakalim and glibenclamide in cardiac and vascular tissues. *Pfluegers Arch* 414: S 190 –, 1989.

11. Brain, S. D., T. J. Williams, J. R. Tippins, H. R. Morris, and I. MacIntyre. Calcitonin gene-related peptide is a potent vasodilator. *Nature* 313: 54 – 56, 1985.

12. Brum, J. M., V. L. W. Go, Q. Sufan, G. lane, W. Reilly, and A. A. Bove. Substance P distribution and effects in the canine epicardial coronary arteries. *Regul Peptides* 14: 41 – 55, 1986.

13. Buckingham, R. E., T. C. Hamilton, D. R. Howlett, S. Mootoo, and C. Wison. Inhibition by glibenclamide of the vasorelaxant action of cromakalim in the rat. *Br J Pharmacol* 97: 57 – 64, 1989.

14. Canty, J. M., Jr, and L. Brownschidle. Coronary flow adjustments to pacing are not affected by inhibiting nitric oxide production in conscious dogs. *Circulation* 86 (Suppl I): 639, 1992.

15. Carretero, O. A., and A. G. Scicli. The kallikrein-kinin system., In Fozzard, H. A., E. Haber, R. B. Jennings, A. M. Katz, and H. E. Morgan (ed.), *The Heart and Cardiovascular System. Scientific Foundations.* 2. New York: Raven Press, 1992, pp 1851 – 1874.

16. Cavero, I., S. Mondot, and M. Mestre. Vasorelaxant effects of cromakalim in rats are mediated by glibenclamide-sensitive potassium channels. *J Pharmacol Ther* 248: 1261 – 1268, 1989.

17. Clapham, J. C., and R. E. Buckingham. The haemodynamic profile of cromakalim in the cat. *J Cardiovasc Pharmacol* 12: 555 – 561, 1988.

18. Clapham, J. C., and S. D. Longman. Haemodynamic differences between cromakalim and pinacidil: comparison with nifedipine. *Eur J Pharmacol* 171: 109 – 117, 1989.

19. Cohen, M. L., and K. D. Kurz. Pinacidil-induced vascular relaxation: comparison to other vasodilators and to classical mechanisms of vasodilation. *J Cardiovasc Pharmacol* 12 (Suppl 2): S5 – S9, 1988.

20. Crossman, D. C., S. W. Larkin, R. W. Fuller, G. J. Davies, and A. Maseri. Substance P dilates epicardial coronary arteries and increases coronary blood flow in humans. *Circulation* 80: 475-484, 1989.

21. Dalsgaard, C.-J., A. Franco-Cereceda, A. Saria, J. M. Lundberg, E. Theodorsson-Norheim, and T. Hökfelt. Distribution and origin of substance P- and neuropeptide Y- immunoreactive nerves in the guinea pig heart. *Cell Tissue Res* 243: 477 – 485, 1986.

22. Dankelman, J., J. A. E. Spaan, H. G. Stassen, and I. Vergroesen. Dynamics of coronary adjustment to a change in heart rate in the anaesthetized goat. *J Physiol (Lond)* 408: 295 – 312, 1989.

23. Dankelman, J., J. A. E. Spaan, C. P. B. VanderPloeg, and I. Vergroessen. Dynamic response of the coronary circulation to a rapid change in its perfusion in the anaesthetized goat. *J Physiol (Lond)* 419: 703 – 715, 1989.

24. Darvish, A., J. Chakraborty, S. L. Britton, and P. J. Metting. Myocardial adenosine production via AMP-specific cytosolic 5'nucleotidase. *FASEB J* 4: 1071 –, 1990.

25. Deussen, A., M. Borst, and J. Schrader. Formation of S-adenylhomocysteine in the heart. I. An index of free intracellular adenosine. *Circ Res* 63: 240 – 249, 1988.

26. DeWeille, J. R., M. Fosset, C. Mourre, H. Schmid-Antomarchi, H. Bernardi, and M. Lazdunski. Pharmacology and regulation of ATP-sensitive $K^+$ channels. *Pfluegers Arch* 414: S80 – S87, 1989.

27. Diez Guerra, F. J., M. Zaidi, P. Bevis, I. MacIntyre, and P. C. Emson. Evidence for release of calcitonin gene-related peptide and neurokinin A from sensory nerve endings *in vivo*. *Neuroscience* 25: 839 – 846, 1988.

28. Dole, W. P. Autoregulation of the coronary circulation. *Prog Cardiovasc Res* 29: 293-323, 1987.

29. Drexler, H., A. M. Zeiher, and H. Just. L-arginine improves endothelial function in the coronary microcirculation of patients with hypercholesterolemia. *Circulation* 84 (Suppl II): 14, 1991.

30. Duncker, D. J., N. S. Van Zon, T. Pavek, M. Crampton, P. Lindstrom, and R. J. Bache. Effect of $K^+_{ATP}$ channel blockade on exercise-induced coronary vasodilation,. *Circulation* 86 (Suool I): 484, 1992.

31. Dzau, V. J., and R. E. Pratt. Renin-angiotensin system., In Fozzard, H. A., E. Haber, R. B. Jennings, A. M. Katz, and H. E. Morgan (ed.), *The Heart and Cardiovascular System. Scientific Foundations*. 2. New York: Raven Press, 1992, pp 1817 – 1849.

32. Engler, R. L., and H. E. Gruber. Adenosine: an autocoid., In Fozzard, H. A., E. Haber, R. B. Jennings, A. M. Katz, and H. E. Morgan (ed.), *The Heart and Cardiovascular System. Scientific Foundations*. 2. New York: Raven Press, 1992, pp 1745 – 1776.

33. Erdos, E. G., and R. A. Skidgel. Neutral endopeptidase 24.11 (enkephlinase) and related regulators of peptide hormones. *FASEB J* 3: 145-151, 1989.

34. Ezra, D., F. R. M. Laurindo, J. Eimerl, R. E. Goldstein, C. C. Peck, and G. Feuerstein. Tachykinin modulation of coronary blood flow. *Eur J Pharmacol* 122: 135 – 138, 1986.

35. Ezra, D., F. R. M. Laurindo, D. S. Goldstein, R. E. Goldstein, and G. Feuerstein. Calcitonin gene-related peptide: a potent modulator of coronary flow. *Eur J Pharmacol* 137: 101 – 105, 1987.

36. Franco-Cereceda, A., H. Henke, J. M. Lundberg, J. B. Petermann, T. Hökfelt, and J. A. Fische. Calcitonin gene-related peptide (CGRP) in capsaicin-sensitive substance P-immunoreactive sensory neurones in animals and man: distribution and release by capsaicin. *Peptides* 8: 399 – 410, 1987.

37. Franco-Cereceda, A., A. Rudehill, and J. M. Lundberg. Calcitonin gene-related peptide but not substance P mimics capsaicin-induced coronary vasodilation in the pig. *Eur J Pharmacol* 142: 235 – 243, 1987.

38. Franco-Cereceda, A., A. Saria, and J. M. Lundberg. Differential release of calcitonin gene-related peptide and neuropeptide Y from isolated heart by capsaicin, ischaemia, nicotine, bradykinin and ouabain. *Acta Physiol Scand* 135: 173 – 187, 1989.

39. Gidday, J. M., J. W. Esther, S. W. Ely, R. Rubio, and R. M. Berne. Time-dependent effects of theophylline on myocardial reactive hyperaemias in the anaesthetized dog. *Br J Pharmacol* 100: 95-101, 1990.

40. Gidday, J. M., H. E. Hill, R. Rubio, and R. M. Berne. Estimates of left ventricular interstitial fluid adenosine during catecholamine stimulation. *Am J Physiol* 254 (Heart Circ Physiol 23): H207-H216, 1988.

41. Greenberg, B., K. Rhoden, and P. Barnes. Calcitonin gene-related peptide (CGRP) is a potent non-endothelium-dependent inhibitor of coronary vasomotor tone. *Br J Pharmacol* 92: 789 – 794, 1987.

42. Grundemar, L., S. E. Jonas, E. D. Högestätt, C. Wahlestedt, and R. Håkanson. Characterization of vascular neuropeptide Y receptors. *Br J Pharmacol* 105: 45 – 50, 1992.

43. Hanley, F. L., M. T. Grattan, M. B. Stevens, and J. I. E. Hoffman. Role of adenosine in coronary autoregulation. *Am J Physiol* 250 (Heart Circ Physiol 19): H558-H566, 1986.

44. Hanley, F. L., L. M. Messina, R. W. Baer, P. N. Uhlig, and J. I. E. Hoffman. Direct measurement of left ventricular interstitial adenosine. *Am J Physiol* 245 (Heart Circ Physiol 14): H327-H335, 1983.

45. He, M., R. D. Wangler, P. F. Dillon, G. D. Romig, and H. V. Sparks. Phosphorylation potential and adenosine release during norepinephrine in guinea pig heart. *Am J Physiol* 253 (Heart Circ Physiol 22): 1987.

46. Headrick, J. P., F. J. Northington, M. R. Hynes, G. P. Matherne, and R. M. Berne. Relative responses to luminal and adventitial adenosine in perfused arteries. *Am J Physiol* 263 (Heart Circ Physiol 32): H1437-H1446, 1992.

47. Hermsmeyer, R. K. Pinacidil actions on ion channel in vascular muscle. *J Cardiovasc Pharmacol* 12 (Suppl 2): S17 – S22, 1988.

48. Holzer, P. Local effector functions of capsaicin-sensitive sensory nerve endings; involvement of tachykinins, calcitonin gene-related peptide and other neuropeptides. *Neuroscience* 24: 739 – 768, 1988.

49. Imamura, Y., K. Muramatu, T. Narishige, and A. Takeshita. ATP-sensitive K channel contributes to hyperemic responses to coronary occlusion in awake dogs. *Circulation* 84 (Suppl II): 275, 1991.

50. Imamura, Y., H. Tomoike, N. Narishige, T. Takahasi, and H. Kasuya. Role of ATP-sensitive potassium channel in control of basal coronary flow in dogs. *Circulation* 84 (Suppl II): 185, 1991.

51. Ishizaka, H., K. Okamura, H. Yamabe, and T. Tsuchiya. Role endothelium-derived nitric oxide in myocardial reactive hyperemia. *Circulation* 84 (Suppl II): 182, 1991.

52. Jiang, C., P. A. Poole-Wilson, and P. Collins. Glibenclamide inhibits in vitro coronary relaxation induced by calcitonin-gene-related peptide (CGRP) but not by vasoactive intestinal peptide (VIP). *Circulation* 84: 218, 1992.

53. Kanatsuka, H., N. Sekiguchi, K. Sata, K. Akai, Y. Wang, K. Ashikawa, and T. Takishima. The role of ATP-sensitive $K^+$ channels in responses of coronary microvessels during brief ischemia and reactive hyperemia. *Circulation* 84 (Suppl II): 184, 1991.

54. Katayama, M., J. A. Nadel, N. W. Bunnett, G. U. Di Maria, M. Haxhiu, and D. B. Borson. Catabolism of calcitonin gene-related peptide and substance P by neutral endopeptidase. *Peptides* 12: 563 – 567, 1991.

55. Kroll, K., F. F. A. Hendriks, and J. J. Schipperheyn. Extracorporeal circulation system for coronary artery perfusion in the closed-chest dog. *Am J Physiol* 236 (Heart Circ Physiol 5): H652-H656, 1979.

56. Liu, G. S., J. Thornton, D. M. Van Winkle, A. W. Stanley, R. A. Olsson, and J. M. Downey. Protection against infarction afforded by preconditioning is mediated by A1 adenosine receptors in rabbit heart. *Circulation* 84: 350 – 356, 1991.

57. Longman, S. D., J. C. Clapham, C. Wilson, and T. C. Hamilton. Cromakalim, a potassium channel activator: a comparison of its cardiovascular haemodynamic profile and tissue specificity with those of pinacidil and nicorandil. *J Cardiovasc Pharmacol* 12: 535 – 542, 1988.

58. Lundberg, J. M., A. Franco-Cereceda, X. Hua, T. Hökfelt, and J. A. Fischer. Co-existence of substance P and calcitonin gene-related peptide-like immunoreactivities in sensory nerves in relation to cardiovascular and bronchoconstrictor effects of capsaicin. *Eur J Pharmacol* 108: 315 – 319, 1985.

59. Macho, P., R. Pérez, J. Huidobro-Toro, and R. J. Domenech. Neuropeptide Y (NPY): a coronary vasoconstrictor and potentiator of catecholamine-induced coronary constriction. *Eur J Pharmacol* 167: 67 – 74, 1989.

60. Mankad, P. S., A. H. Chester, and M. H. Yacoub. 5-hydroxytryptamine mediates endothelium dependent coronary vasodilatation in the isolated rat heart by the release of nitric oxide. *Cardiovasc Res* 25: 244 – 248, 1991.

61. Maxwell, A. J., W. K. Husseini, G. Piedimonte, and J. I. E. Hoffman. Inhibition of neutral endopeptidase augments resting myocardial flow and prevents isoproterenol-induced myocardial hypoperfusion in rats. *Circulation* 88 (Suppl I): I-34, 1993.

62. Maxwell, A. J., W. K. Husseini, G. Piedimonte, and J. I. E. Hoffman. The effects of inhibiting neutral endopeptidase on coronary and systemic hemodynamics in rats. *In preparation* 1994.

63. McEwan, J. R., N. Benjamin, S. Larkin, R. W. Fuller, C. T. Dollery, and I. MacIntyre. Vasodilatation by calcitonin gene-related peptide and by substance P: a comparison of their effects on resistance and capacitance vessels of human forearms. *Circulation* 77: 1072 – 1080, 1988.

64. Minkes, R. K., P. Kvamme, T. R. Higuera, B. D. Nossaman, and P. J. Kadowitz. Analysis of pulmonary and systemic vascular responses to cromakalim, an activator of $K^+_{ATP}$ channels. *Am J Physiol* 260 (Heart Circ Physiol 29): H957 – H966, 1991.

65. Miyashiro, J. K., and E. O. Feigl. Feedforward control of coronary blood flow via coronary β-receptor stimulation. *Circ Res* 73: 252 – 263, 1993.

66. Mulderry, P. K., M. A. Ghatei, K. Rodrigo, J. M. Allen, M. G. Rosenfeld, J. M. Polak, and S. R. Bloom. Calcitonin gene-related peptide in cardiovascular tissues of the rat. *Neuroscience* 14: 947 – 954, 1985.

67. Myers, P. R., P. F. Banitt, R. Guerra Jr, and D. G. Harrison. Role of the endothelium in modulation of the acetylcholine vasoconstrictor response in porcine coronary blood vessels. *Cardiovasc res* 25: 129 – 137, 1991.

68. Nakamura, Y., R. Parent, and M. Lavallee. Opposite effects of substance P on conductance and resistance coronary vessels in conscious dogs. *Am Heart J* 258 (Heart Circ Physiol 27): H565 – H573, 1990.

69. Nelson, M. T., Y. Huang, J. E. Brayden, J. Heschler, and N. B. Standen. Arterial dilations in response to calcitonin gene-related peptide involve activation of $K^+$ channels. *Nature* 344: 770 – 773, 1990.

70. Oda, S., T. Sano, Y. Morishita, and Y. Matsuda. Pharmacological profile of HS-142-1, a novel nonpeptide atrial natriuretic peptide (ANP) antagonist of microbila origin. 2. Restoration by HS-142-1 of ANP-induced inhibition of aldosterone production in adrenal glomerulosa cells. *J Pharmacol Exp Ther* 263: 241 – 245, 1992.

71. Ohyama, Y., K. Miyamoto, Y. Morishita, Y. Matsuda, Y. Saito, N. Minamino, K. Kangawa, and M. H. Stable expression of natriuretic peptide receptors: effects of HS-142-1, a non-peptide ANP antagonist. *Biochem Biophys Res Comm* 189: 336 – 342, 1992.

72. Olsson, R. A., R. Bünger, and J. A. E. Spaan. Coronary circulation., In Fozzard, H. A., E. Haber, R. B. Jennings, A. M. Katz, and H. E. Morgan (ed.), *The Heart and Cardiovascular System. Scientific Foundations.* 2. New York: Raven Press, 1991, pp 1393 – 1425.

73. Patacchini, R., P. Santicioli, M. Astolfi, P. Rovero, G. Viti, and C. A. Maggi. Activity of peptide and non-peptide antagonists at peripheral NK1 receptors. *Eur J Pharmacol* 215: 93 – 98, 1992.

74. Peuss, K. C., G. J. Gross, H. L. Brooks, and D. C. Warltier. Hemodynamic actions of nicorandil, a new antianginal agent in the conscious dog. *J Cardiovasc Pharmacol* 7: 709 – 714, 1985.

75. Piedimonte, G., J. I. E. Hoffman, W. K. Husseini, C. Bertrand, R. M. Snider, M. C. Desai, G. Petersson, and J. A. Nadel. Neurogenic vasodilation in the rat nasal mucosa involves neurokinin₁ tachykinin receptors. *J Pharmacol Exp Therap* 265: 36 – 40, 1993.

76. Piedimonte, G., J. I. E. Hoffman, W. K. Husseini, W. L. Hiser, and J. A. Nadel. Effect of neuropeptides released from sensory nerves on blood flow in the rat airway microcirculation. *Journal of Applied Physiology* 72: 1563-1570, 1992.

77. Piedimonte, G., J. I. E. Hoffman, W. K. Husseini, R. M. Snider, M. C. Desai, and J. A. Nadel. NK₁ receptors mediate neurogenic inflammatory increase in blood flow in rat airways. *J Appl Physiol* 74: 2462 – 2468, 1993.

78. Piedimonte, G., J. A. Nadel, and J. I. E. Hoffman. Neutral endopeptidase modulates neurogenic coronary vasodilation. *Circulation* 88 (Suppl I): I-183, 1993.

79. Piedimonte, G., J. A. Nadel, C. S. Long, and J. I. E. Hoffman. Neutral endopeptidase in the heart: NEP inhibition prevents isoproterenol-induced myocardial hypoperfusion in rats by reducing bradykinin degradation. *In preparation* 1994.

80. Prieto, D., S. Benedito, and N. C. B. Nyborg. Heterogeneous involvement of endothelium in calcitonin gene-related peptide-induced relaxation in coronary arteries from rat. *Br J Pharmacol* 103: 1764 – 1768, 1991.

81. Quast, U., and N. S. Cook. In vitro and in vivo comparison of two K⁺ channel openers, diazoxide and cromakalim, and their inhibition by glibenclamide. *J Pharmacol Exp Therap* 250: 261 – 271, 1989.

82. Regoli, D., and J. J. Barabe. Pharmacology of bradykinin and related kinins. *Pharmacol Rev* 32: 1 – 46, 1980.

83. Richer, C., J. Pratz, P. Mulder, S. Mondot, J. F. Giudicelli, and I. Cavero. Cardiovascular and biological effects of K⁺ channel openers, a class of drugs with vasorelaxant and cardioprotective properties. *Life Sci* 47: 1693 – 1705, 1990.

84. Rouissi, N., F. Nantel, G. Drapeau, N.-E. Rhaleb, S. Dion, and D. Regoli. Inhibition of peptidases: How they influence the biological activities of substance P, neurokinins, kinins and angiotensins in isolated vessels. *Pharmacology* 40: 185 – 195, 1990.

85. Saito, D., C. R. Steinhart, D. G. Nixon, and R. A. Olsson. Intracoronary adenosine deaminase reduces canine myocardial reactive hyperemia. *Circ Res* 49: 1262-1267, 1981.

86. Samaha, F. F., F. W. Heineman, J. Fleming, C. Ince, and R. S. Balaban. ATP-sensitive potassium channel plays a critical role in maintenance of basal coronary tone. *Circulation* 84 (Suppl II): 183, 1991.

87. Smits, G. J., and R. A. Lefebvre. Influence of thiorphan and phosphoramidon on the relaxant effect of vasoactive intestinal polypeptide and non-adrenergic non-cholinergic neurone stimulation in the rat gastric fundus. *Eur J Pharmacol* 200: 361 – 364, 1991.

88. Snider, R. M., J. W. Constantine, J. A. Lowe III, K. P. Longo, W. S. Lebel, H. A. Woody, S. E. Drozda, M. C. desai, F. J. Vinick, R. W. Spencer, and H.-J. Hess. A potent nonpeptide antagonist of the substance P (NK₁) receptor. *Science* 251: 435 – 437, 1991.

89. Standen, N. B., J. M. Quayle, N. W. Davies, J. E. Brayden, Y. Huang, and M. T. Nelson. Hyperpolarizing vasodilators activate ATP-sensitive K⁺ channels in arterial smooth muscle. *Science, Wash DC* 245: 177 – 180, 1989.

90. Sun, L. S., P. C. Ursell, and R. C. Robinson. Chronic exposure to neuropeptide Y determines cardiac α₁-adrenergic responsiveness. *Am J Physiol* 261 (Heart Circ Physiol 30): H969 – H973, 1991.

91. Tay, S. S. W., and W. C. Wong. Immunocytochemical localisation of substance P-like nerves in the cardiac ganglia of the monkey (Macaca fascicularis). *J Anat* 180: 239 – 245, 1992.

92. Tepperman, B. L., and B. J. R. Whittle. Endogenous nitric oxide and sensory neuropeptides interact in the modulation of the rat gastric microcirculation. *Br J Pharmacol* 105: 171 – 175, 1992.

93. Ueeda, M., and R. Olsson. Role of endothelium in coronary reactive hyperemia (RH) in the guinea pig. *Circulation* 84 (Suppl II): 270, 1991.

94. Ursell, P. C., C. L. Ren, A. Alabal, and P. Danilo Jr. Nonadrenergic noncholinergic innervation. Anatomic distribution of calcitonin gene-related peptide-immunoreactive tissue in the dog heart. *Circ Res* 68: 131 – 140, 1991.

95. Weihe, E., M. Reinecke, and W. G. Forssmann. Distribution of vasoactive intestinal polypeptide-like immunoreactivity in the mammalian heart. Interrelation with neurotensin- and substance P-like immunoreactive nerves. *Cell Tissue Res* 236: 527 – 540, 1984.

96. Wharton, J., and S. Gulbenkian. Peptides in the mammalian cardiovascular system. *Experientia* 43: 821 – 832, 1987.

97. Wharton, J., J. M. Polak, G. P. McGregor, A. E. Bishop, and S. R. Bloom. The distribution of substance P-like immunoreactive nerves in the guinea pig heart. *Neuroscience* 6: 2193 – 2204, 1981.

98. Wildey, G. M., K. S. Misono, and R. M. Graham. Atrial natriuretic factor: biosynthesis and mechanism of action., In Fozzard, H. A., E. Haber, R. B. Jennings, A. M. Katz, and H. E. Morgan (ed.), *The Heart and Cardiovascular System. Scientific Foundations.* 2. New York: Raven Press, 1992, pp 1777 – 1796.

99. Yong, T., X. P. Gao, S. Koizumi, J. M. Conlon, S. I. Rennard, W. G. Mayhan, and I. Rubinstein. Role of peptidases in bradykinin-induced increase in vascular permeability in vivo. *Circ Res* 70: 952 – 959, 1992.

100. Zaidi, M., and P. J. R. Bevis. Enhanced circulating levels of neurally derived calcitonin gene related peptide in spontaneously hypertensive rats. *Cardiovasc Res* 25: 125 – 128, 1991.

101. Zaidi, M., P. J. R. Bevis, S. I. Girgis, C. Lynch, J. C. Stevenson, and I. MacIntyre. Circulating CGRP comes from the perivascular nerves. *Eur J Pharmacol* 117: 283 – 284, 1985.

102. Zhang, J., M. J. Somers, and F. R. Cobb. Inhibition of ATP-sensitive potassium channels reduces basal vasomotor tone but not stimulated responses in the coronary vasculature. *Circulation* 86 (Suppl I): 484, 1992.

103. Zhu, Q., G. P. Matherne, R. R. Curnish, C. G. Tribble, and R. M. Berne. Effect of adenosine deaminase on cardiac interstitial adenosine. *Am J Physiol* 263 (Heart Circ Physiol 32): H1322-H1326, 1992.

104. Zusman, R. M. Eicosanoids: prostaglandins, thromboxane, and prostacyclin., In Fozzard, H. A., E. Haber, R. B. Jennings, A. M. Katz, and H. E. Morgan (ed.), *The Heart and Cardiovascular system. Scientific Foundations.* 2. New York: Raven Press, 1992, pp 1797 – 1815.

# 14

# GRADED RESTRICTION OF BLOOD FLOW IN EXERCISING LEG MUSCLES

## A Human Model

Hilding Bjurstedt[1]* and Ola Eiken[2]

[1] Environmental Physiology Laboratory
Department of Physiology and Pharmacology
Karolinska Institute, 171 77 Stockholm, Sweden
[2] Department of Clinical Physiology
Karolinska Institute, Huddinge Hospital
141 86 Huddinge, Sweden

## INTRODUCTION

Much of our knowledge about the effects of acute muscle ischemia on the respiration and central circulation in humans has been derived from studies in which the circulation in muscles has been completely arrested by the use of inflatable cuffs. Notable examples are the classical experiments by Alam & Smirk in the 30s (1937). They blocked the circulation of one forearm and found that the trapping of metabolites within the limb after exercise prevented the postexercise fall in arterial pressure. Later, Asmussen & Nielsen (1964) also used the cuff method during dynamic leg exercise, and presented evidence that both the pressor response and the hyperpnea of exercise may in part originate from stimulation of muscle afferents sensitive to local metabolic changes in ischemic muscle. Yet, complete blocking of the circulation to an exercising limb can only be maintained for short periods and may give rise to pain and other adverse effects with physiological consequences of mixed origin. There has been a need for a method by which the experimenter can induce partial restriction of blood flow in large muscle groups, and control both the degree and the duration of the restriction within wide limits.

A method with this potential was developed in our laboratory some years ago (Eiken & Bjurstedt 1987). We originally designed the method to study the effects of graded muscle ischemia on the respiration and the central circulation, but it has also proved well suited for studies of the effects of long-term muscle ischemia. Because of its usefulness the method and the way it operates will be described here in some detail.

---

* Correspondence to: Dr. H. Bjurstedt, Environmental Physiology Laboratory, Department of Physiology and Pharmacology, Karolinska Institute, S-171 77 Stockholm, Sweden

*Control of the Cardiovascular and Respiratory Systems in Health and Disease*
Edited by C. T. Kappagoda and M. P. Kaufman, Plenum Press, New York, 1995

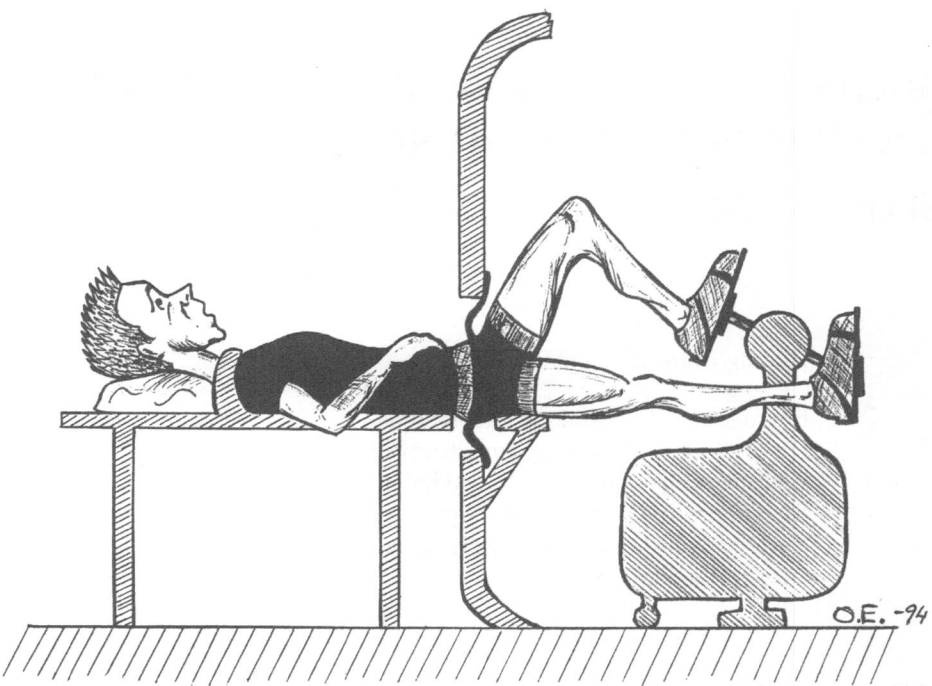

**Figure 1.** Use of pressure chamber for application of Leg Positive Pressure (LPP) during cycle exercise.

## EXPERIMENTAL SET-UP FOR LEG POSITIVE PRESSURE (LPP)

As shown in Fig. 1 the subject is positioned supine in the opening of a pressure chamber, and is hermetically sealed into the chamber at the level of the crotch by using a rubber diaphragm with two holes and short self-sealing sleeves for the thighs. The subject's feet are strapped to the pedals of a cycle ergometer. The chamber pressure can be raised as desired in a controlled fashion. A shoulder support or a harness is used to prevent displacement of the body in the cranial direction when the chamber pressure is raised; this also minimizes the activity of the arm and trunk muscles during cycling. In our initial work a chamber pressure of 50 mm Hg was used, which causes moderate flow restriction in the working leg muscles (see further below). Subsequent studies by others have used stepwise or ramp increases in pressure on exercising limbs up to a maximum of 40 or 60 mm Hg (Joyner 1991, Rowell et al. 1991, Sundberg & Kaijser 1992, Shi et al 1993, Williamson et al. 1993).

## LEG MUSCLE ISCHEMIA AS PRODUCED BY LPP

The effects of exposing the working legs to LPP on muscle perfusion pressure have been reviewed by Eiken (1987). Briefly, the pressure increase is transmitted to the soft tissues of the legs in a nearly 1/1 relationship (cf. Lundvall & Länne 1989). This causes the local

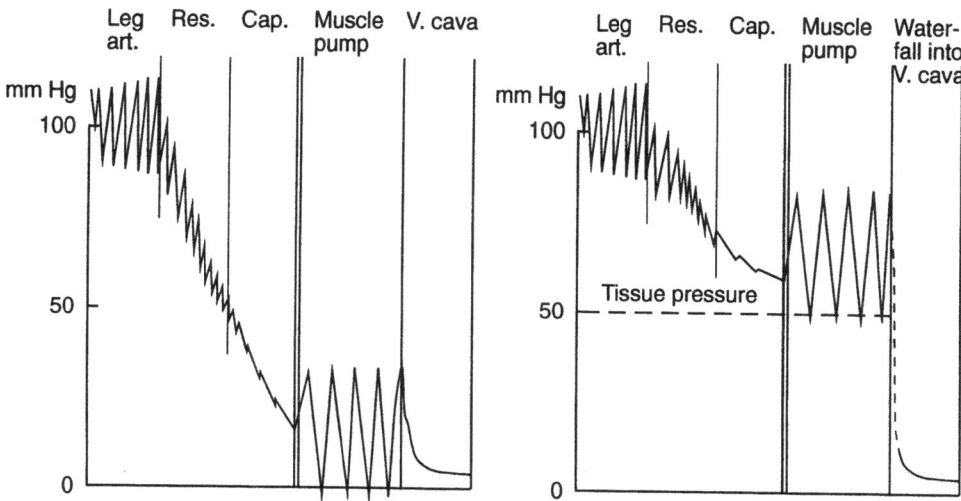

**Figure 2.** Pressure profiles, relative to atmospheric pressure, in the vascular circuits of the legs (arteries, resistance vessels, capillaries and veins) in free-flow exercise (left) and in ischemic exercise (right) induced by application of LPP at 50 mm Hg. LPP elevates the total tissue pressure and thereby the mean venous pressure. As a result, the perfusion pressure (mean arterial-to-mean venous pressure) is reduced by 50 mm Hg.

venous pressure to rise until it just exceeds the elevated tissue pressure and blood flow is resumed (cf. Roddie & Shepherd 1957, Reneman et al. 1980). The arterial pressure, as measured relative to the atmospheric, is not directly affected by the external pressure. As a consequence, the perfusion pressure is reduced.

Fig. 2 gives a schematic representation of pressure profiles in the different segments of the vascular circuits of the legs during dynamic exercise. Two conditions are represented. The left panel shows the normal situation, i. e. without application of external pressure. The pressure profile is complicated by the action of the muscle pump, and a simple ohmic resistance does not exist across the contracting muscle as it does in the resting state. The figure shows the pressure oscillations in the muscle veins relative to the outside atmospheric pressure. After each contraction the veins are empty, and venous valves prevent a back flow. The pressure in small muscle veins momentarily falls to zero and can become negative for a brief period due to the elastic recoil of the vein walls (Laughlin 1987).

In the right panel of Fig. 2 the exercising legs are exposed to positive pressure of 50 mm Hg. The venous pressure swings still appear in the pressure profile, but the mean venous pressure is elevated. Consequently the arterial-mean venous pressure difference, i.e. the perfusion pressure, is reduced. At the extreme right is shown the drop in venous pressure that occurs in the transition zone from a higher to a lower tissue pressure. This pressure drop represents a pressure discontinuity in a hydraulic sense, and does not form part of the driving pressure in the muscles (cf. Duomarco & Rimini 1954). The phenomenon can be likened to a waterfall, where the height of the fall has no influence on the flow upstream of the fall. This means that the pressure in the downstream caval and central veins has no influence on the blood flow in the leg muscles.

In resting legs, LPP-induced impairment of muscle blood flow may be somewhat greater than would be expected from the lowered perfusion pressure *per se,* since the venous flow resistance may be increased by collapse of the veins (cf. Nielsen 1983). Even small increases in tissue pressure tend to collapse postcapillary vessels so that their resistance to flow is slightly increased (cf. Dahn et al. 1967, Reneman et al. 1980). Judging from animal experiments (Reneman et al. 1980), the diameter of the arterioles is if anything increased, indicating that arterioles are not flow-limiting even though the transmural pressure in these vessels is decreased by the raised tissue pressure. Circulatory arrest appears to occur primarily at capillary level (Reneman et al. 1989). Dahn et al. (1967) found that with increasing external pressure, blood flow ceases completely when the pressure exceeds diastolic pressure.

In rhythmically exercising legs, by contrast, LPP acts on a hemodynamic picture which is marked by vasodilatation of the resistance vessels and the activity of the muscle pump. Vasodilatation due to intramuscular buildup of metabolites and/or myogenic autoregulation may tend to counteract restrictions in blood flow caused by increases in tissue pressure, and reflexly induced increases in arterial pressure (see further below) may operate in the same direction.

Actual measurements of blood flow in leg muscles during submaximal exercise using the constant-infusion dye-dilution method (Sundberg & Kaijser 1992) have shown that LPP at 50 mm Hg reduces flow by an average of 16 percent. The reduction increases somewhat with the work load, and during strenuous exercise amount to about 20 percent. These flow measurements have enabled the same authors to calculate the resistance in the vascular beds of the exercising legs which showed no increase during application of LPP at 50 mm Hg. Thus, in rhythmically exercising legs, LPP-induced restriction of the blood flow occurs solely by reducing the perfusion pressure.

Judging from experiments in animals, it appears likely that the maldistribution of blood flow within exercising muscles (cf. Pendergast et al. 1985) is exaggerated by LPP, since blood flow heterogeneity in skeletal muscles increases with arterial flow restriction (Tyml & Mikulash 1988).

In our original study (Eiken & Bjurstedt 1987) a square-wave positive pressure of 50 mm Hg was employed. Subsequent investigations of the effects of acute exposure to positive pressure have used square-wave, stepwise or ramp increases in pressure up to 40 or 60 mm Hg (Joyner 1991, Rowell et al. 1991, Sundberg & Kaijser 1992, Williamson et al. 1993).

## RESPONSES TO ACUTE ISCHEMIC EXERCISE

To study how work performance and the normal exercise-induced changes in respiration and central circulation are influenced by LPP we used a standard incremental work rate exercise protocol (Eiken & Bjurstedt 1987). The subjects worked on the cycle ergometer at a pedalling rate of 60 rev per min, the load starting at zero and being increased every 4 min in steps of 50 W up to 100 W and thereafter in steps of 30 W until the subject was unable to maintain the pedalling rate for the required 4-min period. The preceding load was defined as peak load.

Two conditions of exercise were tested and compared, (1) exercise under normal conditions, and (2) exercise with muscle blood flow in the legs reduced by LPP at 50 mm Hg. The two conditions will here be termed *free-flow exercise* and *flow-restricted* or *ischemic exercise*, respectively.

Flow-restricted exercise was found to impair work performance considerably. With incremental-load exercise, the endurance time was reduced from an average of 30 min in free-flow exercise to about 18 min in ischemic exercise. The peak load attained was lowered by about 40 percent from an average of 35 W in free-flow exercise. Local leg fatigue was invariably reported as the main reason for discontinuing exercise in the ischemic condition.

Using the incremental work rate exercise protocol, we measured and plotted respiratory, metabolic and cardiovascular variables against work load. Some of our results are shown in Fig. 3 and 4, in which data obtained with the legs exposed to LPP at 50 mm Hg, i.e. in ischemic exercise, are compared with data from free-flow exercise.

A key finding was that the point of exhaustion was attained at a considerably lower blood lactate level in ischemic than in free-flow exercise (Fig. 3). Muscular fatigue is known to be closely linked to intra-muscular lactate accumulation (Karlsson 1971). In cycle ergometry, a positive relationship is known to exist between intramuscular and blood lactate levels, so that at exhaustion blood lactate shows similar values whether exercise is performed in hyperoxia, normoxia or hypoxia (Linnarsson et al. 1974). In ischemic exercise, however, the normal exercise-induced gradient between intramuscular and blood lactate was greatly

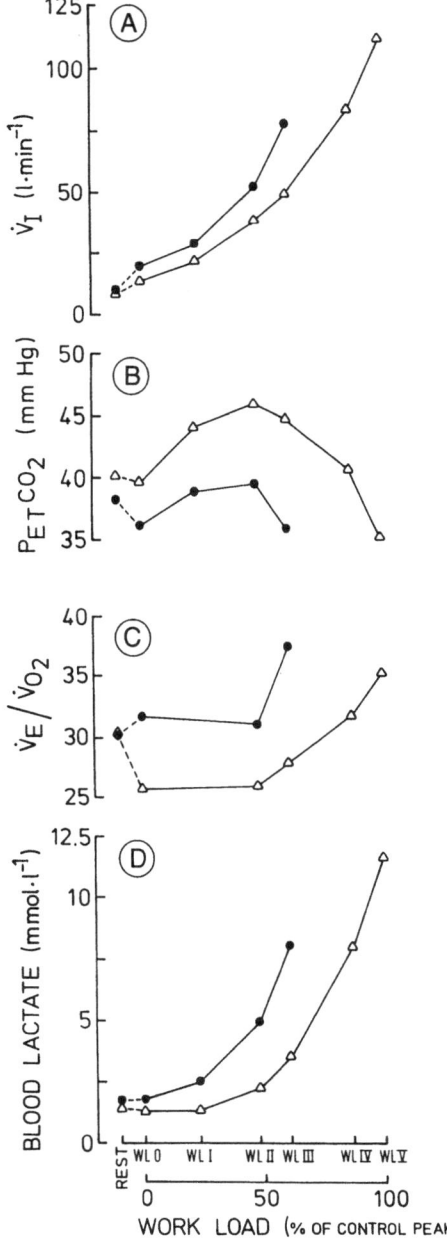

**Figure 3.** Inspired minute ventilation (A), end-tidal $P_{CO_2}$ (B), ventilatory equivalent for oxygen (C), and blood lactate (D) as functions of external work in ischemic (●) and free-flow (△) exercise, all data referring to steady state. On the abscissa, work loads are given as percentages of the peak load attained in the free-flow condition, with WL 0-V representing 0, 23, 48, 61, 87 and 100 % of this load and WL III indicating the highest load that could be managed during ischemic exercise. Values are means, n=8. Note that the point of exhaustion was attained at a considerable lower blood lactate level in ischemic than in free-flow exercise, and that the ventilatory minute volume was no less than about 30 liter/min greater than in free-flow exercise at the corresponding work load, with a marked lowering of end-tidal $P_{CO_2}$, indicating relative hyperventilation.

exaggerated because of reduced elimination of intramuscular lactate (Eiken 1987). The lowered blood lactate concentration in ischemic exercise at the point of exhaustion must therefore be ascribed to intramuscular 'trapping' of lactate. A corollary to this is that the common notion of blood lactate as a determinant of aerobic exercise tolerance is no longer applicable when comparing ischemic and free-flow exercise conditions.

As shown in Fig. 3, ventilation increased much more in the ischemic than in the free-flow condition. At the same time end-tidal $P_{CO_2}$ was consistently and markedly lowered in ischemic exercise, indicating a condition of a relative hyperventilation. The behavior of the ventilatory equivalent for oxygen indicates that ischemic exercise caused ventilation to increase out of proportion to the overall oxygen uptake. These findings tally with recent results of Williamson et al. (1993), who used a lower body positive pressure chamber to induce muscle ischemia during cycle exercise. Thus, with flow-restriction and associated intramuscular 'trapping' of metabolites, ventilation behaved as if the work rate was considerably higher than it actually was. We believe that this hyperpneic response to ischemic exercise was directly related to an increase in intramuscular metabolite concentration, and that the hyperpnea could not have been the result of stimulation by metabolically generated blood-borne factors. Instead, there appears to be an additional ventilatory drive, a reflex component which conceivably originates from within the exercising ischemic muscle.

Using the same protocol with incremental work-load cycle exercise we also studied the effects of ischemic exercise on the central circulation. Fig 4 shows that the pressure response was clearly exaggerated in ischemic exercise. At peak load for ischemic exercise

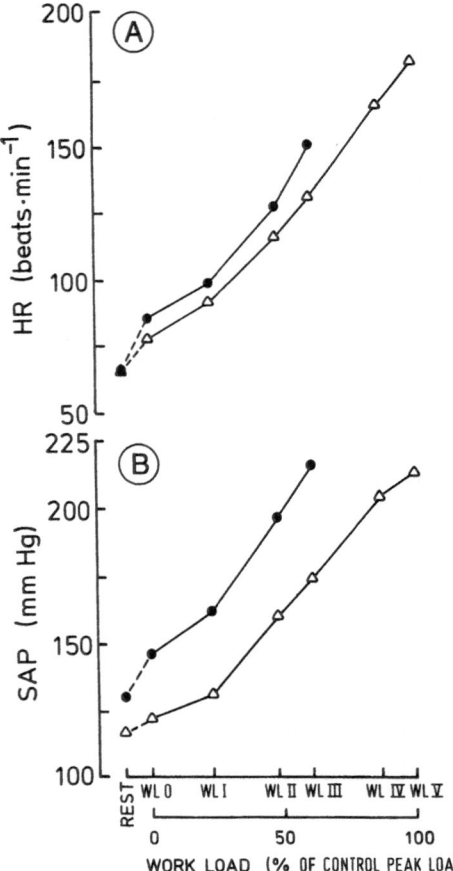

**Figure 4.** Heart rate (A) and systolic arterial pressure (B) as functions of external work load in ischemic (●) and free-flow exercise (△). Abscissa, see explanation in Fig. 3. Values are means, n=7. Note that the arterial pressure - here represented by the systolic pressure - rose markedly higher in ischemic than in free-flow exercise at corresponding work loads.

mean arterial pressure was calculated to be about 30 mm Hg higher than at the corresponding work load in the free-flow condition. Also the heart rate response was somewhat stronger during ischemic exercise at corresponding work loads.

The pressor and other cardiovascular responses that were observed during ischemic exercise in our experiments are presumably akin to those commonly thought to be mediated by "the muscle chemoreflex". It is well established that the signals originate in intramuscular chemosensors responding to exercise-induced accumulation of metabolites and other agents, and to a certain extent also to mechanical stimuli. The impulse traffic is mediated by afferents within group III and IV (McCloskey & Mitchell 1972, Kaufman & Rybicki 1987). Our observations from ischemic exercise suggest that a ventilatory drive may be part of the same reflex.

It must be emphasized, that the results obtained by the use of our model were caused by an intervention that impeded free-flow conditions in the working muscles. From our experiments, we have no way of knowing, whether or to which extent the respiratory and cardiovascular effects seen during ischemic exercise are due to signals that are tonically active under normal free-flow dynamic exercise.

Nor do we know if or to what extent a central command component might have been involved in these effects, even though it appears that the effects on the heart rate - commonly thought to reflect the central command drive - were generally modest.

Some of our reflex control data can be viewed as an extension of previous observations using cuffs. We would like to emphasize, however, that the LPP method offers certain advantages in the study of responses to acute ischemic exercise. Complete blocking of the circulation to exercising limbs is avoided, and since the degree of flow restriction and its duration can be varied at will, the method readily lends itself to closer studies of the nature of the local and systemic responses that accompany acute, flow-restricted dynamic exercise.

## ISCHEMIC TRAINING

To our knowledge, experimental reduction of blood flow in human muscles undergoing training has not previously been employed to elucidate the possible impact of long-term local ischemia on training-induced improvements in work performance. The following account deals with the application of the LPP method to induce muscle ischemia in repeated exposures to high intensities of aerobic exercise, i.e. in the process of physical conditioning.

It is well known that the local aerobic capacity of human skeletal muscle improves in response to endurance training. It is also known that patients suffering from intermittent claudication caused by impaired circulation to the legs commonly show promptly improved exercise capacity in response to endurance training (Larsen & Lassen 1966, Alpert et al. 1969, Zetterquist 1970, Dahllöf et al. 1974, Saltin 1981). However, the stimulus by which training triggers adaptation is poorly understood.

To study the effects of local muscle ischemia on the adaptive process of endurance training, the LPP method was applied in a series of investigation recently summarized in a doctoral thesis by Sundberg (1994). By employing this method it became possible to functionally imitate, in healthy individuals, the therapeutic use of walking training. One central hypothesis was that ischemia in working muscles acts as a stimulus to the metabolic and structural adaptive changes that occur in response to endurance training.

In the investigations to be reported here, healthy subjects underwent endurance training on a cycle ergometer four times a week over a period of 4 - 5 weeks. All investigations employed *one-legged exercise,* one leg undergoing training with ischemic exercise, the contralateral leg being trained under free-flow conditions. This regimen was used in all studies on training with the LPP method to make it possible to distinguish, in the same subject, the effects of repeated muscle ischemia from those of muscle training *per se*. The two training modes will here be referred to as *ischemic* and *free-flow training,* respectively.

All training sessions lasted for 45 min, and the work load chosen was the highest tolerable that the individual subject could maintain under ischemic conditions, with the understanding that the whole 45-min session had to be completed. This work load was usually about 50-60 percent of the peak load attained by the individual in a standard incremental work rate test, the average for the group being 70 W. The same work-load protocol was then followed for the contralateral leg under free-flow conditions. In this way, both legs produced the same mechanical power and work throughout the study.

The ischemic training regimen was invariably experienced by the subjects as extremely strenuous with periods of ischemic muscle pain occurring frequently during the training sessions, whereas the free-flow training (contralateral leg) was experienced as very light.

When work performance was tested before and after a 4-week training regimen using a standard incremental work rate scheme, it was observed that ischemic training is markedly more effective in improving endurance performance than training under free-flow conditions. Time to exhaustion was prolonged by about 20 percent, and peak oxygen uptake increased by 18 percent (Sundberg et al. 1993).

It should also be noted, that the improvement in endurance performance in the ischemically trained leg, as tested in incremental work rate exercise, was found to be exaggerated when the test was repeated under ischemic conditions, indicating that there is a specificity in response to ischemic training. The results of these experiments clearly show that local muscle ischemia acts as an additive stimulus to endurance training.

This raises the question as to what mechanisms might underlie the enhanced training effect by ischemia? It was thought that some partial answers might be obtained by using the LPP method to investigate certain metabolic and structural changes that occur in human skeletal muscle undergoing ischemic endurance training. Accordingly, biopsies were taken from the vastus lateralis muscle in the ischemic and the control leg by a percutaneous technique before and after 4 - 5 weeks of one-legged training.

Parallel to the improvements in work performance, the biopsy analyses showed that ischemic training produces well-defined adaptive changes of both metabolic and structural kinds (Kaijser et al. 1990, Esbjörnsson et al. 1993). It was found that ischemic training changes the muscle metabolic profile in a direction facilitating aerobic metabolism. It increases the metabolic capacity of muscle as evidenced by a higher activity of the enzyme citrate synthase, a marker of the capacity for oxygen utilization. Proliferation of capillaries is normally a major factor, permitting increased oxygen extraction after endurance training and improving conditions for diffusion of oxygen. It was observed that ischemic training increases the local density of capillaries per fiber in the ischemic leg, but not in the control leg, presumably because of the relatively low work rates employed in the training regimen. It should be remembered at this point that the work rate used was experienced as extremely heavy for the ischemic leg, but as very light for the control leg. There are also changes in fiber type: ischemic training leads to an increased percentage of slow-twitch (type-1) fibers, and also to larger cross-sectional areas of all fiber types (Esbjörnsson et al. 1993). The higher percentage of type-1 fibers may have been a factor behind the observed higher citrate synthase activity in the ischemic leg.

A somewhat unexpected finding was obtained in testing the effects of ischemic training on muscle force development. It was found that the maximal dynamic strength development in the quadriceps muscle of the ischemic leg was significantly reduced as compared with the strength performance of the control leg. Biopsies showed that the proportion of slow-twitch fibers was higher, and the mean slow-twitch fiber area larger in the ischemic than in the control leg. The relative preponderance of slow-twitch fibers in the ischemically trained muscle might have been a contributing factor in the reduction of muscle strength (Eiken et al. 1991).

The mechanisms by which ischemia in working muscles acts as a stimulus to the metabolic and structural adaptive changes in endurance training are a virgin field open for continued investigations. We believe the LPP method holds promise as a tool for continued investigations of local and systemic effects resulting from prolonged, iterative ischemic training.

# REFERENCES

ALAM, M. & F.H. SMIRK. 1937. Observations in man upon a blood pressure raising reflex arising from the voluntary muscles. J. Physiol. 89: 372-383.

ALPERT, J.S., O.A. LARSEN & N.A. LASSEN. 1969. Exercise and intermittent claudication. Blood flow in calf muscle during walking studied by the Xenon-133 clearance method. Circ. 39: 353-359.

ASMUSSEN, E. & M. NIELSEN. 1964. Experiments on nervous factors controlling respiration and circulation during exercise employing blocking of the blood flow. Acta Physiol. Scand. 60: 103-111.

DAHLLÖF, A.-G., P. BJÖRNTORP, J. HOLM & T. SCHERSTEN. 1974. Metabolic activity of skeletal muscles in patients with peripheral arterial insufficiency. Effects of physical training. Eur. J. Clin. Invest. 4: 9-15.

DAHN, I., N.A. LASSEN & H. WESTLING. 1967. Blood flow in human muscles during external compression or venous stasis. Clin. Sci. 32: 467-473.

DUOMARCO, J.L. & R. RIMINI. 1954. Energy and hydraulic gradients along systemic veins. Am. J. Physiol. 178: 215-220.

EIKEN, O. 1987. Responses to dynamic leg exercise in man as influenced by changes in muscle perfusion pressure. Acta Physiol. Scand. 131. Suppl. 566: 1-37.

EIKEN, O. & H. BJURSTEDT. 1987. Dynamic exercise in man as influenced by experimental restriction of blood flow in the working muscles. Acta Physiol. Scand. 131: 339-345.

EIKEN, O., C.J. SUNDBERG, M. ESBJÖRNSSON, A. NYGREN & L. KAIJSER. 1991. Effects of ischemic training on force development and fibre-type distribution in human skeletal muscle. Clin. Physiol. 11: 41-49.

ESBJÖRNSSON, M., E. JANSSON, C.J. SUNDBERG, O. EIKEN, A. NYGREN & L. KAIJSER. 1993. Muscle fibre types and enzyme activities after training with local leg ischemia. Acta Physiol. Scand. 148: 233-241.

JOYNER, M.J. 1991. Does the pressor response to ischemic exercise improve blood flow to contracting muscles in humans? J. Appl. Physiol. 71: 1496-1501.

KAIJSER, L., C.J. SUNDBERG, O. EIKEN, A. NYGREN, M. ESBJÖRNSSON, C. SYLVEN & E. JANSSON. 1990. Muscle oxidative capacity and work performance following training under local leg ischemia. J. Appl. Physiol. 69: 785-787.

KARLSSON, J. 1971. Lactate and phosphagen concentrations in working muscle of man with special reference to oxygen deficit at the onset of work. Acta Physiol. Scand. Suppl. 358.

KAUFMAN, M.P. & K.J. RYBICKI. 1987. Discharge properties of group III and IV muscle afferents: their responses to mechanical and metabolic stimuli. Circ.Res. 61 (Suppl. I), 60-65.

LARSEN, O.A. & N.A. LASSEN. 1966. Effect of daily muscular exercise in patients with intermittent claudication. Lancet. 2: 1093-1096.

LAUGHLIN, M.H. 1987. Skeletal muscle blood flow capacity: role of muscle pump in exercise hyperemia. Am. J. Physiol. 253 (Heart Circ. Physiol. 22), H993-H1004.

LINNARSSON, D., J. KARLSSON, L. FAGRAEUS & B. SALTIN. 1974. Muscle metabolites and oxygen deficit with exercise in hypoxia and hyperoxia. J. Appl. Physiol. 36: 399-402.

LUNDVALL, J. & T. LÄNNE. 1989. Transmission of externally applied negative pressure to the underlying tissue. A study on the upper arm of man. Acta Physiol. Scand. 136: 403-409.

McCLOSKEY, D.I. & J.H. MITCHELL. 1972. Reflex cardiovascular and respiratory responses originating in exercising muscle. J. Physiol. (Lond.) 224: 173-186.

NIELSEN, H.V. 1983. External pressure-blood flow relations during limb compression in man. Acta Physiol. Scand. 119: 253-260.

PENDERGAST, D.R., J.A. KRASNEY, A. ELLIS, B. McDONALD, C. MARCONI & P. CERRETELLI. 1985. Cardiac output and muscle blood flow in exercising dogs. Resp. Physiol. 61: 317-326.

RENEMAN, R.S., D.W. SLAAF, L. LINDBOM, G.J. TANGELDER & K.E. ARFORS, 1980. Muscle blood flow disturbances produced by simultaneous elevated venous and total muscle tissue pressure. Microvasc. Res. 20: 307-318.

RODDIE, I.C. & J.T. SHEPHERD. 1957. Evidence for critical closure of digital resistance vessels with reduced transmural pressure and passive dilatation with increased presssure. J. Physiol. 136: 498-506.

ROWELL, L.B., M.V. SAVAGE, J. CHAMBERS & J.R. BLACKMON. 1991. Cardiovascular responses to graded reductions in leg perfusion in exercising humans. Am. J. Physiol 261: H1545-H1553.

SALTIN, B. Physical training in patients with intermittent claudication. In: COHEN, I.S., M.B. MOCK & I. RINGQUIST (eds.). Physical conditioning and cardiovascular rehabilitation, chapt. 13, p 181-196. John Wiley and Sons, New York. 1981.

SHI, X., C.G. CRANDALL & P.B. RAVEN. 1993. Hemodynamic responses to graded lower body positive pressure. Am. J. Physiol. 265: H69-H73.

SUNDBERG, C.J. 1994. Exercise and training during graded leg ischaemia in healthy man. Acta Physiol. Scand. 150. Suppl. 615: p 1-50.

SUNDBERG, C.J., O. EIKEN, A. NYGREN & L. KAIJSER. 1993. Effect of one-legged ischemic training on dynamic muscle performance and peak oxygen uptake in man. Acta Physiol. Scand. 148:13-19.

SUNDBERG, C.J. & L. KAIJSER. 1992. Effects of graded restriction of perfusion on circulation and metabolism in the working leg; quantification of a human ischaemia-model. Acta Physiol. Scand. 146: 1-9.

TYML. K. & K. MIKULASH. 1988. Evidence for increased perfusion heterogeneity ins skeletal muscle during reduced flow. Microvasc. Res. 35: 316-324.

WILLLIAMSON, J.W., P.B. RAVEN, B.H. FORESMAN & B.J. WHIPP. 1993. Evidence for an intramuscular ventilatory stimulus during dynamic exercise in man. Resp. Physiol. 94: 121 - 135.

ZETTERQUIST, S. 1970. The effect of active training on the nutritive blood flow in exercising ischemic legs. Scand. J. Clin. Lab. Invest. 25: 101-111.

# 15

# CARDIOVASCULAR REFLEXES FROM VENTRICULAR AND CORONARY RECEPTORS

Roger Hainsworth

Institute for Cardiovascular Research
Research School of Medicine
University of Leeds
Leeds, LS2 9JT United Kingdom

## INTRODUCTION

The cardiac ventricles are innervated by both myelinated and nonmyelinated fibers and these run in both vagal and sympathetic nerves (Hainsworth, 1991). There are far more nonmyelinated than myelinated afferents, and the innervation of the left ventricle greatly exceeds that of the right (Coleridge, Coleridge and Kidd, 1964; Paintal, 1955). Despite the abundant physiological evidence for a rich afferent innervation, no specific sensory nerve endings have yet been identified.

It has proved to be technically difficult to apply discrete physiological stimuli to cardiac receptors and there have been many studies which have deduced the reflex responses which originate in the heart by observing the effects of procedures such as vagotomy or changes in venous return to the heart. This has led to the use of the unfortunate term of "cardiopulmonary receptors". These are considered to be attached to sensory nerves from the heart and lungs, which run in the vagus nerves, and activity in which causes vasodilatation and bradycardia. The flaw with this concept is that the heart and lungs do not contain a homogeneous population of receptors but many different groups and these are activated by different events. For example, atrial receptors are stimulated by atrial distension caused by increased cardiac filling, and their reflex responses are increases in heart rate and urine flow (Linden and Kappagoda, 1982; Linden chapter). Pulmonary arterial receptors are also stimulated by increases in venous return and pulmonary blood flow, but their reflex effects are to cause vasoconstriction (Ledsome and Kan, 1977). Ventricular receptors, as will be discussed below, can be stimulated by chemical events, mechanical stimuli, or both and stimulation of them results in bradycardia and vasodilatation. However, as will be explained below, there is really no convincing evidence that they have any role at all in normal physiological regulation. In view of the wide variation in the reflexes elicited from stimulation of relexogenic areas in the heart and lungs, use of the term "cardiopulmonary receptors" is misleading, causes confusion, and is best avoided.

*Control of the Cardiovascular and Respiratory Systems in Health and Disease*
Edited by C. T. Kappagoda and M. P. Kaufman, Plenum Press, New York, 1995

In this chapter, I will describe the reflexogenic areas in the ventricle and coronary arteries, the stimuli which are responsible for exciting them, and the reflex responses which result. I will also attempt to assess the likely physiological significance of these reflexes.

## REFLEXES FROM CHEMOSENSITIVE RECEPTORS

It has been known for a very long time that powerful depressor reflexes may result from chemical stimulation of receptors in the heart. These responses were first described by Von Bezold and Hirt in 1867. Dawes (1947) injected veratridine into the cavity of the ventricle or directly into branches of the coronary arteries and showed that reflex responses occured following very much smaller injections into the coronary branches supplying the left ventricle compared with the doses that were needed to elicit the reflex by intraventricular injection. He also showed that injection into the right ventricle or the coronary arteries supplying the right ventricle had little or no effect. The distribution of the main reflexogenic area to the left ventricle was also shown in other studies in which the agents were applied by means of pledgelets to the epicardial surface of the heart (Sleight, 1964). Several other chemicals given by intracoronary injection or epicardial application have also been shown to be capable of eliciting the reflex. These include capsaicin, nicotine, acetylstrophanthidine, ouabain, bradykinin and prostaglandins.

Injection or application of the stimulant chemicals results, after a short latency, in an intense burst of activity in vagal afferent nerves, particularly in nonmyelinated fibers (Fig 1). The reflex responses to this stimulation are bradycardia and hypotension. The bradycardia is predominantly vagal and, in some particularly responsive preparations, there is likely to be a period of asystole. The mechanism for the hypotension is complex. Accompanying the bradycardia is a decrease in the cardiac output (Zucker et al. 1981, Barron and Bishop, 1982).

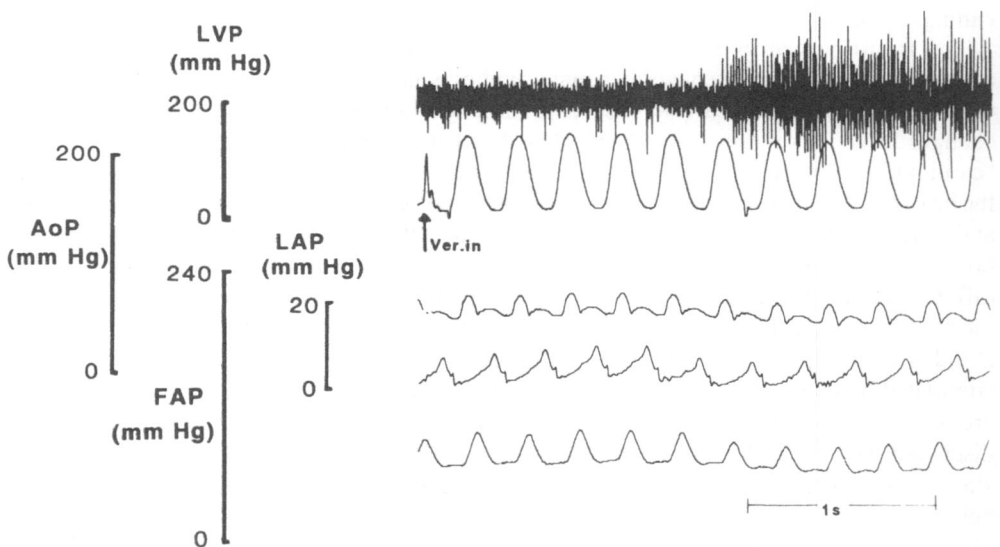

**Figure 1.** Response of nonmyelinated vagal afferents to veratridine. At the arrow 25 μg veratridine was injected into the left ventricle. Traces of action potentials in slip of vagus nerve, left ventricular pressure (LVP), aortic blood pressure (AoP), left atrial pressure (LAP), and femoral arterial blood pressure (FAP). The latency of the response (about 1.5s) can be attributed to the time between ventricular injection and the drug entering the coronary circulation. (From ref. 20)

Vascular resistance decreases in the limb (Fig 2) (McGregor et al. 1986), abdominal circulation (Hainsworth et al. 1986), renal circulation (Thames, 1979) and coronary circulation (Clozel et al. 1990). Capacitance vessels also reflexly dilate (Tutt et al. 1986) and this also contributes to the hypotension. Studies of the responses to repeated injections of chemical stimulants are complicated by the phenomenon of tachyphylaxis. This effect can be minimised if the interval between injections is maintained above about 10 min.

The physiological significance of the coronary chemoreflex is uncertain. It is unlikely that chemosensitive afferents would normally be stimulated by vegetable extracts such as the veratrum alkaloids or capsaicin. However, there are endogenous chemicals, including bradykinin, prostaglandins and thromboxanes, which have been shown to excite chemosensitive afferents and to cause reflex depressor responses. These chemicals may be released in humans during myocardial ischemia and infarction (Berger et al. 1977; Hirsch et al. 1981). The circulatory effects of coronary arterial occlusion are complex due to possible additive, opposing and interacting effects caused by a decrease in cardiac pumping and stimulation of either or both vagal depressor and sympathetic pressor reflexes. In conscious humans, the reflex effects are also complicated by psychological reactions to a painful and frightening event. It seems that obstruction of the circumflex branch of the left coronary artery, which results in ischemia of the inferior and lateral walls of the left ventricle and leads predominantly to excitation of a vagal depressor reflex, tends to cause reflex bradycardia and hypotension (Thames et al. 1978). Occlusion of the anterior descending branch, however, probably excites mainly sympathetic afferents and this tends to increase blood pressure and may predispose to arrhythmias. It is claimed that the pressor responses are seen best in conscious preparations and after removing the depressor reflex by vagotomy (Brown and Malliani, 1971; Lombardi et al. 1984).

The coronary chemoreflex may also be elicited following acute administration of various drugs which stimulate cardiac nonmyelinated afferents and cause reflex hypotension (Semple et al. 1988). Also, intracoronary injections of iodine containing contrast medium

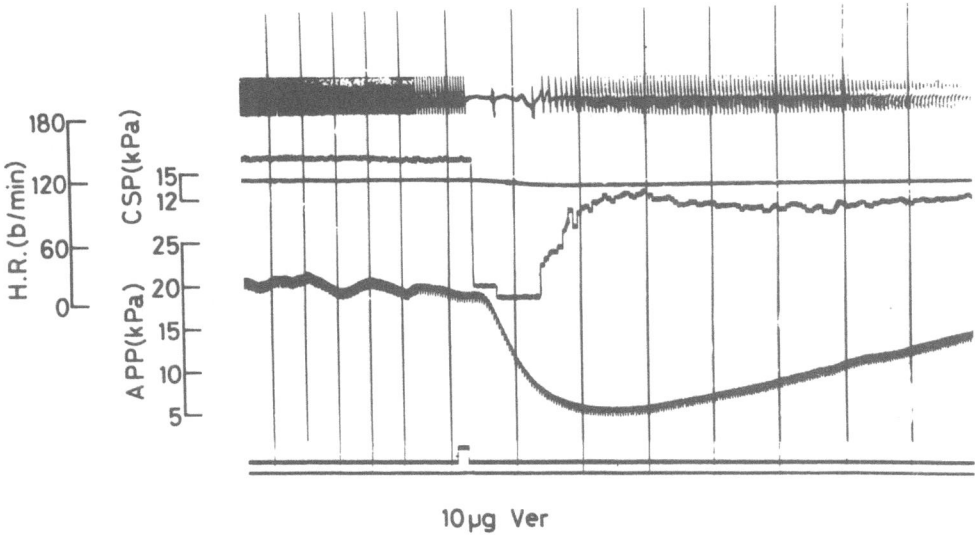

10 μg Ver

**Figure 2.** Reflex responses to injection of veratridine (10μg) into aortic root. Traces of E.C.G., heart rate (HR), carotid sinus pressure (CSP), aortic perfusion pressure (APP) perfused at constant flow. Vertical lines are at 10s intervals. Pressures in kPa, 1kPa = 7.5 mmHg. Note that after a short latency veratridine results in profound bradycardia and vasodilatation (decrease in perfusion pressure at constant flow). (From ref.36).

may cause acute hypotension and bradycardia by a similar mechanism (Perez-Gomez and Garcia-Aguado, 1977, Arrowood et al 1989).

# REFLEXES FROM VENTRICULAR MECHANORECEPTORS

The vast majority of ventricular afferent nerves consist of nonmyelinated fibers. Nevertheless, there have been reports of some myelinated fibers which were thought to end in the ventricle (Coleridge et al. 1964; Gupta and Thames, 1983). Indeed Gupta and Thames (1983) claimed to have recorded from a surprisingly large number of myelinated ventricular mechanoreceptor afferents. However, in our experience it is not always easy to be certain of the exact location of ventricular endings and it is possible that some of the nerves described by Gupta and Thames may actually have terminated either in the adjacent atria or in the coronary arteries. We recently recorded from several myelinated afferents which appeared to be ventricular in origin in that they responded to ventricular probing and to aortic occlusion, but when we changed ventricular and atrial pressures independently, the nerves were shown actually to be attached to atrial receptors (Moore, 1994). Many apparent ventricular receptors attached to myelinated fibers may actually be coronary arterial receptors and in our recent study (Drinkhill et al. 1993) we observed that *all* the myelinated fibers which responded to changes in aortic root and ventricular pressures were almost certainly coronary receptors.

Nonmyelinated vagal afferents do terminate in the left ventricle and some of these respond to mechanical events. Some ventricular afferents can be excited by both mechanical and chemical stimuli. In our recent study (Drinkhill et al. 1993) we recorded from 11 nonmyelinated afferents with conduction velocities of 0.15-1.5m/s and mean resting discharge rates of 2.3 impulses/s. Of these fibers, 5 increased their discharge to increases in ventricular pressure but not to ventricular or aortic injections of veratridine, 3 responded only to veratridine and 3 to both stimuli. These findings are broadly in agreement with the findings of Coleridge and colleagues (Baker et al. 1979; Kaufman et al. 1980), that ventricular afferents form a spectrum between those that are solely mechanosensitive, through those that are both mechano- and chemosensitive, to afferents displaying only chemosensitivity.

In most studies of the behavior of ventricular afferents, fibers were usually selected on the basis that they had a spontaneous, often cardiac related, discharge. However, according to Thames et al. (1977), most ventricular afferents in closed chest animals are normally silent. The behavior of ventricular receptors to increases in pressure is often erratic. There is usually no increase in activity in response to moderate changes in ventricular systolic pressures. We recently noted that pressures in excess of 225 mmHg were necessary to increase the discharge and even then the increase in activity was usually not sustained. Others have claimed that a more potent stimulus is an increase in the force of ventricular contraction or diastolic ventricular distension, although it has been noted that, in absence of ventricular contraction, distension does not stimulate the receptors, (Sleight 1964, Thoren, 1977). A more detailed account of the behavior of ventricular afferents is given in the preceding chapter.

Localisation of a mechanical stimulus to the left ventricle is technically quite difficult. Studies of the effects of cutting or stimulating cardiac nerves do not provide useful quantitative information. Another approach has been to occlude the ascending aorta (Oberg and Thoren, 1973; Mark et al. 1973; Zucker et al. 1986). However, this not only increases ventricular pressure but also alters the pressure distending several other reflexogenic areas, particularly baroreceptors in the aorta itself, carotid sinuses and coronary circulation. Daly and Verney (1927) inserted both ventricles inside a cardiometer and noted that distension, caused by applying a subatmospheric pressure to the cardiometer, resulted in a small bradycardia and vasodilatation. This was an ingenious approach but the stimulus is not really physiological and may not even be localised to the ventricle. There have been a number of other attempts to stimulate ventricular mechanoreceptors, using preparations involving cardiopulmonary bypass in which either ventricular inflow was increased, outflow obstructed, or the ventricle distended by a balloon (Aviado and Schmidt, 1959; Chevalier et al. 1974; Hoka et al. 1988; Kostreva et al. 1979; Ross et al. 1961; Salisbury et al. 1960; Zelis et al. 1977). Usually these interventions caused either no effect on heart rate or a small and often transient bradycardia. Other responses were a decrease in vascular resistance and dilatation of capacitance vessels (Hoka et al. 1988; Tutt et al. 1988). The earlier work, therefore, suggests that ventricular mechanoreceptors may behave like arterial baroreceptors in that both result in responses of vasodilatation and possibly, bradycardia. However, interpretation of the importance of these responses is unclear because the stimuli tended to be poorly localised unphysiological, or both.

In our recent study (Al-Timman et al. 1993) we devised a preparation which was intended to allow us to apply a localized and discrete stimulus to ventricular receptors. The pressure distending the aortic and carotid baroreceptors and the atrial receptors was controlled and we also controlled mean pressure in the coronary arteries. We found, using this technique, that an increase in ventricular systolic pressure did result in vasodilatation but that the response was much smaller that that occuring when pressure in the coronary arteries increased by a similar amount. Futhermore, increases in ventricular pressure using this technique did actually stimulate coronary arterial mechanoreceptors possibly through mechanical forces transmitted from the ventricle to the coronary arteries or from changes in coronary arterial pulse pressure (Drinkhill et al. 1993).

It seems likely, therefore, from the results of our experiments and an analysis of results from the experiments of others, that ventricular mechanoreceptors are probably unimportant in circulatory control. Electrophysiological studies showed that they were stimulated only by extremes of stimuli and, even then, their response tended to be erratic and transient. The responses obtained from reflex studies tended to be small and in most cases could be attributed to the poor localisation or to the extreme and unphysiological nature of the stimuli applied.

# REFLEXES FROM CORONARY MECHANORECEPTORS

The coronary arteries are richly innervated. Indeed, it has been claimed that no other artery in the body has a richer innervation (Woolard, 1926). Many vagal branches terminate in the adventitia of the coronary arteries, although no specific nerve endings have yet been described and it is unknown whether the nerves described were afferent or efferent.

The variable effects of total occlusion of a major coronary artery have already been described and any reflex responses are probably mainly the result of locally released chemicals due to ischemia. Obstruction to outflow from the coronary sinus, when it is sufficient to cause coronary sinus pressure to exceed 50mmHg, results in bradycardia and hypotension (Muers and Sleight, 1972). However, it is unclear whether this effect is a reflex response to distension of coronary veins or coronary arteries or whether it is an effect of congestion or ischemia.

Brown (1965) wedged a cannula into a coronary artery of the cat and recorded the effects of transient changes in coronary pressure on the activity in afferent slips of vagus nerve. He noted that increases or decreases in coronary arterial pressure influenced the discharge rate in presumed coronary mechanoreceptors. Brown's work provided strong evidence for the existence of coronary arterial mechanoreceptors, but it remained uncertain

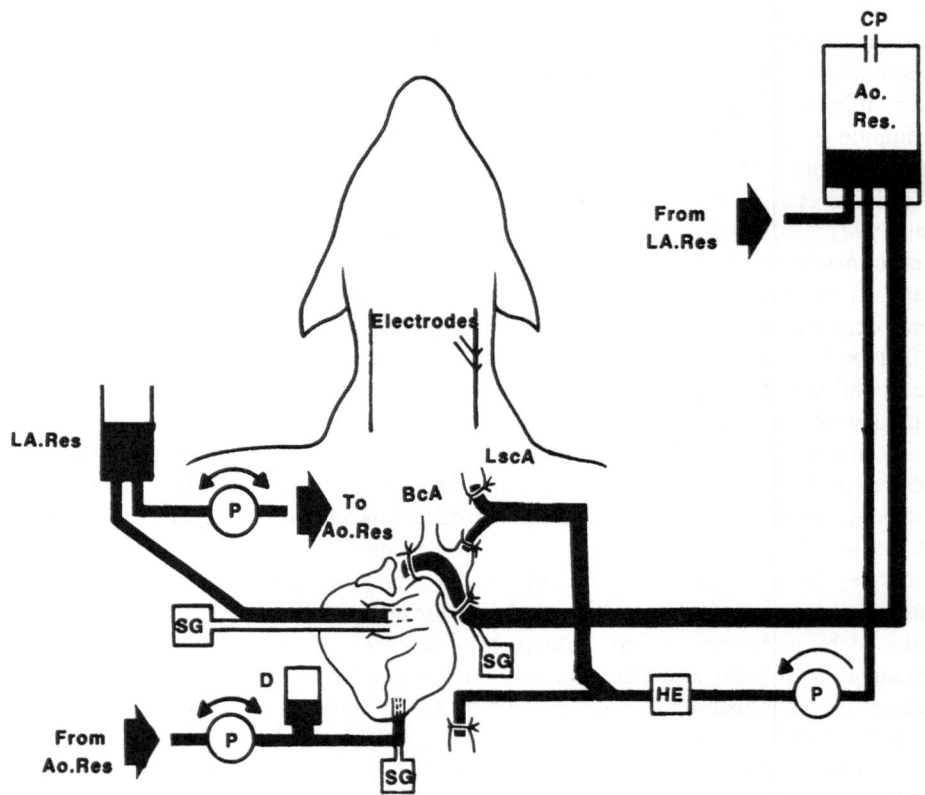

**Figure 3.** Preparation used to separate pressure stimulus applied to aortic root plus coronary arteries from that to the left ventricle. The pressure applied to the aortic reservoir is transmitted to the aortic root and coronary circulation. Cannulae in the left atrium, connected to a reservoir, and in the left ventricle through its apex and linked to the aortic reservoir, allow control of ventricular pressures. Abbreviations: CP, constant pressure, Ao. Res., aortic reservoir; LA.Res, left atrial reservoir; P, pump; BcA, brachiocephalic artery; LscA, left subclavian artery; SG, strain gauge manometer; D, damping chamber; HE, heat exchanger. (From Ref. 20).

whether they were actually a distinct population of receptors, or whether they were merely ventricular afferents which happened to be near to a coronary vessel. We recently reinvestigated this question using a preparation in anesthetized dogs which allowed us to apply controlled pressure stimuli to both the ventricle and the coronary arteries together or to either region separately with the pressure to the other region controlled.

Fig. 3 shows the preparation that we used for separating the stimulus to the coronary and ventricular receptors. The pressure to the coronary arteries is controlled by use of a cannula tied into the ascending aorta. Control of ventricular pressure was achieved by use of a partial bypass between the left atrium and a reservoir connected to the aorta root cannula, and a further cannula inserted directly into the cavity of the ventricle through which blood could be pumped rapidly into or out of the ventricle. We recorded from 10 vagal afferent strands which discharged in phase with ventricular contraction. An example is shown in Fig. 4. This shows a single fiber preparation of a myelinated vagal afferent which increases its discharge in response to an increase in pressure in the aortic root combined with an increase in ventricular systolic pressure (combined test). When only ventricular pressure is increased (although this also causes an increase in aortic pulse pressure, but not in mean pressure) the effect on the discharge is small. When only aortic root and coronary arterial

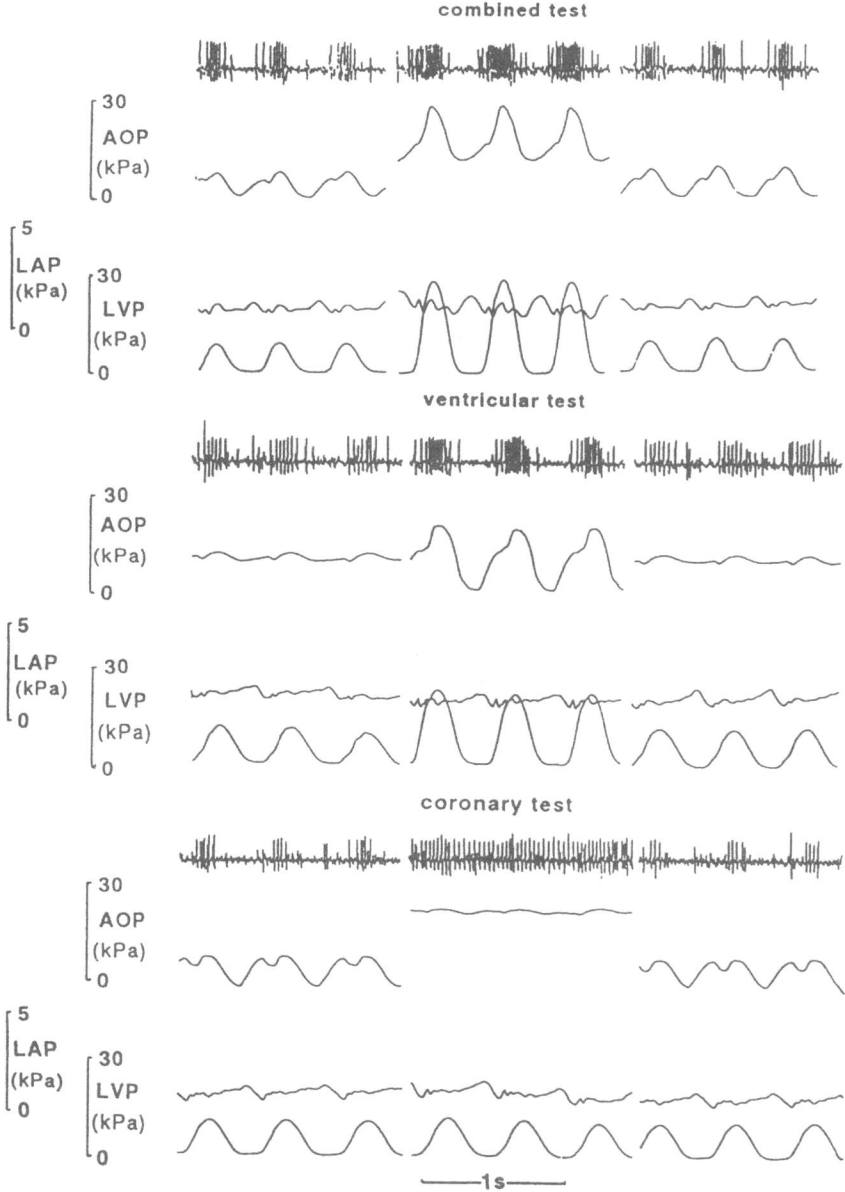

**Figure 4.** Responses of an afferent vagal unit to combined changes in aortic root plus coronary arterial pressure and ventricular systolic pressure (combined test), changes in ventricular systolic pressure at constant mean aortic pressure (ventricular test), changes in aortic root plus coronary arterial pressure at constant ventricular pressure (coronary test). Traces of aortic root pressure (AOP), left atrial pressure (LAP) and left ventricular pressure (LVP). Note that the coronary test increases afferent neural activity more than the LV test and by nearly the same amount as the combined test. Note also that the pulsatility of the discharge follows that of the aortic root. (From ref 20).

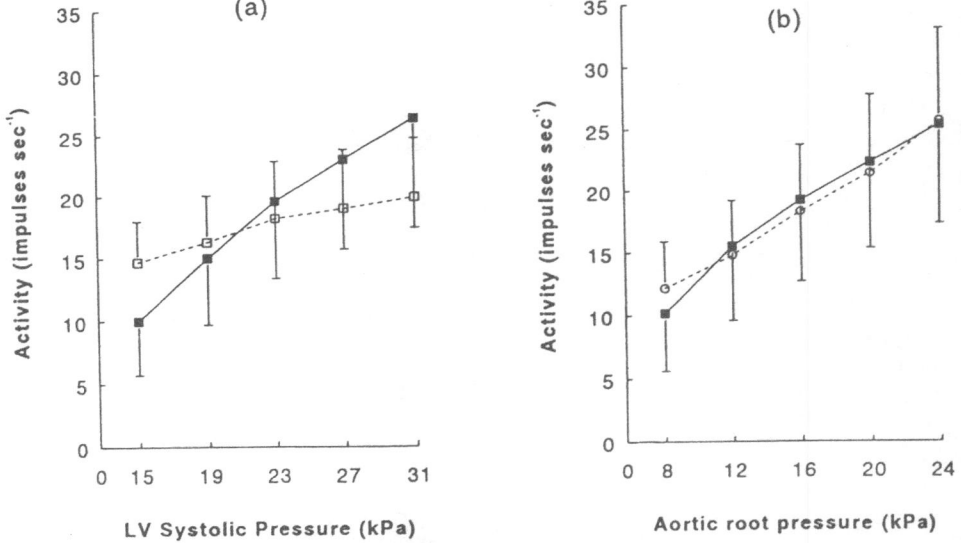

**Figure 5.** Responses of activity in six vagal afferent fibers to step changes in pressure in both aortic root plus coronary pressure and left ventricular pressure (■), changes only in ventricular pressure (□), and change only in aortic plus coronary pressure (○). Results show means ± SE of activity. Combined pressure changes are plotted as the ventricular pressures in (a) and as the associated aortic pressure in (b). Note that when only aortic plus coronary pressures were changed, the responses were nearly the same as those to the combined test. (From ref. 20).

pressures were increased, and ventricular pressure held constant, there was a large increase in discharge frequency and the pulsatility of the discharge closely followed that of the coronary pulse pressure. The greater sensitivity of the fibers to changes in coronary pressure than to changes in ventricular pressure is also seen from Fig. 5 in which the response to graded changes in coronary pressure were almost identical to the responses when both coronary and ventricular pressures changed simultaneously.

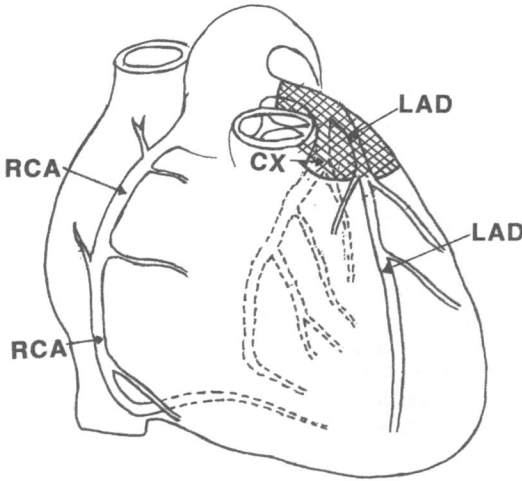

**Figure 6.** Location of receptive fields of mechanoreceptors attached to myelinated vagal afferents. Abbreviations: LAD, left anterior descending coronary artery; CX, circumflex artery; RCA, right coronary artery. All receptors were localized to the left coronary artery and the proximal parts of its main branches. (From Ref. 20).

All the coronary mechanoreceptors from which we recorded were localized to the region of the left coronary artery and its divisions into the circumflex and anterior descending branches (Fig. 6). This is in agreement with the findings of Brown (1965). All fibers had conduction velocities in excess of 3.2m/s and were, therefore, considered to be myelinated. All fibers were also tested for chemosensitivity by injection of veratridine either intraventricularly or into the aortic root. Unlike half of our nonmyelinated ventricular afferents (6

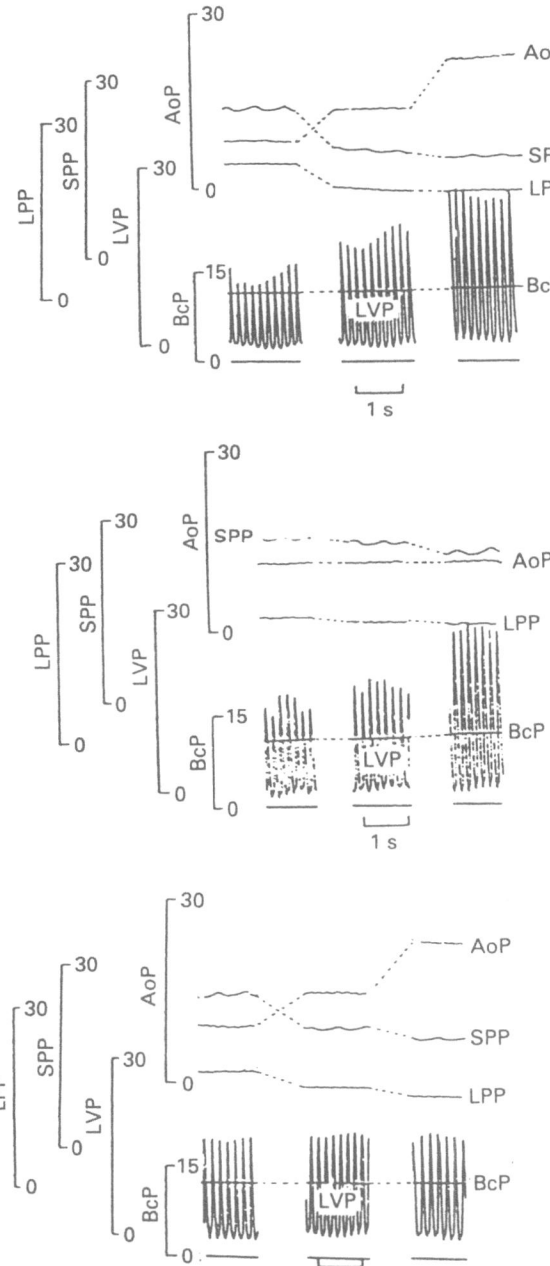

**Figure 7.** Comparison in one dog of the effects of combined changes in ventricular pressure and aortic root plus coronary pressures (top traces), with effects of changes only in ventricular pressure (center), and only aortic plus coronary pressure (bottom). Traces of aortic root pressure (AoP), systemic and limb perfusion pressures (SPP, LPP), brachiocephalic artery perfusion pressure (BcP, =pressure to aortic and carotid baroreceptors). Note that the decreases in perfusion pressures to increases in aortic and coronary pressures were much greater than these to ventricular pressure and similar in magnitude to those to the combined stimulus. (From ref 3).

of 11) and the mechanoreceptors described by Brown (1965), none of our coronary arterial receptors showed chemosensitivity. The discrepancy between our results and Brown's is puzzling. One possibility is that there may be a species difference as Brown studied cats and we studied dogs. This explanation is quite plausible as there is evidence from several studies for differences between cats and dogs in the cardiovascular response to injections of various chemicals (Dawes and Comroe, 1955).

We consider the receptors to be coronary arterial baroreceptors. They respond to a wide range of coronary arterial pressures although, not surprisingly, they are also influenced to some extent by ventricular contraction. Further points to be emphasized are that none of the nonmyelinated afferents responded to physiological pressure changes and none of the myelinated afferents from which we recorded were more sensitive to ventricular pressure changes than to changes in coronary arterial pressure. This leads us to suggest that the only mechanosensitive afferents from the heart which are likely to have a normal physiological role are actually coronary arterial baroreceptors.

Having established that coronary arterial baroreceptors show discharge characteristics similar to those of arterial baroreceptors elsewhere, we need to consider whether their reflex responses are also similar. We carried out a study of the reflex vascular responses to changes in ventricular and coronary arterial pressures, using a similar approach to that of

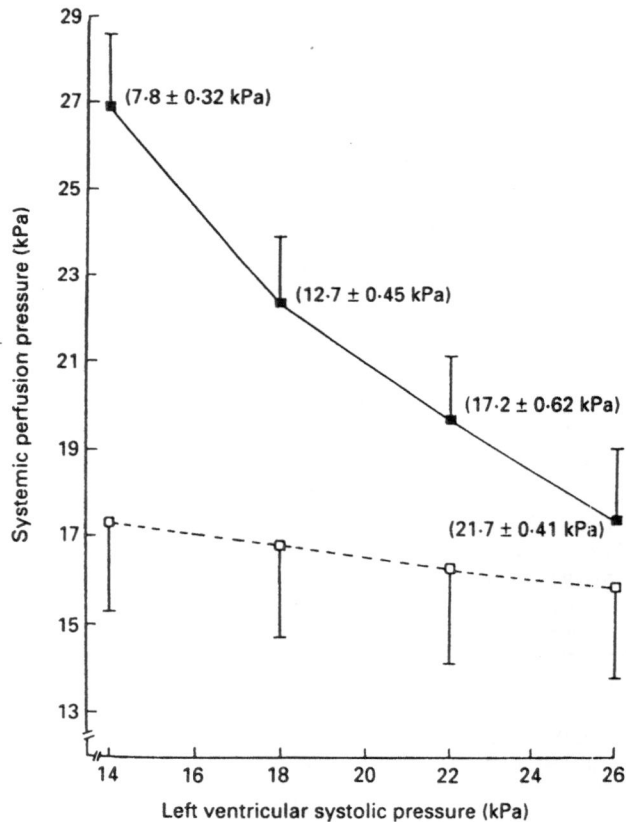

**Figure 8.** Reflex vascular responses to graded changes in aortic plus coronary pressure and ventricular pressure (■) and only in ventricular pressure (□). Numbers in parentheses denote concomitant changes in aortic root pressure. Note the very small response when only ventricular pressure was changed compared to that when aortic plus coronary pressure was also changed. (From Ref 3).

the electrophysiological study. In a preparation in which we had independent control of coronary and ventricular pressures, we determined reflex vascular responses in a vascularly isolated perfused hind-limb, as well as in the remainder of the systemic circulation. The responses from one dog are shown in Fig 8. The top traces show that increases in both aortic root (coronary) pressure and ventricular pressure (combined test) resulted in decreases in systemic and limb perfusion pressures which, at constant flows, denote vasodilatation. Very much smaller responses occured when only the ventricular pressure was changed with coronary pressure controlled. However when only the aortic root pressure was changed, the response was little different from that to the combined stimulus. The data from stepwise changes in pressure, shown in Fig 8, also shows that the responses to the changes in aortic root pressure were much greater than those to changes only in ventricular pressure, confirming that the responses did indeed arise from stimulation of coronary mechanoreceptors.

Although coronary mechanoreceptors function in much the same way as other arterial baroreceptors, there do appear to be some differences. Unlike the aortic and carotid baroreceptors, the coronary receptors do not cause consistent changes in the heart rate (Challenger et al. 1987, Tutt et al. 1988). We (Crisp et al. 1989) also determined effects on respiration by recording efferent phrenic nerve activity and noted that increases in aortic root and coronary arterial pressure, unlike increases in carotid sinus pressure, consistently resulted in a decrease in respiratory drive (e.g. Fig 9). The main role of baroreceptors, however, is to control arterial blood pressure, and this is achieved mainly through effects on resistance blood vessels and here there is no qualitative difference between the various groups of receptors. Also, not only do all groups of baroreceptors cause dilatation of resistance vessels, but they also dilate capacitance vessels. This was shown by Tutt et al. (1988) who perfused the systemic circulation at constant flow and drained it at constant venous pressure, conditions which are required so that changes in vascular volume can be a measure of changes in capacitance (Hainsworth, 1986, 1990), and showed that increases in coronary arterial pressure caused dilatation of capacitance vessels.

We have recently noted that coronary arterial baroreceptors differ from other baroreceptors in that the speed of the increase in vascular resistance following reflex

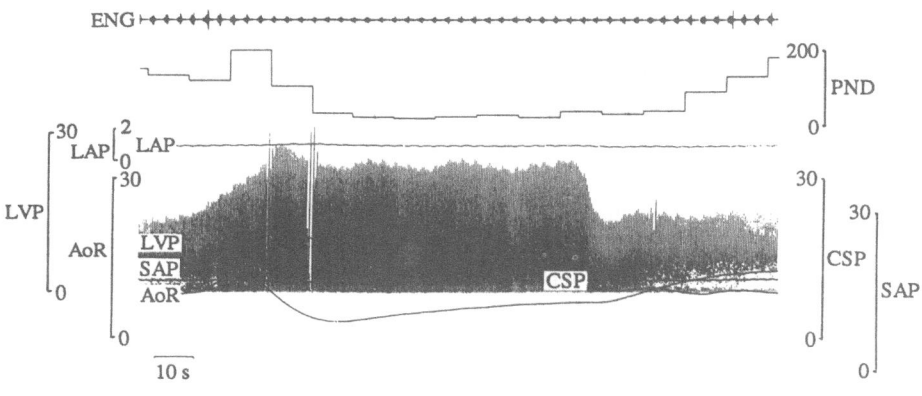

**Figure 9.** Effects of changes in aortic root plus coronary arterial pressure and left ventricular systolic pressure on respiratory activity. Traces of phrenic nerve electroneurogram (ENG), histogram of phrenic nerve discharge (PND) as action potentials/10s, left atrial pressure (LAP), left ventricular pressure (LVP), systemic arterial perfusion pressure perfused at constant flow (SAP), carotid sinus pressure (CSP), and aortic root and coronary arterial pressure AoR. All pressures are in kPa (1kPa = 7.5mmHg). Traces show increases in aortic root and ventricular pressures resulting in a decrease in respiratory activity as well as vasodilatation. (From ref 16).

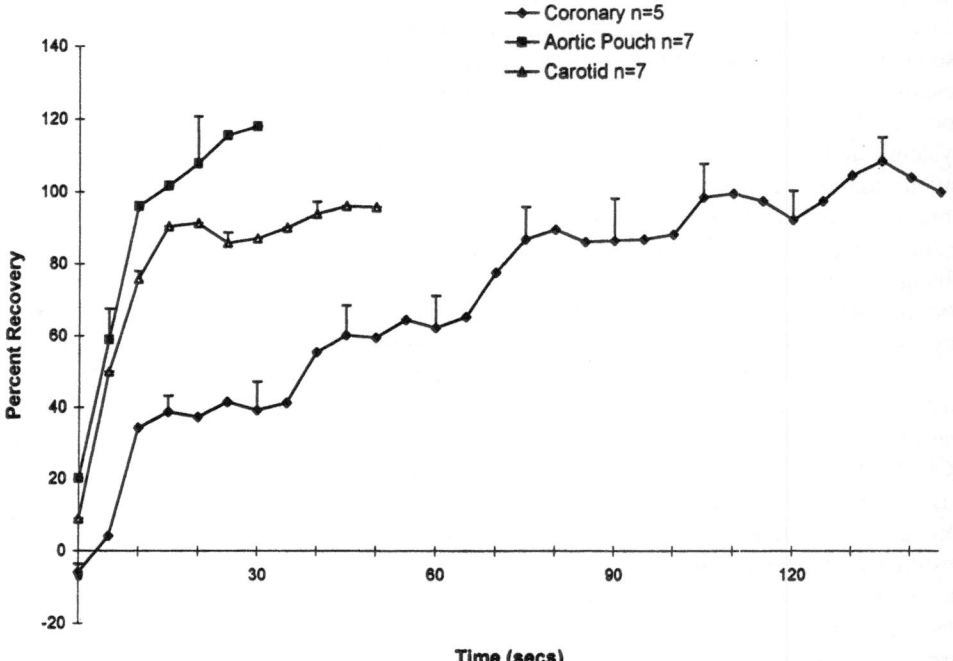

**Figure 10.** Time courses of vasoconstriction in response to step decreases in aortic arch, carotid sinus and coronary arterial pressures. Each baroreceptor pressure was independently elevated from threshold to saturation level and, after 2 min, decreased back to threshold and the recovery of the perfusion pressure was plotted. Note the very rapid response to decreases in aortic and carotid pressures, but that the response to the change in coronary pressure required about 2 min.

inhibition is much slower (McMahon et al. 1994). This effect is seen in Fig.10. The pressure distending each baroreceptor region was raised to near the saturation level and held for 2 min before being reduced back to threshold. As can be seen, after decreasing aortic and carotid pressures, the perfusion pressure recovered within seconds near to its previous level. However, the response to decreasing the coronary pressure occured much more slowly and took about 2 min to recover fully. The time required for recovery was related to the duration for which the pressure stimulus had been applied and, after an 8 min stimulus, recovery required even longer. The reason why the responses to unloading the coronary receptors are so slow is unknown, but it is not likely to be a property of the receptors or the afferent limb of the reflex, and probably is either a central effect or a property of the efferent pathway. A central effect seems to be the more likely mechanism since reflex studies indicate that there is likely convergence of the two reflexes on to the same sympathetic motoneurones.

We tested for interactions between the coronary baroreceptor reflex and the carotid sinus reflex by determining the effects of changes in pressure in one area while the pressure in the other was held at different constant levels (Vukasovic et al. 1989). These experiments showed that a high pressure in one area resulted in smaller responses to a change in pressure in the other. However, over the linear range of the reflexes there was no difference in slope (Fig 11). This findings are consistent with the view that the two reflexes summate in a simple arithmetic way and that the response is limited when activity in the efferent sympathetic neurones approaches zero.

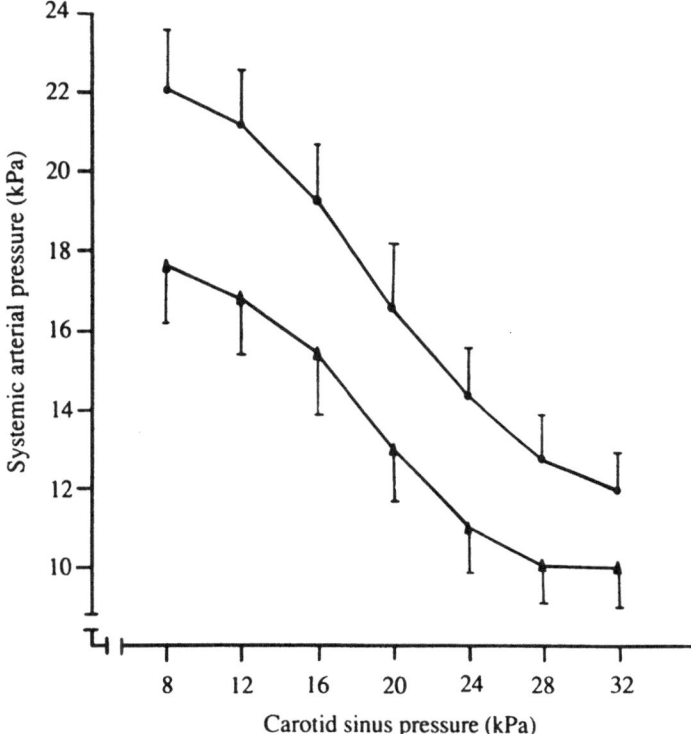

**Figure 11.** Reflex vascular responses to graded chages in carotid sinus pressure with aortic root pressure held at 7.6 ± 0.4 (●) or 15.7 ± 1.0 kPa (▲). Values are means ± SE from 9 dogs. The high aortic root (including coronary and ventricular) pressure resulted in a decrease in perfusion pressure (vasodilatation), but no change in the maximum slope of the plot. The greater overall response at the low aortic pressure is due to non-linearities at the extremes of the curve and not to a change in peak sensitivity of the reflex. (From ref 57).

## INFLUENCE OF VENTRICULAR FILLING AND INOTROPIC STATE ON RESPONSES TO CHANGES IN AORTIC ROOT AND LEFT VENTRICULAR PRESSURE

Nonmyelinated ventricular mechanoreceptors are reported to be stimulated by increases in ventricular inotropic state (e.g. Sleight, 1979). They are also said to be stimulated by increases in diastolic filling although some may also be excited at low volumes (Thoren chapter). The combination of a nearly empty ventricle and a positive inotropic change provides an effective stimulus for some receptors (Oberg and Thoren, 1972) and this effect is believed by some to be the important initiating mechanism for the vasovagal reaction (e.g. Abboud, 1989). We (Al-Timman and Hainsworth, 1992) tested this hypothesis, using anesthetized dogs, in which we determined the effects of changes in ventricular filling using a partial ventricular bypass, and changes in inotropic state effected by crushing and then stimulating the left cardiac sympathetic nerves, on the responses to changes in left ventricular systolic pressure. These experiments showed that, provided aortic root pressure did not change, sympathetic stimulation, sufficient to cause an increase in ventricular dP/dt max averaging 72%, had little effect on vascular resistance (Fig 12). This indicates that, assuming this stimulus does excite ventricular mechanoreceptors, these receptors must have little or no effect on vascular resistance. These experiments actually confirmed our earlier results

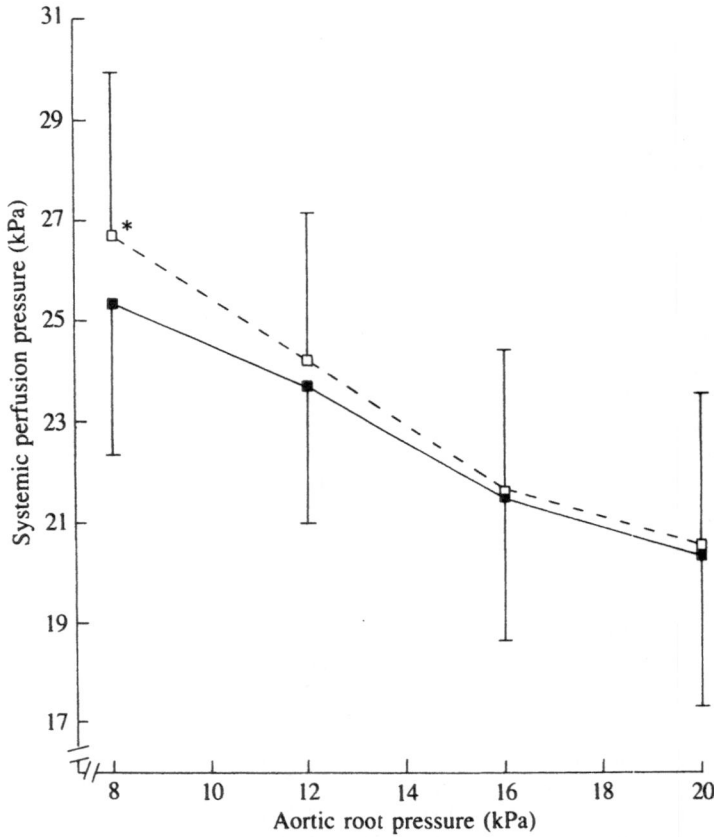

**Figure 12.** Effect of stimulation of efferent cardiac sympathetic nerves on reflex vascular responses to changes in aortic root pressure. Sympathetic stimulator off: ■, on: □ . * P<0.05 compared to sympathetic off. Note that apart from a small vasoconstriction at the lowest aortic pressure, sympathetic stimulation had no effect on vascular resistance at any aortic root pressure. There was no suggestion at any aortic root pressure that cardiac sympathetic stimulation caused vasodilatation. Results of means ± SE (n=5). (From ref 2).

that changes in inotropic state brought about pharmacologically by administration of propranolol or dobutamine also did not cause reflex responses (Tutt et al. 1988).

Changes in ventricular filling, brought about by pumping blood either into or out of the left atrium, also did not cause vasomotor responses provided that coronary arterial pressure was constant. Furthermore, even the combination of sympathetic stimulation and ventricular bypass, conditions which were thought to be particularly effective in stimulating ventricular receptors and possibly be responsible for initiating vasovagal reactions, also did not cause reflex effects. In these experiments, changes in inotropic state or in venous filling caused quite large changes in aortic root and coronary arterial pressures at constant ventricular systolic pressures. This effect can be attributed to changes in the flow of blood from the ventricle into the aorta. If coronary pressure is not controlled, then reflex responses were indeed obtained. However, the combination of ventricular bypass (which decreases coronary pressure at constant ventricular pressure) and sympathetic stimulation (which increases coronary pressure) results in no change in ventricular systolic pressure or in aortic and coronary pressures, and little or no reflex response (Fig. 13).

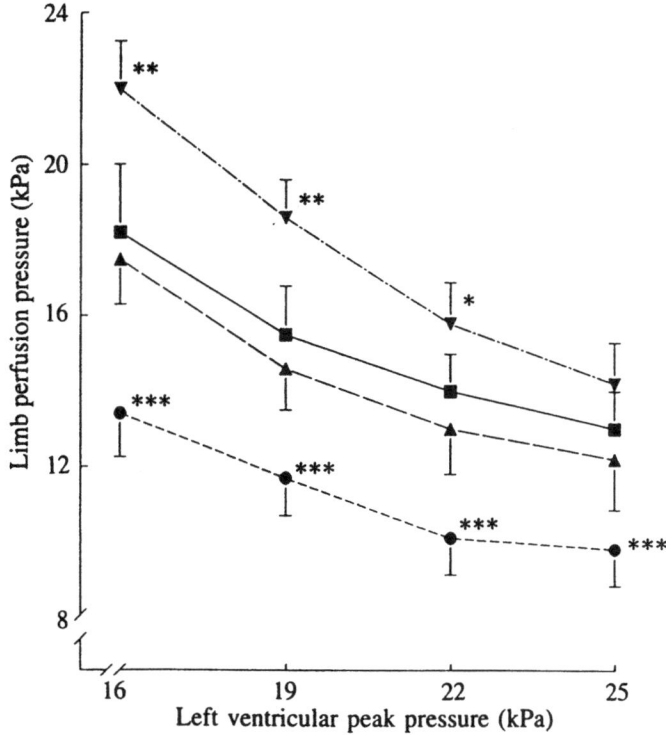

**Figure 13.** Effects on vascular resistance in a perfused hind-limb of changes in left ventricular systolic pressure during normal ventricular filling and no sympathetic stimulation (■); ventricular bypass, i.e. low filling (●); sympathetic stimulation (▼); and both bypass and stimulation (▲). Note that, at any ventricular pressure, sympathetic stimulation causes vasoconstriction, bypass causes dilatation, but the combination has no effect. Note also that these changes are almost certainly due to the concomitant changes in coronary pressure at any ventricular pressure (sympathetic stimulation decreases and bypass increases coronary pressure). These results, however, do confirm that sympathetic stimulation or bypass alone or even together do not potentiate the responses to changes in either ventricular or coronary pressure. (From ref. 2).

The absence of a reflex responce to the combination of ventricular bypass and cardiac sympathetic stimulation implies that ventricular receptors are likely to be unimportant in the initiation of vasovagal syncope.

There are also other pieces of evidence against ventricular receptors being involved in the vasovagal reaction. In dogs, the response is not prevented by chronic cardiac denervation or acute blockade (Shen et al 1990; 1991). Also, in people cardiac transplantation, which results in denervated ventricles, does not render the subject immune to vasovagal reactions (Fitzpatrick et al. 1993).

The vasovagal reaction may be associated with the various humoral changes which occur immediately before syncope (see Ludbrook, 1993). Endogenous opioids in particular ß endorphins may be involved and high doses of naloxone in animals may prevent the response (Schadt and Ludbrook, 1991). Also, it has recently been reported that in people, before the onset of vasovagal syncope there is an increase in the plasma level of opioids (Wallbridge et al. 1994). Other mechanisms, including a possible role for serotinergic receptors (Kosinski et al. 1994) may also be involved. However, further research is needed to determine exactly how the vasovagal reaction is initiated.

# SUMMARY AND CONCLUSIONS CONCERNING THE ROLES OF VENTRICULAR AND CORONARY RECEPTORS

Ventricular receptors are distributed throughout the left ventricle and most, if not all, are attached to nonmyelinated nerve fibers. The receptors may be chemosensitive, mechanosensitive or both. Chemosensitive afferents are classically excited by exogenous chemicals such as veratridine, although endogenous chemicals such as bradykinin and prostaglandins, which are released during ischemia, also excite these nerves. The reflex responses can be very powerful, resuting in profound bradycardia and hypotension. A normal physiological role for these receptors seems unlikely although it is probable that they contribute to the changes occuring in some pathological states.

Ventricular mechanoreceptors, some of which may also exhibit chemosensitivity, are excited by increases in ventricular systolic pressure, but only when the pressure increases to extreme levels. They also appear to react to increases in inotropic state and increases, and possibly also to decreases, in ventricular filling. It seems that ventricular mechanoreceptors do not show the same intense response as is seen in the chemosensitive afferents following chemical stimulation and probably as a consequence of this their reflex responses are also weak and probably of little importance. Previous assertions that they are involved in the vasovagal reaction can probably now be discounted.

The existence of coronary arterial baroreceptors has been suspected for about 30 years. This has now been confirmed and they have been shown to respond to pressure changes in much the same way as the well known carotid and aortic baroreceptors. There are, however, some interesting differences. Coronary baroreceptors, at least in the dog, do not control the heart rate, although they do influence respiratory activity. Another intriguing difference is that when vascular resistance has been inhibited reflexly by perfusing coronary receptors at a high pressure, it takes several minutes for the vascular resistance to increase when coronary pressure is again lowered. The implications of this are uncertain, but it is conceivable that, whereas carotid baroreceptors are involved in the responses to rapid changes in pressure, coronary baroreceptors may be more concerned with the regulation of the long-term level of arterial blood pressure.

# REFERENCES

1. ABBOUD, F.M. Ventricular syncope: is the heart a sensory organ? (Editorial). N. Eng. J. Med. 320; 390-392, 1989.
2. AL-TIMMAN, J.K.A., AND R. HAINSWORTH. Reflex vascular responses to changes in left ventricular pressures, heart rate and inotropic state in dogs. Exp. Physiol. 77: 455-469, 1992.
3. AL-TIMMAN, J.K.A., M.J. DRINKHILL, AND R. HAINSWORTH. Reflex responses to stimulation of mechanoreceptors in the left ventricle and coronary artery in anaesthetized dogs. J. Physiol. Lond. 472: 769-784, 1993.
4. ARROWOOD, J.A. P.K. MOHANTY, J.M. HODGSON, M.E. DIBNER-DUNLAP, AND M.D. THAMES. Ventricular sensory endings mediate reflex bradycardia during coronary arteriography in humans. Circulation 80; 1293-1300, 1989.
5. AVIADO, D.M., JR., AND C.F. SCHMIDT. Cardiovascular and respiratory reflexes from the left side of the heart. Am. J. Physiol. 196: 726-730, 1959.
6. BAKER, D.G., H.M. COLERIDGE, AND J.C.G. COLERIDGE. Vagal afferent C fibres from the ventricle. In: Cardiac Receptors, edited by R. Hainsworth, C. Kidd, and R.J. Linden. Cambridge, UK: Cambridge Univ. Press, 1979, p. 117-137.
7. BARRON, K.W., AND V.S. BISHOP. Reflex cardiovascular changes with veratridine in the conscious dog. Am. J. Physiol. 242. 11: H810-817, 1982.
8. BERGER, H.J., B.L. ZARET, L. SPEROFF, L.S. COHEN, AND S. WOLFSON. Regional cardiac protaglandin release during myocardial ischemia in anesthetized dogs. Circ. Res. 38: 566-571, 1976.

9. BROWN, A.M. Mechanoreceptors in or near the coronary arteries. J. Physiol. Lond. 177: 203-214, 1965.

10. BROWN. A.M. The depressor reflex arising from the left coronary artery of the cat. J. Physiol. Lond. 184: 825-836, 1966.ó

11. BROWN, A.M. AND A. MALLIANI. Spinal sympathetic reflexes initiated by coronary receptors. J. Physiol. Lond. 12: 685-705, 1971.

12. CHALLENGER, S., K.H. McGREGOR, AND R. HAINSWORTH. Peripheral vascular responses to changes in left ventricular pressure in anaesthetized dogs. Q. J. Exp. Physiol. 72: 271-283, 1987.

13. CHEVALIER, P.A, K.C. WEBER, G.W. LYONS, D.M. NICOLOFF, AND I.J. FOX. Hemodynamic changes from stimulation of left ventricular baroreceptors. Am. J. Physiol, 227: 719-728, 1974.

14. CLOZEL, J.P., T.E. PISARRI, H.M. COLERIDGE, AND J.C.G. COLERIDGE. Reflex coronary vaso-dilatation evoked by chemical stimulation of cardiac afferent vagal C fibres in dogs. J. Physiol. Lond. 428: 215-232, 1990.

15. COLERIDGE, H.M., J.C.G. COLERIDGE, AND C. KIDD. Cardiac receptors in the dog, with particular reference to two types of afferent ending in the ventricular wall. J. Physiol. Lond. 174: 323-339, 1964.

16. CRISP, A.J., S.M. TUTT, K.H. McGREGOR, AND R. HAINSWORTH. The effects of changes in left ventricular pressure on respiratory activity in anaesthetized dogs. Q. J. Exp. Physiol. 74: 291-300, 1989.

17. DALY, I.B AND E.B. VERNEY. The localization of receptors involved in the reflex regulation of the heart rate. J. Physiol. Lond. 62: 330-340, 1927.

18. DAWES, G.S. Studies on veratrum alkaloids. VII. Receptors areas in the coronary arteries and elsewhere as revealed by the use of veratridine. J. Pharmacol. Exp. Ther. 89: 325-342, 1947.

19. DAWES, G.S., AND J.H. COMROE J.R. Chemoreflexes from the heart and lungs. Physiol. Rev. 34: 167-201, 1954.

20. DRINKHILL, M.J., J. MOORE, AND R. HAINSWORTH. Afferent discharges from coronary arterial and ventricular receptors in anaesthetized dogs. J. Physiol. Lond. 472: 785-800, 1993.

21. FITZPATRICK, A.P., N. BANNDER, A. CHENG, M. YACOUB, AND R. SUTTON. Vasovagal rections may occur after orthotropic heart transplantation. J. Am. Coll. Cardiol. 21: 1132-1137, 1993.

22. GUPTA, B.N., AND M.D. THAMES. Behavior of left ventricular mechanoreceptors with myelinated and nonmyelinated afferent vagal fibers in cats. Circ. Res. 52: 291-301. 1983.

23. HAINSWORTH, R. Vascular capacitance: its control and importance. Rev. Physiol. Biochem. Pharmacol. 105: 101-173, 1986.

24. HAINSWORTH, R. The importance of vascular capacitance in cardiovascular control. N.I.P.S. 5: 250-254, 1990.

25. HAINSWORTH, R. Reflexes form the heart. Physiol. Rev. 71: 617-658, 1991.

26. HAINSWORTH R., K.H. McGREGOR, AND R. FORD. Effects of veratridine injected into the aortic root on resistance and capacitance in the abdominal circulation in anaesthetized dogs. Q.J. Exp. Physiol. 71: 589-599, 1986.

27. HIRSCH, P.D., L.D. HILLIS, AND W. B. CAMPBELL. Release of prostaglandins and thromboxane from the coronary circulation in patients with ischemic heart disease. N. Engl. J. Med. 304: 685-691, 1981.

28. HOKA, S., Z.J. BOSNJAK, D. SIKER, R.J. LUO. AND J. KAMPINE. Dynamic changes in venous outflow by baroreflex and left ventricular distension. Am. J. Physiol. 254 R204-R211, 1988.

29. KAUFMAN, M.P., D.G. BAKER, AND H.M. COLERIDGE. Stimulation by bradykinin of afferent vagal C-fibers with chemosensitive ending in the heart and aorta of the dog. Circ. Res. 46: 476-484, 1980.

30. KOSINKI, D.J., B.P. GRUBB, AND P.N TEMESY-ARMOS. The use of serotonin re-uptake inhibitors in the treatment of neurally mediated cardiovascular disorders. J. Serotonin. Res. 1: 85-90, 1994.

31. KOSTREVA, D.R., F.A. HOPP. E.J. ZUPERKU, AND J.P. KAMPINE. Apnea, tachypnea, and hypotension elicited by cardiac vagal afferents. J. Appl. Physiol. 47: 312-318, 1979.

32. LEDSOME, J.R. AND K. KAN. Reflex changes in hind limb and renal vascular resistance in response to distension of the isolated pulmonary arteries of the dog. Circ. Res. 40: 64-72, 1977.

33. LINDEN, R.J. AND C.T. KAPPAGODA. Atrial Receptors. Cambridge, UK. Cambridge Univ. Press, 1982.

34. LOMBARDI, F., C. CASALONE, P.D. DELLA BELLA, G. MALAFATTO, M. PAGANI, AND A. MALLIANI. Global versus regional myocardial ischemia: differences in cardiovascular and sympathetic responses in cats. Cardiovasc. Res. 18: 14-23, 1984.

35. LUDBROOK, J. Haemorrhage and Shock. In: Cardiovascular reflex control in health and disease, edited by R.Hainsworth and A.L. Mark. London, U.K.: W.B. Saunders Co, 1993, P 463-490.

36. MARK, A.L., F.M. ABBOUD, P.G. SCHMID, AND D.G. HEISTAD. Reflex vascular responses to left ventricular outflow obstruction and activation of ventricular baroreceptors in dogs. J. Clin. Invest. 52: 1147-1153, 1973.

37. McGREGOR, K.H., R. HAINSWORTH, AND R. FORD. Hind limb vascular responses in anaesthetized doges to aortic root injection of veratridine. Q. J. Exp. Physiol 71: 577-587, 1986.

38. McMAHON, N.C., AND M.J. DRINKHILL. Reflex vascular responses to stimulation of carotid, aortic and coronary baroreceptors in anaesthetized dogs. J. Physiol. Lond. Proc. Aberdeen meeting, 1994.

39. MOORE, J.P. Responses to stimulation of coronary mechanoreceptors. PhD thesis, University of Leeds, 1993.

40. MUERS, M.F. AND P. SLEIGHT. The reflex cardiovascular depression caused by occlusion of the coronary sinus in the dog. J. Physiol. Lond. 221: 259-282, 1972.

41. OBERG, B., AND P. THOREN. Studies on left ventricular receptors, signalling in non-medullated vagal afferents. Acta Physiol. Scand. 85: 145-163, 1972.

42. OBERG, B., AND P. THOREN. Increased activity in left ventricular receptors during hemorrhage or occlusion of caval veins in the cat. A possible cause of the vaso-vagal reaction. Acta Physiol. Scand. 85: 164-173, 1972.

43. PAINTAL, A.S. The study of ventricular pressure receptors and their role in the Bezold reflex. Q. J. Exp. Physiol. 40: 348-363, 1955.

44. PEREZ-GOMEZ, F. AND A. GARCIA-AGUADO. Origin of ventricular reflexes caused by coronary arteriography. Br. Heart J. 39: 967-973, 1977.

45. ROSS, J., JR., C.J. FRAHM, AND E BRAUNWALD. The influence of intracardiac baroreceptors on venous return, systemic vascular volume and peripheral resistance. J. Clin. Invest. 40: 563-572, 1961.

46. SALISBURY, P.F., C.E. CROSS, AND P.A. RIEBEN. Reflex effects of left ventricular distension. Circ. Res. 8: 530-535, 1960.

47. SCHADT, J.C., AND J. LUDBROOK. Hemodynamic and neurohumoral responses to acute hypovolemia in consious mammals. Am. J. Physiol. 260: H305-H313, 1991.

48. SEMPLE, P.F., P. THOREN, AND A.F. LEVER. Vasovagal reactions to cardiovascular drugs: the first dose effect. J. Hypertens. 6: 601-606, 1988.

49. SHEN, Y-T., D.R. KNIGHT, J.X. THOMAS, AND S.F. VATNER. Relative roles of cardiac receptors and arterial baroreceptors during hemorrhage in concious dogs. Circ. Res. 66: 397-405, 1990.

50. SHEN, Y-T, A.W. COWLEY, AND S.F. VATNER. Relative roles of cardiac and arterial baroreceptors in vasopressin regulation during hemorrhage in conscious dogs. Circ. Res. 68: 1422-1436, 1991.

51. SLEIGHT, P. A cardiovascular depressor reflex form the epicardium of the left ventrical of the dog. J. Physiol. Lond. 173:321-343, 1964.

52. SLEIGHT, P. Possible physiological stimuli for ventricular receptors and their significance in man. In: Cardiac Receptors, edited by R. Hainsworth, C. Kidd and R.J Linden. Cambridge, U.K. Cambridge University Press, 1979, p 241-258.

53. THAMES, M.D. Reflex suppression of renin release by ventricular receptors with vagal afferents. Am. J. Physiol. 233: H181-H184, 1977.

54. THAMES, M.D. Acetylstrophanthidin-induced reflex inhibition of canine renal sympathetic nerve activity mediated by cardiac receptors with vagal afferents. Circ. Res. 44: 8-15, 1979.

55. THOREN, P. Characteristics of left ventricular receptors with nonmedullated vagal afferents in cats. Circ. Res. 40: 231-237, 1977.

56. TUTT, S.M., J.K.A. AL-TIMMAN, AND R. HAINSWORTH. reflex responses of vascular resistance in anaesthetized dogs to independent changes in ventricular systolic pressure and cardiac inotropic state. Q. J. Exp. Physiol 73: 801-804, 1988.

57. TUTT, S.M., K.H. McGREGOR, AND R. HAINSWORTH. Reflex vascular responses to changes in left ventricular pressure in anaesthetized dogs. Q. J. Exp. Physiol. 73: 425-437, 1988.

58. VON BEZOLD, A.V., AND L. HIRT. Uber die physiolischen Wirkungendes essigsauren Veratrins. Unter Physiol. Lab. Wurzberg 1: 75-156, 1867.

59. VUKASOVIC, J.L., S.M. TUTT, A.J. CRISP, AND R. HAINSWORTH. The influence of left ventricular pressure on the vascular responses to changes in carotid sinus pressure in anaesthetized dogs. Q. J. Exp. Physiol. 74: 735-746, 1989.

60. WALLBRIDGE, D.R., H.E. MACINTYRE, C.E. GRAY, M.A. DENVIR, K.G. OLDROYD, A.P. RAE, AND S.M. COBBE. Increase in plasma β endorphins preceeds vasodepressor syncope. Brit. Heart. J. 71: 446-448, 1994.

61. WOOLLARD, H.H. The innervation of the heart. J. Anat. 60: 345-373, 1926.

62. ZELIS, R., M. LOTYSCH, M. BRAIS, C.L. PENG, E. HURLEY, AND D.T. MASON. Effects of isolated right and left ventricular stretch on regional arteriolar resistance. Cardiovasc. Res. 11: 419-426, 1977.

# 16

# MECHANISMS OF RELEASE OF ATRIAL NATRIURETIC FACTOR *IN VIVO*

J. R. Ledsome[*] and K. A. King

Department of Physiology
Faculty of Medicine
The University of British Columbia
Vancouver, B.C., Canada V6T 1Z3

## ABSTRACT

The aim of the study was to define the relationship between the concentration of immunoreactive atrial natriuretic factor (ANF) in the plasma and atrial stretch, atrial pressure and atrial wall stress during changes in blood volume within the physiological range. Advantage was taken of the potentiation of the release of ANF, in response to blood volume expansion, in the anaesthetized rabbit, by section of the carotid sinus, aortic depressor and vagus nerves. Mean arterial pressure and right and left atrial pressures were measured. Right and left atrial dimensions were measured by sonomicrometry. Blood volume was expanded by 20% and then decreased at 1% of the blood volume per minute for 40 min, before and after section of the nerves. Plasma ANF did not change significantly in response to either the increase or decrease in blood volume in the presence of intact nerves. After the nerves were sectioned, blood volume expansion significantly increased IR-ANF (34.6 +/- 12 to 260.6 +/- 89.8 pg/ml). The relationship between ANF and the changes in left atrial dimensions (extension ratio) was exponential. A significant linear correlation was found between ANF and left atrial pressure and left atrial wall stress. Over a range of atrial pressures from 1-10 cm $H_2O$ the atrial wall was extremely distensible, but there were only minor changes in ANF. At higher atrial pressures there was little further extension of the atrial wall but there were much larger increases in plasma ANF. In reponse to changes in blood volume within the physiological range cardiovascular reflexes normal maintain atrial dynamic function within a narrow range and there is little stimulus for the release of ANF. In conditions in which the cardiovascular reflexes are unable to maintain atrial function constant then there will be changes in the release of ANF and there will be an exponential relationship between atrial wall stretch and plasma ANF.

---

[*] Correspondence: Dr. J.R. Ledsome, Department of Physiology, University of British Columbia, 2146 Health Sciences Mall, Vancouver, B.C. Canada V6T 1W5; Tel: 604-822-2318. FAX: 604-822-6048

*Control of the Cardiovascular and Respiratory Systems in Health and Disease*
Edited by C. T. Kappagoda and M. P. Kaufman, Plenum Press, New York, 1995

175

# INTRODUCTION

The diuretic and natriuretic actions of atrial natriuretic factor (ANF), as well as its effects on the renin-angiotensin-aldosterone system, suggest that this hormone may play a role in the regulation of body fluid volume. This view is supported by observations that volume expansion stimulates ANF release [1,13,18,21]. Alterations in blood volume over the physiological range (+/- 20%) have been shown to elicit only small changes in plasma ANF in anesthetized rabbits [5]. However, after section of the vagus and aortic depressor nerves and in the presence of a constant carotid sinus pressure of 100 mmHg, alteration of blood volume by 20% produced much larger changes in plasma ANF [5]. The potentiation of ANF release in response to blood volume changes after nerve section could not be attributed to increased mean arterial (MAP) or right atrial pressure (RAP) [5]. It is unlikely to be explained by a direct neural effect on the heart as we were unable to demonstrate any influence of the autonomic nervous system on the release of ANF in anaesthetized rabbits [24]. Numerous humoral agents have been shown to influence the release of ANF *in vitro* [25] but their role *in vivo* has not been established. Before postulating a possible neural or humoral mechanism it was necessary to determine whether changes in left atrial function, not measured in the previous experiments, could have contributed to the potentiation of the release of ANF.

Many studies have indicated that at least in the case of blood volume expansion, it is atrial stretch rather than atrial pressure itself which determines the release of ANF [3,4,7,12,16,26,30]. There may be relatively small changes in atrial stretch during severe acute volume loading [29] in the intact animal. This partly due to the exponential nature of the relationship between atrial pressure and atrial diameter and partly because the atrial volume is high in the closed chest, conscious animal thus placing the atrium on the flat upper portion of the pressure volume curve. During hemorrhage there may be larger changes in wall stretch but only small changes in plasma ANF [27]. In these conditions changes in plasma ANF are linearly correlated with atrial wall stress rather than wall stretch. Smaller and slower changes in blood volume, over a physiological range, produce only small changes in atrial dynamics and plasma ANF [12] and the relationship between atrial stretch, atrial pressure, atrial wall stress and release of ANF has not been fully defined over a large range of atrial pressures.

We examined the effects of alterations in blood volume over a physiological range (+/- 20%) in anaesthetized rabbits before and after section of the vagus, aortic depressor and carotid sinus nerves. The resulting changes in immunoreactive ANF (IR-ANF), right and left atrial pressures (RAP, LAP) and right and left atrial dimensions (RAD, LAD) were measured and right and left atrial wall stress (RAS, LAS) was calculated. After nerve section there was a wide range of changes in left atrial pressures and dimensions and there were large changes in plasma IR-ANF. This provided an opportunity to examine the relationship between plasma ANF and atrial pressure, atrial stretch and atrial wall stress over this wide range of pressures and dimensions.

# METHODS

Eight male New Zealand White rabbits (3.2 +\- 0.1 kg) were anaesthetized with a mixture of alpha-chloralose (100 mg/kg) and urethane (1 g/kg) which was injected into an ear vein. A cannula was inserted into the trachea and the animals were artificially ventilated with a mixture of room air and 100% $O_2$. Cannulae (PE 190, Becton Dickinson & Co. NY) placed in the right jugular vein and right femoral artery were connected to pressure transducers (Model P23Db; Statham Instruments Co., Puerto Rico) and MAP and RAP were recorded on an ultraviolet recorder (Visicorder, Model 1508, Honeywell, Denver, CO). Intratracheal pressure was recorded using a pressure transducer attached to a needle inserted in the inspiratory line. The electrocardiogram, from which heart rate (HR) was obtained, was monitored from leads attached to the right leg and chest. Cannulae were placed in the left femoral artery (PE 190) for the infusion and withdrawal of blood, and in the bladder (PE 240) to drain the urine. A continuous infusion of 0.45% saline was delivered at 1 ml/min by a constant flow infusion pump (Masterflex Model 173-20-25, Cole-Parmer, Chicago, IL) into a cannula placed in the right femoral vein to maintain plasma osmolality at 290-300 mosm/kg. Blood samples (< 1 ml) were taken during the preliminary surgery in order to measure pH, $PCO_2$, and $PO_2$, using a blood gas analyzer (Model 165/2, Corning Glass Works, Medfield, MA). The rate of the respiratory pump was altered and $NaHCO_3$ (1 M) was administered as necessary to maintain $PaCO_2$ within the range of 25 - 30 mm Hg and pH within the range 7.35-7.55 units, values similar to those measured previously in conscious rabbits. Esophageal temperature was maintained at 37°C by heating bars in the surgical table. Ligatures were placed around the cervical vagus nerves, the aortic depressor nerves and the carotid sinus nerves to aid later section. The aortic depressor nerves were isolated at their junctions with the superior laryngeal nerves and the carotid sinus nerves at their junctions with the glossopharyngeal nerves.

The chest was opened along the midline. Left atrial pressure was recorded from a cannula inserted into the tip of the left atrial appendage. The systems recording atrial pressures had a flat frequency response (± 5%) to better than 50 Hz [11]. Manometers were calibrated in stepwise fashion using mercury and saline manometers. Atrial pressures were referred to zero atrial pressure which was recorded as the pressure at the cannula tip, placed at the level of the middle of the body of the atrium, free in air, at the end of the experiment. Right and left atrial dimensions were measured using a sonomicrometer (Triton Technology Inc., San Diego, CA). Two transducers (2 mm) were attached externally with glue (Instant Krazy Glue, Krazy Glue Inc., Chicago, IL), one on each side of the right atrium at the junctions with each of the two cranial venae cavae. A transducer (0.5 mm) was also glued on each side of the left atrial appendage in order to measure LAD. The placement of the transducers has previously been described in more detail [14]. The system was calibrated with an internal quartz clock, and the pulse was monitored with an oscilloscope (Model #5513, Tektronix Inc., Beaverton, OR) to check the triggering stability. In addition to mean atrial pressure, peak "a" and "v" wave pressures and the corresponding atrial dimensions were measured at end-expiration. Systolic atrial wall stress was calculated by multiplying peak "a" wave pressure by the atrial dimension corresponding to the peak of the "a" wave and dividing by 4. This corresponds to the formula (P x r)/(2 x h) (where P = pressure, r = radius), which provides an indication of the force per unit cross-sectional area (wall stress) of heart muscle [2]. It proved to be impractical to measure changes in atrial wall thickness (h) during contraction, therefore wall thickness was measured after cardiac arrest using potassium chloride, and was found to be close to 1 mm. Since the unstressed wall thickness was used, the stress calculated is the Lagrangian stress rather than the Eulerian stress [9]. Diastolic atrial wall stress was determined using the values for "v" wave pressure and dimension. Atrial wall strain was calculated as the natural strain [17] or extension ratio ( dl/l; where l was taken to be the diameter of the atrial appendage at the lowest peak "v" wave pressure in each animal). The average lowest "v" wave pressure was -1.2 cm $H_2O$ close to the tip of the atrial appendage making l a reasonable approximation for $l_0$; it is not possible to measure true $l_0$ in the intact beating heart [17]). Changes in peak "v" wave pressure were calculated as changes from this lowest "v" wave pressure.

Blood from an anaesthetized donor rabbit was used to expand blood volume and to replace blood sampled from the experimental rabbit. Blood obtained by cardiac puncture from the donor rabbit was filtered through gauze, exchanged with the blood of the experimental rabbit and kept at 37°C in a water bath. Blood from the donor rabbit had a plasma IR-ANF content of 57.7 +\- 10.9 pg/ml, which is higher than basal plasma ANF presumably due to the method of blood removal. The plasma IR-ANF concentration of the blood after exchange was 16.2 +\- 2.3 pg/ml, which suggests that the higher ANF content of the donor blood does not significantly alter the plasma IR-ANF of the experimental rabbit.

On completion of the surgical preparation, a further dose of anaesthetic (10% of the original dose) was given, and an equilibration period of approximately 15 min was allowed before the protocol was started.

## Experimental Protocol

Two pretreatment blood samples (3 ml) were obtained 5 min apart from the femoral artery. All blood samples were replaced with equal volumes of the previously exchanged donor blood. Mean arterial pressure, HR, RAP, LAP, RAD and LAD were recorded immediately before each blood sample was obtained. A volume of donor blood equivalent to 10% of the blood volume (approximately 20 ml) was then injected into a femoral artery, over 1 min. The blood volume of the New Zealand White rabbit has been previously found to be 60 ml/kg [6]. Five min later, an identical volume of blood was injected over 1 min, so that blood volume was expanded by 20%. Blood samples were obtained immediately and 5 min and 10 min after the injection of blood was completed. Blood was then withdrawn by means of a roller pump at a rate of 1% blood volume per min for 40 min, until blood volume was decreased by 40% (to 20% less than basal blood volume), or until MAP reached 40 mmHg. Blood samples were obtained every 5 min. Blood volume was then restored to basal by injecting a volume of blood equivalent to 10% of blood volume over 1 min, followed by an identical injection 5 min later. The aortic depressor nerves, the carotid sinus nerves and the vagus nerves were then sectioned, and after 10 min, the protocol was repeated. Satisfactory section of the aortic depressor and carotid sinus nerves was confirmed by a significant increase in MAP and by an increase in MAP in response to carotid artery occlusion before, but not after, section of the nerves.

## Treatment of Blood Samples

Arterial blood samples were withdrawn into prechilled syringes, then placed in chilled test tubes containing EDTA (Vacutainer, Becton Dickinson Canada Inc., Mississauga, Ont.). The blood samples were centrifuged in a refrigerated centrifuge (Silencer, Model H-102NA, Western Scientific, San Francisco, CA) at 2500 x G for 10 min. The plasma was removed, extracted within 48 hours of sampling, and stored at -20°C for subsequent analysis of IR-ANF levels by radioimmunoassay (RIA).

## Radioimmunoassay of IR-ANF

Plasma IR-ANF was measured according to a previously described method [31] using mouse monoclonal anti-alpha ANF serum in a dilution of 1:11,250,000. The plasma was extracted as described previously [23]. The upper limit of RIA sensitivity (20% depression of maximum binding) was 2.4 +\- 0.1 pg/tube (n=8), while the value for 50% depression of maximum binding was 7.9 +\- 0.3 pg/tube (n=8). The interassay error (CV %) was 10% (n=8) for one quality control sample, 7.1% (n=7) for a second sample, and 4.2% (n=7) for a third sample.

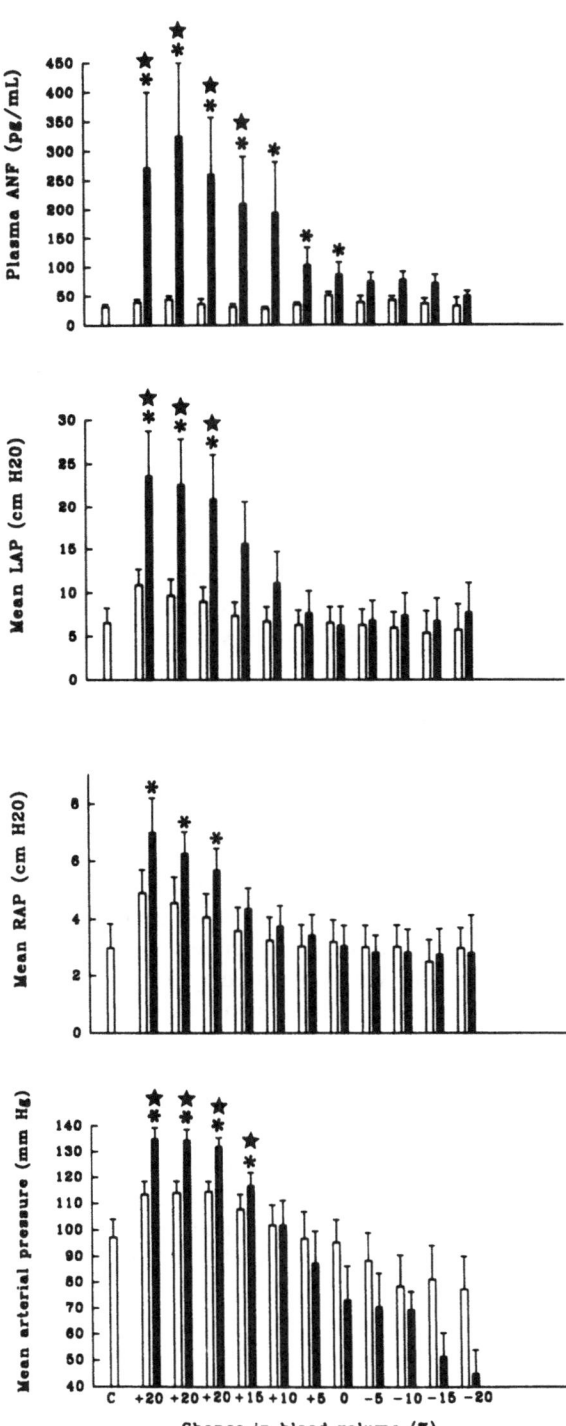

**Figure 1.** The effect of changing blood volume on mean arterial pressure (MAP), mean left atrial pressure (LAP), mean right atrial pressure (RAP) and plasma ANF, before and after section of the carotid sinus, aortic depressor and vagus nerves. Values represent the mean of 8 experiments, +/- SEM. The open bars are values obtained with the nerves intact, the solid bars after nerve section. The control values before volume expansion are labelled **C**. Blood volume was expanded by 20% of estimated blood volume and 10 min later decreased at a rate of 1%/min. ★ significantly different from the control value; * significantly different from the value observed at the same blood volume before nerve section.

## Statistical Analysis

All values represent the mean +/- SEM. The data obtained before and after nerve section were analyzed using analysis of variance with Duncan's multiple range test to assess the significance of individual differences. Comparison of values at the same blood volumes, before and after nerve section, was made using a paired t-test. Because the ANF values did not fall within a normal distribution the Wilcoxon signed-ranks test was used. A value of $p < 0.05$ was selected as the criterion for statistical significance. Correlation and regression analysis (NCSS, Kaysville, Utah.) was carried out to determine the relationships between pressure and diameter of the atria and also which variables were positively correlated with plasma IR-ANF.

## RESULTS

At the start of the experiments the MAP was 99 +/- 7 mm Hg and plasma IR-ANF was 32 +/- 2.8 pg/ml. Ten minutes after volume expansion by 20% of the blood volume MAP was 117 +/- 4 mm Hg and plasma ANF was 44 +/- 8.3 pg/ml; these changes did not reach statistical significance. The cardiovascular responses to decreasing blood volume (BV) from +20% BV to -20%BV, before and after section of the carotid sinus, aortic depressor and vagus nerves are shown in Fig. 1. Before nerve section, there were no statistically significant changes in MAP over the range of blood volume examined. After blood volume had been restored to normal the MAP was 102 +/- 9 mm Hg and this was increased to 145 +/- 8 mm Hg immediately after nerve section. However, the increase in MAP was transient and 10 min later MAP had decreased to 119 +/- 12 mm Hg. Volume expansion by 20% BV then significantly increased MAP to 142 +/- 4 mm Hg. After nerve section, MAP was significantly greater than before nerve section at blood volumes greater than +15% BV. There were no statistically significant changes in heart rate (not shown) in response to alterations in blood volume or nerve section. There were no changes in right atrial pressure before nerve section, but at 20% volume expansion RAP was higher after, than before nerve section (Fig. 1). There were no significant changes in mean LAP before nerve section, but in the presence of 20% volume expansion LAP was significantly greater after nerve section than before nerve section. There were no significant changes in plasma IR-ANF in response to changing BV, before nerve section (Fig. 1). Plasma IR-ANF was not significantly changed by nerve section, but after nerve section increased from 34.6 +/- 12 pg/ml to 261 +/- 89.9 pg/ml when BV was increased by 20%. The values of plasma IR-ANF immediately after, 5 min after or 10 min after volume expansion were not different from each other. Plasma IR-ANF levels were significantly greater after nerve section at blood volumes equal to or greater than the original blood volume.

The large increase in LAP after nerve section appeared to be related to an increase in afterload, since the largest changes in left atrial pressure were seen in those rabbits in which the MAP remained high after nerve section. There was a wide variation in LAP after nerve section and volume expansion and this provided an opportunity to examine the dynamics of atrial wall function and their relationship to the release of ANF over a wide range of LAP. The stress/strain relationship for the left atrial appendage is shown in Fig. 2, which shows the data points for all animals. There was an exponential relationship between peak "v" wave left atrial pressure (atrial filling) and wall strain (extension ratio), measured at the same time as the peak of the "v" wave. The data obtained before nerve section covers a smaller range of atrial pressures but there was no indication of a difference in atrial distensibility before and after nerve section. When the data obtained after nerve section were plotted as change in pressure divided by extension ratio (elastic stiffness), against change in "v" wave pressure (Fig. 2), there was an excellent linear correlation ($r = 0.85$, $n = 64$, $p < 0.0001$), confirming the exponential nature of the relationship between pressure and exten-

**Figure 2.** Stress-strain relationships of the left atrial appendage. Lower panel shows the change in peak "v" wave pressure plotted against left atrial wall strain (extension ratio). Data points were obtained by decreasing blood volume from +20% at a rate of 1%/min and taking measurements every 5 min. Open circles are values obtained with nerves intact; closed circles are values obtained after nerve section. Data are from 8 experiments. The continuous line is the best-fit exponential calculated using the data after nerve section. In the upper panel the change in "v" wave pressure is plotted against left atrial elastic stiffness (change in pressure divided by extension ratio (dp/(dl/l)). The continuous line is the linear regression line for these data and the broken lines are the 95% confidence limits for the line. The slope of the line (elasticity constant) is 1.1.

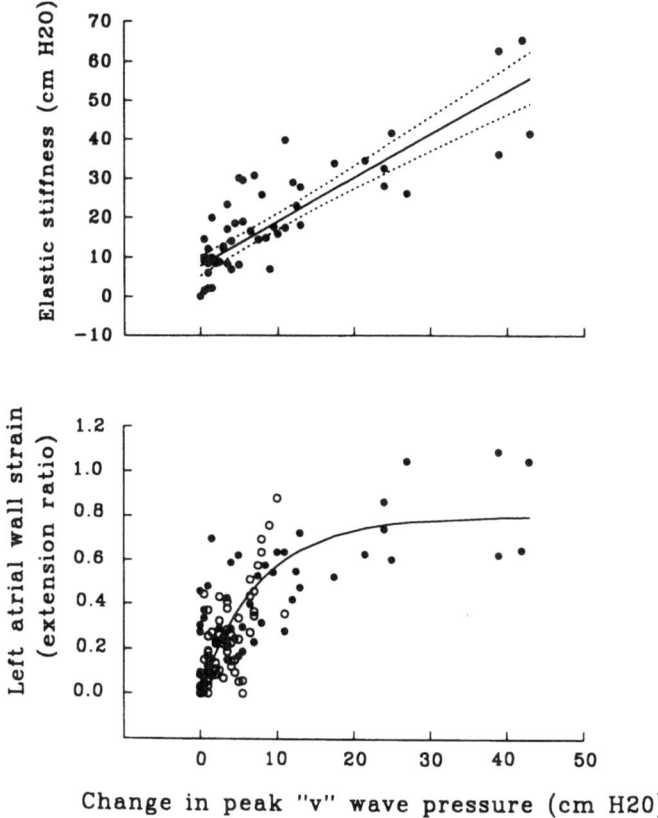

sion ratio. The value of the slope, or the "stiffness constant" of the tissue [9, 17] was calculated for each animal and the mean slope was 1.3 ± 0.2 (the slope for the pooled data shown in Fig 2. was 1.1). Similar calculations for the stiffness constant at the peak of the "a" wave showed a linear correlation betwen pressure and extension ratio in each animal, and the mean slope was 1.2 ± 0.2. Directionally similar but smaller changes were observed in right atrial pressure and dimensions in response to changes in blood volume.

Before nerve section, none of the variables examined in this study were significantly correlated with plasma IR-ANF and indeed there were no significant changes in ANF. However, after nerve section, there were signifcant changes in ANF and the relationship between plasma IR-ANF and wall strain (extension ratio) was exponential (Fig. 3) and this relationship was emphasized by the linear relationship between the logarithm of plasma IR-ANF and the extension ratio (r = 0.65, p< 0.0001). There was a significant positive linear correlation between plasma IR-ANF and changes in left atrial "v" wave pressure (r = 0.77, p < 0.0001) or diastolic left atrial wall stress (r = 0.76, p < 0.0001) (Fig. 4). The changes in left atrial "a" wave pressure and systolic left atrial wall stress (not shown) were similar to those of the diastolic or "v" wave measurements and had a similar relationship to IR-ANF.

## DISCUSSION

Previous work showed that when the vagus nerves and the aortic depressor nerves were cut and the carotid sinus pressure was maintained constant at 100 mm Hg then the

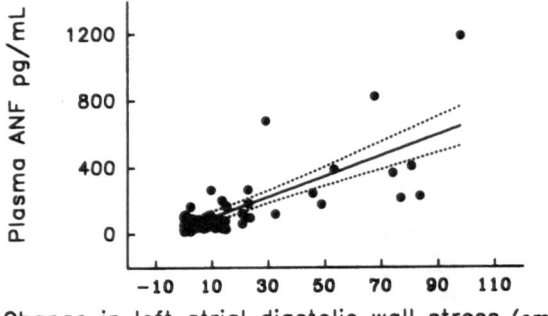

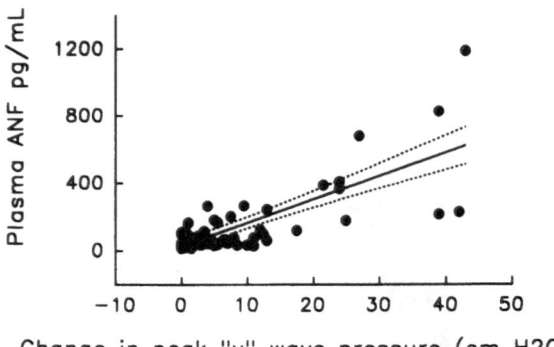

Figure 3. Relationships between plasma IR-ANF and changes in peak left atrial "v" wave pressure and left atrial wall stress. Lower panel shows changes in left atrial "v" wave pressure and IR-ANF. Upper panel shows the changes in left atrial diastolic wall stress and IR-ANF. Conventions as in Fig. 2.

increases in plasma IR-ANF induced by volume expansion were potentiated [5]. That the potentiation of the response was not due to time was shown by the previous experiments in which the vagus nerves were sectioned with the aortic nerves either intact or sectioned; potentiation of the release of ANF was present only in animals with the aortic nerves

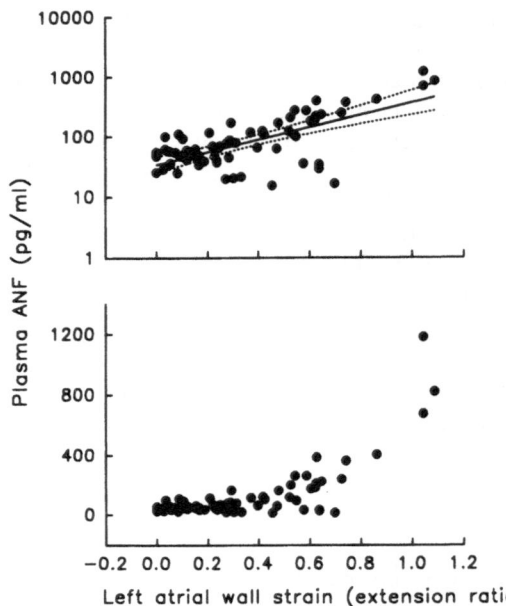

Figure 4. Relationships between plasma IR-ANF and atrial wall strain (extension ratio). In the lower panel plasma IR-ANF is plotted against left atrial wall strain, in the upper panel plasma IR-ANF is plotted on a logarithmic scale against atrial wall strain. Conventions as in Fig.2.

previously sectioned. The results presented here confirm that when the major cardiovascular reflexes were eliminated by section of the vagus, aortic depressor and carotid sinus nerves that the increase in plasma ANF associated with volume expansion was indeed potentiated. These data emphasize the importance of the cardiovascular reflexes in the normal adaptation to acute changes in blood volume.

In rabbits in which the cardiovascular reflexes were intact, although there were small changes in MAP, mean RAP and mean LAP with changes in blood volume of +/- 20%, none of these changes reached statistical significance (Fig. 1). At the same time alteration of the BV produced no significant change in plasma IR-ANF. In contrast, a number of previous studies have demonstrated a small but significant increase in plasma ANF in response to similar increases in blood volume in rats [1, 21], dogs [19, 22] and rabbits [5]. Also, in some but not all experiments hemorrhage of a similar degree has been shown to evoke a small but significant reduction in plasma ANF in rats [18, 32] rabbits [5] and humans [8]. The discrepancy between the results of this study and previous experiments may have arisen because the protocol we used for BV expansion was to expand the blood volume in stages and to reduce blood volume gradually over a period of 40 minutes. In the presence of active cardiovascular reflexes, the cardiovascular changes in response to the blood volume change were buffered and there were no significant changes in cardiovascular variables. Examination of the stimulus/response characteristics between LAP, and plasma IR-ANF (Fig. 3) shows that with only small changes in LAP there would be no significant stimulation of the release of ANF. The failure of the volume change to induce larger changes in LAP and plasma IR-ANF may have been influenced by a lowered effective blood volume and low effective atrial pressure (atrial pressure minus intrathoracic pressure) in these thoracotomized, instrumented animals.

After the vagus, aortic depressor and carotid sinus nerves were sectioned expansion of the BV by 20% caused a large increase in MAP and mean LAP. Mean arterial pressure decreased in an almost linear fashion as BV was decreased from +20% to -20% BV. In contrast mean LAP decreased from the value at +20% BV until normal BV was reached but then did not decrease further with a further decrease in BV (Fig.1). Thus the relationship between MAP and LAP was non-linear. Stewart et al. [28] have recently reported a linear relationship between MAP and LAP when MAP was increased by pressor agents and then allowed to fall. However, their protocol did not include decreasing blood volume below normal and close inspection of their data (their Fig. 1) indicates that there may have been a non-linear relationship, although linear regression lines were drawn through the data. It appears that in the absence of the cardiovascular reflexes, volume expansion by 20% of the blood volume caused an acute left ventricular failure secondary to the increase in both afterload and blood volume. Right atrial pressure showed only a small increase with volume expansion and a gradual decrease with volume reduction.

The large variation in LAP caused by volume expansion, after nerve section, provided a model in which it was possible to examine atrial dynamic behaviour and ANF release, over a wide range of left atrial pressures. In our analysis we have presented data related to diastolic function since it is diastolic events which govern atrial stretch. Changing blood volume is associated with similar changes in both diastolic and systolic atrial function [14] and similar conclusions would be reached using the systolic data from the present experiments. Because the relationships between atrial pressure and diameter and blood volume show hysteresis we chose to examine the pressure-diameter relationship of the atria during gradual decreases in blood volume after expanding the blood volume by 20%. Pressure and diameter each show a similar hysteresis with respect to blood volume and the pressure-diameter relationship is not significant different during inflation and deflation, at least at the rates of change of blood volume used here (14)

The stress/strain relationship of the left atrial wall, in diastole, is plotted in Fig. 2 and demonstrates an exponential relationship between changes in peak "v" wave (filling) pressure and the extension ratio. There was a 50-60% increase in atrial diameter when atrial pressure increased by 10 cm $H_2O$; further increases in atrial pressure produced much smaller increments in atrial diameter. The use of a single measure of the diameter of the atrial appendage to provide the extension ratio is clearly an approximation and any calculation of wall stress, in terms of lengthening of the myocytes, will be complicated by the irregular shape of the atrium, the non-uniformity of fibre direction and the variable wall tension dependant on the Laplace relationship [20]. If the stress-strain relationship of a biological material has an exponential form, then the elastic stiffness, expressed as change in stress divided by change in strain ($dP/(dl/l)$), is a linear function of the stress [9, 17]. The elastic stiffness of the atrial appendage did not appear to vary significantly between animals, or with section of afferent nerves. The slope of the linear relationship shown in Fig. 2 is known as the stiffness constant, and allows comparisons between elastic structures of different size and shape. Fung [9] regarded the stiffness constant as the most important parameter to characterize the stress-strain relationship or elastic stiffness of the tissue. The average stiffness constant for the atrial appendage of 1.4, found in the current experiments, is some 30 times smaller than values found for rat or dog ventricle [17]. It is also lower than the value of 2.5 found for cat papillary muscle by Parmley and Sonnenblick [20]. We are not aware of previous calculations from measurements of atrial diameter of the stiffness constant for atrial muscle *in vivo*. The high distensibility of the atrial appendage is consistent with the data of Stone et al. [30] who showed an increase in diameter of the right atrial appendage of 50%, with an increase in RAP of 11 cm $H_2O$, in the dog, with the pericardium opened. It should be noted that in the latter experiments, when the pericardium was closed there were only small changes in RAD; the pericardium was widely opened in our experiments. We also found that the elastic stiffness of the atrial appendage was not significantly changed during systole. This is in contrast to the finding of Stewart et al. [27] who found that the diastolic compliance was 10 to 30 times that of the systolic compliance. It is true that atrial volume is lower during systole compared to diastole at any given atrial pressure but our data do not indicate that much greater pressure changes are needed for a given change in atrial volume during systole compared to diastole. This marked difference in our conclusions can probably be attributed to the ways in which compliance was calculated

Atrial natriuretic factor is released into the plasma by atrial distension [13, 15] and it is widely accepted that it is stretch rather than pressure itself that determines the release of ANF [3,4,7,12,16,26,30]. We analysed the data from the current experiments to determine the relationship between plasma ANF and atrial pressure, atrial stretch (extension ratio) and atrial wall stress. There was an exponential relationship between the extension ratio of the atrial wall (diastolic atrial stretch) and plasma ANF (Fig. 3). At low atrial pressures the changes in dimensions for a given change in atrial pressure are larger and contribute more to the change in wall stress than at high atrial pressures, when there is a smaller contribution from the changes in dimensions. This means that with changes in atrial pressure of up to 10 cm $H_2O$ there were considerable changes in extension with relatively little change in ANF; major increases in ANF were seen only with much larger increases in atrial pressure which were associated with only small further changes in extension. These changes were observed with blood volume changes within the physiological range ($\pm$ 20%) but only in the absence of modulating cardiovascular reflexes. A similar relationship requiring very high atrial appendage wall stresses before there were marked increases in plasma ANF was shown in conscious dogs [27] subjected to a large and rapid volume expansion with saline. Since in our experiments the relationship between atrial pressure and extension was also exponential, there was a good linear relationship between both diastolic left atrial wall stress and the change in plasma IR-ANF and also between left atrial diastolic pressure and ANF (Fig. 4).

The similarity is to be expected, since the changes in LAP during atrial filling are much greater than the changes in dimensions and therefore the change in pressure dominates the calculation of wall stress. The mechanism by which a change in atrial wall stress is transduced to cause a release of ANF remains unknown.

Stewart et al. [29] have recently concluded that V-wave atrial wall stress accounts for 60% of the change in plasma ANF in response to volume loading and A-wave wall stress 40%. This was based on a study in which heart rate was held constant by ventricular pacing leading to a fixed A-wave pressure during volume loading. Our results are in agreement with these observations that both diastolic and systolic stress increase during volume loading.

It is concluded that the potentiation of the release of ANF, in response to changes in blood volume, by section of the carotid sinus, aortic and vagus nerves, can be accounted for in terms of changes in the dynamics of the left atrial wall. The results provide no evidence for neural modulation of the release of ANF. The results illustrate the importance that the cardiovascular reflexes normally play in maintaining atrial dynamics relatively constant during changes in blood volume within a physiological range. It is only when these reflexes fail to prevent a significant degree of atrial distension that there are increases in plasma ANF. Although wall stretch may be the primary stimulus to the release of ANF the relationship with plasma ANF is exponential and the linear relationship between atrial wall stress and plasma ANF makes atrial wall stress and/or atrial filling pressure more convenient indexes of the stimulus to the release of ANF.

## ACKNOWLEDGEMENTS

The authors are grateful to Jeannie Sharp-Kehl and Carolyn Redekopp for excellent technical assistance. The authors wish to thank Dr. F.J. Morich for the monoclonal antibody, which was a gift to Dr. N. Wilson. This work was supported by The Medical Research Council of Canada and The Heart and Stroke Foundation of B.C. and Yukon.

## REFERENCES

1. Anderson JV, Christofides ND, Bloom SR. Plasma release of atrial natriuretic peptide in response to blood volume expansion. J Endocrinol 109: 9-13, 1986.
2. Badeer HS. Contractile tension in the myocardium. Am Heart J 66:432-434, 1963.
3. Brouwer RML, Wenting GJ, Zijlstra F, Balk AHMM, Mochtar B, Derkx FHM, Bruin RJ, Weimar W, Schalekamp MADH. Atrial wall stress rather than pressure *per se* might be responsible for the increased secretion of atrial natriuretic factor after heart transplantation. J Hypertension 6:S330-S332, 1988.
4. Christensen G, Ilebekk A, Aakeson I, Kiil F. The release mechanism for atrial natriuretic factor during blood volume expansion and tachycardia in dogs. Acta Physiol Scand 134:263-270, 1988.
5. Courneya CA, Wilson N, Ledsome JR. Plasma vasopressin and atrial natriuretic factor in response to blood volume changes in the anaesthetized rabbit. Can J Physiol Pharmacol 67:344-352, 1989.
6. Courneya CA, Wilson N, Ledsome JR. Carotid sinus pressure, blood volume, and vasopressin in the anaesthetized rabbit. Can J Physiol Pharmacol 67:1386-1390, 1989.
7. Edwards BS, Zimmerman RS, Schwab TR, Heuplein DM, Burnett JC. Atrial stretch, not pressure, is the principal determinant controlling the acute release of atrial natriuretic factor. Circulation Res 62:191-195, 1988.
8. Finn WL, Gordon RD, Tunny TJ, Klemm SA, Hamlet SM. Effects of volume expansion and contraction on plasma levels of atrial natriuretic peptide in man. Clin Exper Pharmacol Physiol 15:311-315, 1988.
9. Fung YCB. Elasticity of soft tissues in simple elongation. Am J Physiol 216:1532-1544, 1967.
10. Hall C, Sanderud J, Risoe C. Is there a pericardial restriction to the cardiac secretion of atrial natriuretic peptide. Eur. Surg. Res. 25: 155-161, 1993.
11. Hansen AT. Pressure measurement in the human organism. Acta Physiol Scand 19 suppl 68:1-230, 1949.

12. King KA Ledsome JR. Atrial dynamics, atrial natriuretic factor, tachycardia and blood volume in anaesthetized rabbits. Am J Physiol 261:H22-H28, 1991.
13. Lang RE, Tholken H, Ganten D, Luft FC, Ruskoaho H, Unger T. Atrial natriuretic factor - a circulating hormone stimulated by volume loading. Nature 314:264-266, 1985.
14. Ledsome JR, King KA. Atrial dynamics and release of atrial natriuretic factor in vivo. Can J Physiol Pharmacol 69:1507-1513, 1991.
15. Ledsome JR, Wilson N, Courneya CA, Rankin AJ.Release of atrial natriuretic peptide by atrial distension. Can J Physiol Pharmacol 63:739-742, 1985.
16. Mancini GBJ, McGillem MJ, Bates ER, Weder B, Deboe SF, Grekin RJ. Hormonal responses to cardiac tamponade: inhibition of release of atrial natriuretic factor despite elevation of atrial pressures. Circulation 76:884-890, 1987.
17. Mirsky I. Elastic properties of the myocardium: a quantitative approach with physiological and clinical applications, in BERNE RM (ed): Handbook of Physiology. Bethesda, MA, Am Physiol Soc, Publishers, Section 2, Vol 1, pp 497-531, 1979.
18. Morris M, Cain M, Chalmers J. Complementary changes in plasma atrial natriuretic peptide and antidiuretic hormone concentrations in response to volume expansion and haemorrhage: studies in conscious normotensive and spontaneously hypertensive rats. Clin Exper Pharmacol Physiol 14:283-289, 1987.
19. Nishida Y, Miyata A, Morita H, Hosomi H. Physiological factors of atrial natriuretic polypeptide release and its neural regulation in conscious dogs. Jap Circulation J 52:1425-1429, 1988.
20. Parmley WW, Sonnenblick EH. Series elasticity in heart muscle. Circulation Res 20:112-123, 1967.
21. Petterson A, Hedner J, Ricksten S-E, Towle AC, Hedner T. Acute volume expansion as a physiological stimulus for the release of atrial natriuretic peptides in the rat. Life Sciences 38:1127-1133, 1986.
22. Pichet R, Gutkowska J, Cantin M, Lavallee M. Hemodynamic and renal responses to volume expansion in dogs with cardiac denervation. Am J Physiol 254:F780-F786, 1986
23. Rankin AJ, Ledsome JR, Keeler R, Wilson N. Extracted and non-extracted atrial natriuretic peptide in rabbits during tachycardia. Am J Physiol 253:R696-R700, 1986.
24. Rankin AJ, Wilson N, Ledsome JR. Effects of autonomic stimulation on plasma immunoreactive atrial natriuretic peptide in the anesthetized rabbit. Can J Physiol Pharmacol 65:532-537, 1987.
25. Ruskoaho H. Atrial natriuretic peptide: synthesis, release, and metabolism. Pharmacol. Rev. 44:479-602, 1992
26. Schiebinger R, Linden J The influence of resting tension on immunoreactive atrial natriuretic peptide secretion from the perfused rat heart. Circulation Res 59: 105-109, 1986.
27. Stewart JM, Odea DJ, Shapiro GC, Patel MB, McIntyre JT, Gewitz MH, Hoegler CT, Shapiro JT Zeballos GA Hintze TH. Atrial compliance determines the nature of passive atrial stretch and plasma atrial natriuretic factor in the conscious dog. Cardiovasc. Res. 25: 784 - 792, 1991.
28. Stewart JM, Wang J, Singer A, Zeballos GA, Ochoa M, Patel MB, Gewitz MH, Hintze TH. Regulation of plasma ANF after increases in afterload in conscious dogs. Am J Physiol 259: H1736-H1742, 1990.
29. Stewart JM, Wang J, Zeballos GA, Ochoa M, Shustek M, Hintze TH. Role of tachycardia and V wave wall stress in the release of ANF during volume loading. Am. J. Physiol. 264: H217-H223, 1993
30. Stone JA, Wilkes PRH, Keane PM, Smith ER, Tyberg JV. Pericardial pressure attenuates release of atriopeptin in volume-expanded dogs. Am J Physiol 256:H648-H654, 1989.
31. Wilson N, Ledsome JR, Keeler R, Rankin AJ, Wade JP, Courneya CA. Heterologous radioimmuno-assay of atrial natriuretic polypeptide in dog and rabbit plasma. J Immunoassay 7:73-96, 1986.
32. Yoshida K, Kawano Y, Hirata Y, Yoshimi H, Kuramochi M, Ito K, Omae T. Regulation of plasma atrial natriuretic peptide and the cardiopulmonary baroreflex in the rat. Jap Circulation J 51:1310-1314, 1987.

# 17

# EXERCISE PRESSOR REFLEX
## Studies on the Effect of Skeletal Muscle Fiber Type and Spinal Cord Transmission

L. Britt Wilson and Jere H. Mitchell*

UT Southwestern Medical Center
Harry S. Moss Heart Center
5323 Harry Hines Boulevard
Dallas, Texas 75235-9034

## INTRODUCTION

Static skeletal muscle contraction increases arterial blood pressure, heart rate, myocardial contractility, and sympathetic nerve activity. These changes occur when humans or conscious cats perform static exercise, and can also be produced by evoking static contractions using anesthetized animals (14). Two potential mechanisms for these cardiovascular changes have been proposed: 1) a reflex arising from the contracting muscles, and 2) a descending input from higher brain centers. The first theory, termed the "exercise pressor reflex," proposes that the cardiovascular changes are produced reflexly by a contraction induced activation of muscle afferents (1,12,15). The second theory proposes that the cardiovascular changes are elicited by signals arising from central areas that recruit motor units and also activate cardiovascular neurons located in the medulla and/or spinal cord (central command) (6,11,27). Currently, both mechanisms are thought to be important in mediating the cardiovascular responses during exercise. However, the relative contributions of these mechanisms to the cardiovascular changes produced by skeletal muscle contraction and how they are integrated needs further study.

The experiments described in this chapter will focus only on the reflex cardiovascular responses evoked by static contraction of skeletal muscle. More specifically, we will discuss studies concerning the effect of skeletal muscle fiber type on the exercise pressor reflex as well as the possible neurotransmitter/neuromodulators that are released at the first synapse in the dorsal horn of the spinal cord in this reflex.

---

*Phone: (214) 648-3424; Fax: (214) 648-3566

*Control of the Cardiovascular and Respiratory Systems in Health and Disease*
Edited by C. T. Kappagoda and M. P. Kaufman, Plenum Press, New York, 1995

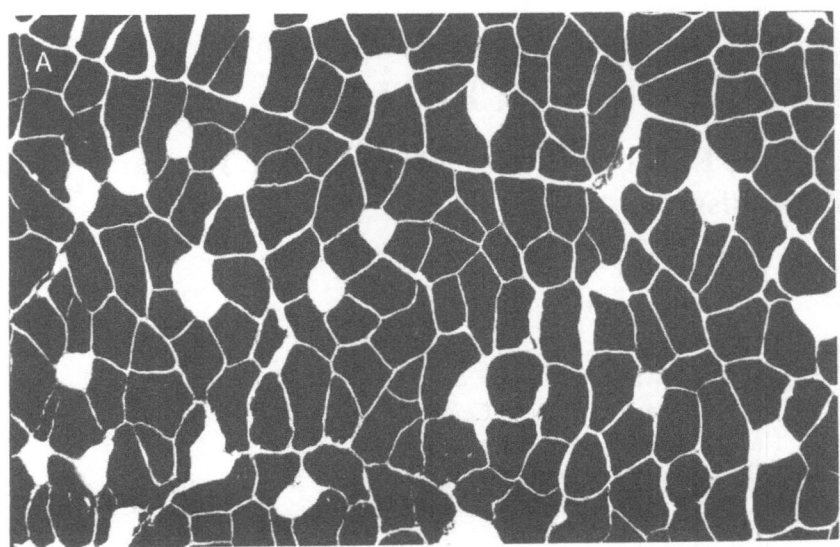

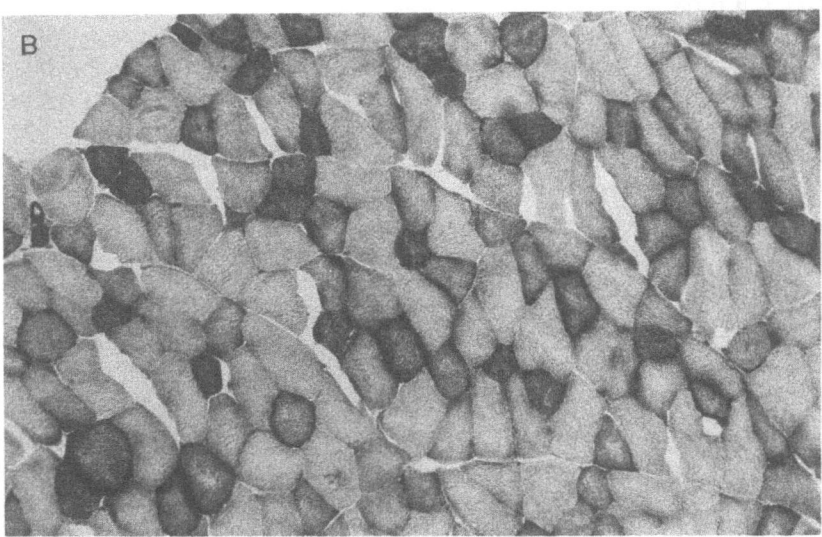

**Figure 1.** Light micrographs of the gastrocnemius muscle of a rabbit. Panels A and C show the black and white staining pattern of the myofibrillar actomyosin ATPase, which characterizes the Type I and Type II fibers. Panel A - Muscle from the control, unstimulated (glycolytic) muscle, with a predominance of fast twitch, Type II fibers. Panel C - Muscle from the contralateral, stimulated (oxidative) muscle which shows an increased number of fibers with light brown, tan, and white stain indicating reduced amounts of myofibrillar actomyosin ATPase. Panel B - Section of NADH stained muscle tissue from the glycolytic hindlimb with the variegated staining pattern typical of untrained muscle. Most cells are lightly stained indicating a highly glycolytic capacity. Panel D - Muscle from the contralateral stimulated muscle stained for NADH, showing a uniform, dense stain for the majority of fibers indicating a greater oxidative potential for these cells than those depicted in panel B. (Reproduced with permission. *J. Appl. Physiol.* 79:1744-1752, November, 1995.) [For color representation see insert at the end of the chapter.]

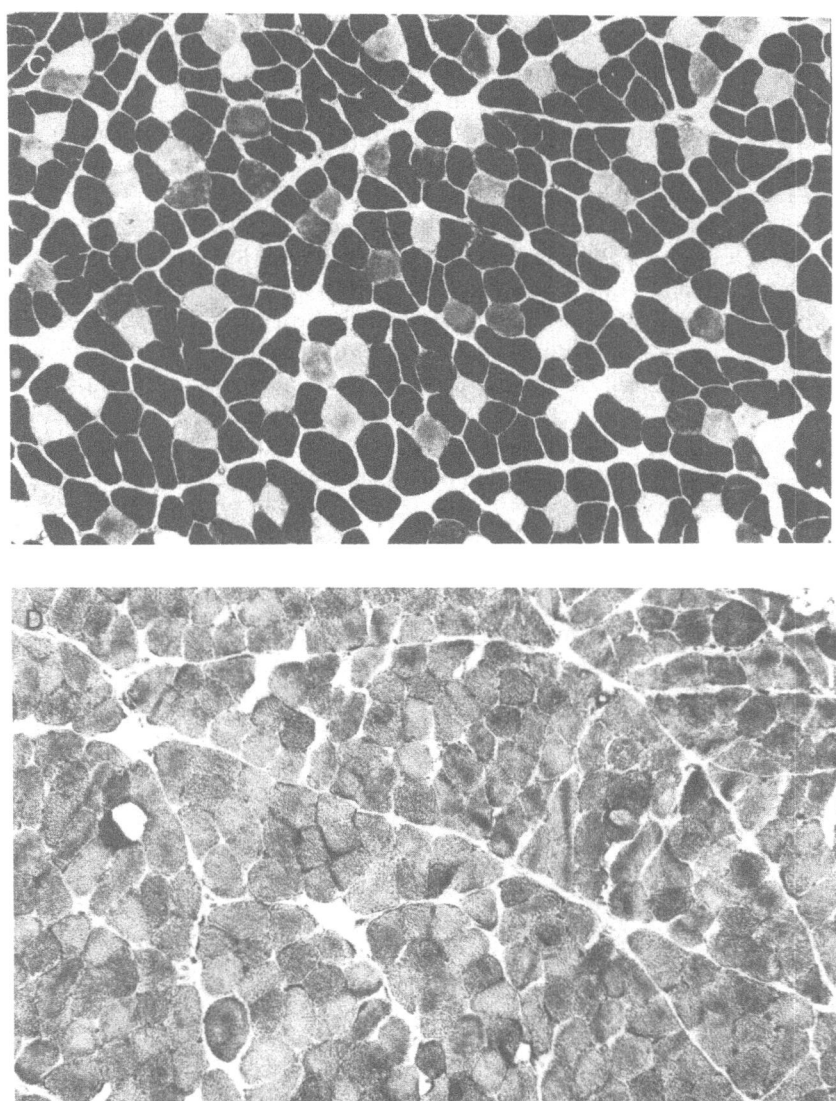

# EFFECT OF SKELETAL MUSCLE FIBER TYPE

An important question concerning the exercise pressor reflex is whether different types of muscle fibers evoke divergent reflex cardiovascular and sympathetic responses. In general, mammalian skeletal muscle contains three different fiber types (2): 1) type I, slow-twitch (highly oxidative, low glycolytic), 2) type IIa, fast-twitch (moderately oxidative and glycolytic), and 3) type IIb, fast-twitch (highly glycolytic, low oxidative). Type I and IIa muscle contains numerous mitochondria and capillaries, hence the high oxidative capacity. These fiber types are resistant to fatigue, minimal lactate is produced during the contraction, and the rate of tension development is slow in Type I fibers, but fast in type IIa (2,16). In contrast, type IIb muscle contains few mitochondria and capillaries, and energy production occurs primarily via glycolysis. This fiber type is quick to fatigue, lactate release is prominent during the contraction, and the rate of tension development is rapid (2,16).

Studies have suggested that the pressor response evoked by isometric skeletal muscle contraction is mediated predominantly by fast-twitch fibers (4,5,17,18). Using anesthetized cats, Petrofsky et al. showed that isometric contraction of the medial gastrocnemius muscle (primarily fast-twitch) increases arterial blood pressure while contraction of the soleus muscle (slow-twitch) fails to evoke a pressor response (17). Furthermore, only a very small pressor response to gastrocnemius contraction was observed when only the slow-twitch fibers of this muscle were contracted (18). The authors suggested that a greater release of $K^+$, as measured in venous blood, from the fast-twitch muscle was the mechanism for the greater pressor response. Injecting $K^+$ into the arterial supply of skeletal muscle activates group III and IV muscle afferents and increases arterial pressure (21,22,28). An alternative possibility, although not mutually exclusive, is a greater production of lactic acid by the fast-twitch, glycolytic muscle fibers. Lactic acid also activates muscle afferents and increases arterial blood pressure (20). In contrast to the studies of Petrofsky et al., it was shown that isometric contraction of the soleus muscle increases arterial blood pressure in decerebrate cats (8). Thus, contraction of slow-twitch muscle may increase blood pressure if anesthesia is removed.

In rabbits, chronic electrical stimulation changes the skeletal muscle from a fast-twitch, glycolytic to a slow-twitch, oxidative muscle (19,23). This model has been used recently to further elucidate the effects of skeletal muscle fiber types on the cardiovascular response to static muscle contraction (30). From this study the effects of chronic stimulation on the histochemistry of rabbit gastrocnemius muscle are shown in Figure 1. The upper two panels are the control muscle, while the lower two panels represent gastrocnemius muscle

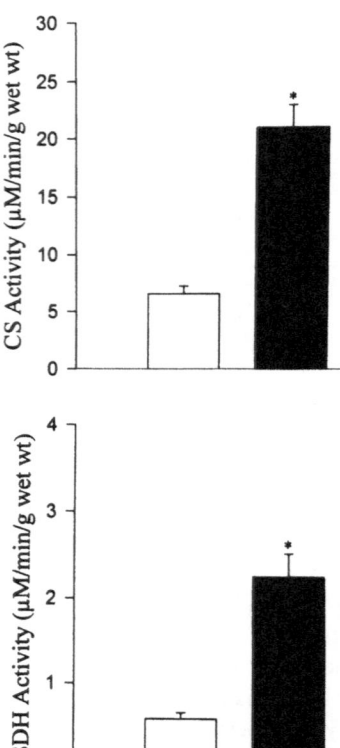

**Figure 2.** The specific activity of citrate synthase (CS; upper graph) and succinate dehydrogenase (SDH; lower graph) from gastrocnemius muscles of adult rabbits that were stimulated (oxidative; closed bars) for 21 days as compared with the contralateral, unstimulated (glycolytic; open bars) gastrocnemius muscles from the same animals. *-denotes significant difference compared to glycolytic muscle. Numbers are mean±SE. (n=14). [Reproduced with permission. *J. Appl. Physiol.*, 79:1744-1752, November, 1995.]

from the contralateral leg that had undergone 21 days of chronic stimulation. The upper left panel, which was stained for actomyosin ATPase, shows a marked predominance of fast twitch (Type II) fibers. The upper right panel, which was stained for NADH, shows that most of these fibers have low oxidative capacity. The lower left panel shows that there has been a decrease in myofibrillar actomyosin ATPase with chronic stimulation. More importantly, the right lower panel demonstrates that after chronic stimulation there is a marked enhancement of oxidative capacity.

The biochemical analysis of samples taken from the control and the stimulated gastrocnemius muscle of the rabbit are shown in Figure 2. Citrate synthase (CS) and succinate dehydrogenase (SDH) activity were measured as an indicator of oxidative capacity. Chronic stimulation markedly increased the oxidative capacity of the muscle since CS increased by $250\pm43\%$ and SDH by $324\pm65\%$.

The heart rate (HR) and arterial blood pressure (AP) responses to induced static exercise in the control (glycolytic) leg and the stimulated (oxidative) leg are shown in Figure 3. In the upper panel both HR and AP transiently decreased and then increased above

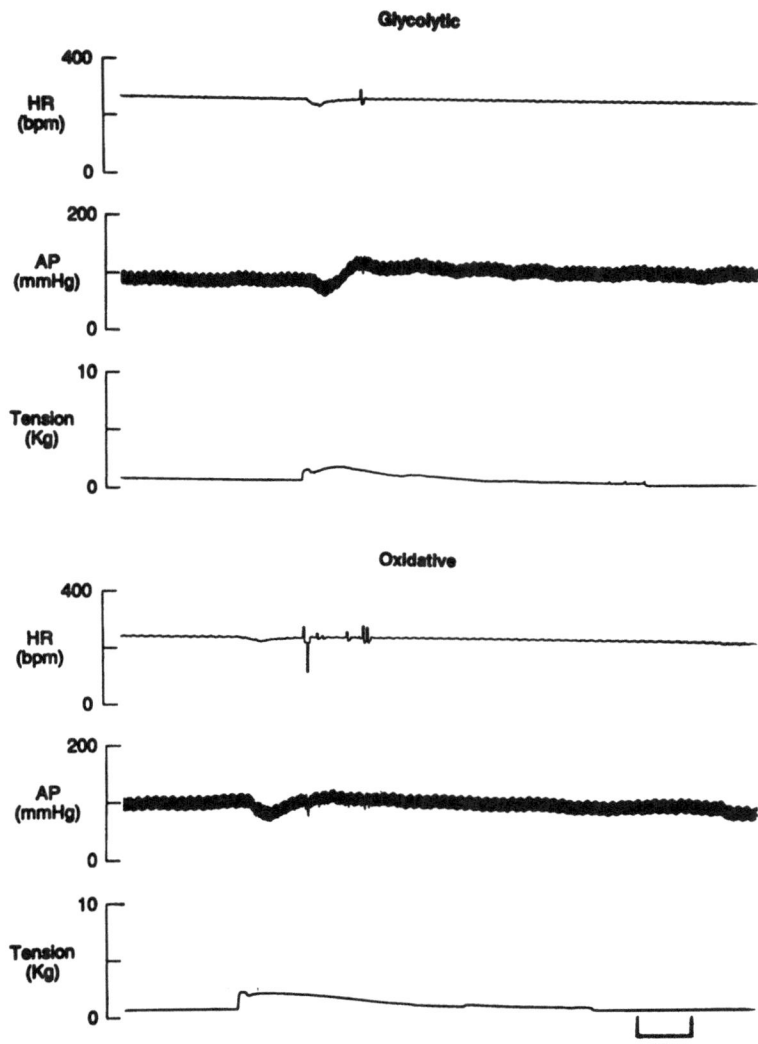

**Figure 3.** Original recordings from one rabbit showing the heart rate and arterial blood pressure responses to static contraction of the glycolytic (upper tracings) and oxidative (lower tracings) muscles.

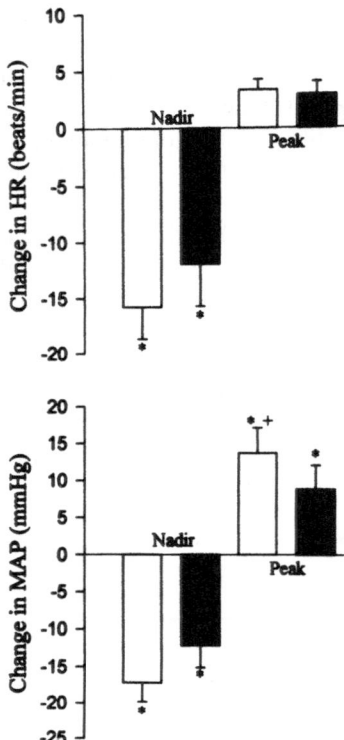

**Figure 4.** These graphs illustrate the nadir and peak increases in HR (upper graph) and MAP (lower graph) in response to static contraction of the glycolytic (open bars) and oxidative (closed bars) muscles. *denotes significant change from baseline. +denotes significant difference compared to oxidative muscle. Numbers are mean±SE. (n=12). [Reproduced with permission. *J. Appl. Physiol.*, 79:1744-1752, November, 1995.]

resting values in the glycolytic leg. This response is the same as that reported in rabbits (24,25,31). In the lower panel, HR and AP transiently decreased and then increased only slightly above resting values.

The average values for the nadir and peak HR and AP changes from 12 rabbits are shown in Figure 4. There was a tendency for the nadir to be greater from contraction of the glycolytic muscle as compared to the oxidative muscle. The important finding, however, was that the AP peak was greater when the glycolytic muscle was contracted than it was when the oxidative muscle was contracted.

## SPINAL CORD TRANSMISSION

As indicated above, group III and group IV muscle afferents comprise the initial arm of the exercise pressor reflex. The majority of these muscle afferents synapse in the superficial layers of the dorsal horn of the spinal cord (13), ultimately activating the cells of ascending tracts that project to the medulla of the brain. Although projection neurons to the medulla are activated by the muscle afferents, the number of synapses that are involved in this activation is unknown. Further, the dorsal horn contains a vast network of nerve fibers and cell bodies (29). Thus, although the dorsal horn represents the first synapse of the exercise pressor reflex, it is quite likely that this region serves as a site of integration of the neural signals arising from the contracting muscle.

Anatomical studies have demonstrated a large number of putative neurotransmitters within the dorsal horn (3,26,29). Substance P is one neuropeptide whose concentration is quite abundant in the superficial region of the dorsal horn (7); work from our laboratory

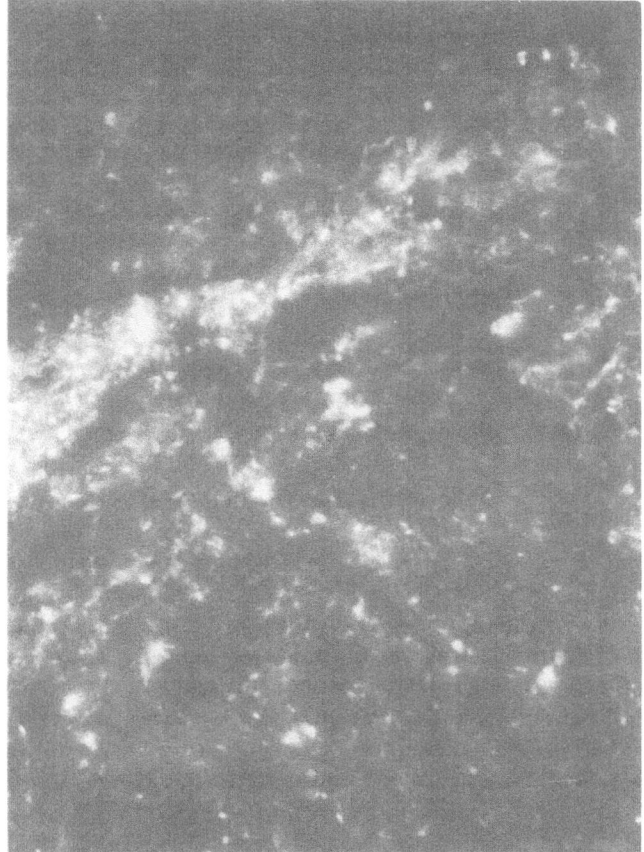

**Figure 5.** Immunofluorescence micrograph of the dorsal horn of the spinal cord of a cat incubated with antiserum to substance P. The bright green is the fluorescence of FITC conjugated to the secondary antibody, thus indicating the presence of substance P. [For color representation see insert at the end of the chapter.]

verifying this is shown in Figure 5. In this picture, the marked green fluorescence demonstrates the presence of substance P.

The intrathecal injection of an antagonist (9) or an antibody (10) to substance P markedly decreases the blood pressure and heart rate responses to induced muscle contraction in the cat. It has also been shown that the microinjection of a substance P antagonist into the dorsal horn region of the spinal cord markedly attenuates the cardiovascular response to induced muscle contractions (34). Thus, the results of these studies suggest that the release of substance P in the dorsal horn plays a role in mediating the exercise pressor reflex.

More recently, studies have shown that substance P is released into the dorsal horn during a static contraction in cats (32,33). In these studies, a laminectomy was performed and the spinal roots, $L_6$, $L_7$ and $S_1$ were identified. The preparation is shown in Figure 6. The $L_6$ and $S_1$ dorsal and ventral roots were sectioned, and the $L_7$ ventral root was sectioned near the cord. The peripheral ends of the $L_7$ and $S_1$ ventral roots were placed on stimulating electrodes. A microdialysis probe was inserted into the $L_7$ dorsal horn region. During induced static contraction by electrical stimulation of the ventral roots, fluid from the perfusion of the microdialysis probe was collected and the concentration of substance P was measured

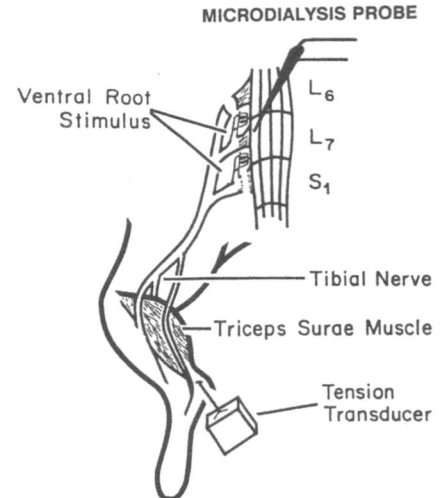

**Figure 6.** A depiction of the experimental preparation used to quantitate substance P release in the dorsal horn in response to a static muscle contraction. The microdialysis probe was inserted into the midpoint (rostro-caudal) of the $L_7$ dorsal horn.

using a radioimmunoassay (32). During an induced muscle contraction substance P was released as shown in Figure 7. The release of substance P after insertion of the microdialysis probe into the $L_7$ region of the dorsal horn is shown by the first three bars. After substance P reached a control value, a muscle contraction was induced. During the period of the contraction, the release of substance P increased significantly. After the contraction (control 1 and control 2) substance P release returned to control values.

It has also been shown that the release of substance P, as well as the heart rate and blood pressure responses, during an induced muscle contraction are influenced by the developed tension (33). These results are shown in Figure 8. The tension time index was

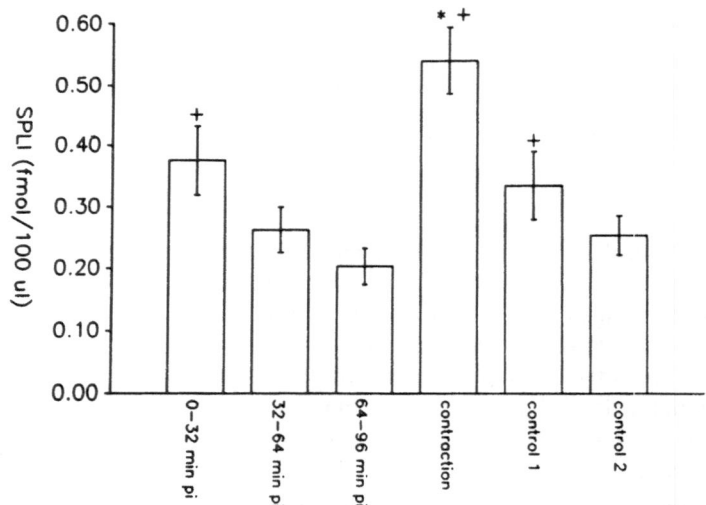

**Figure 7.** The concentration of SP-LI in the dialysate from the microdialysis probe before, during, and after a 5-8 min static contraction. Abbreviations are SP-LI—substance P-like immunoreactivity; pi—Post insertion of the probe. * indicates significantly different compared to all other points; + indicated significantly different compared to 64-96 min pi (third bar). Numbers are mean±S.D. (n=6). [Reprinted from the *J Physiol* 460:79-90, 1993.]

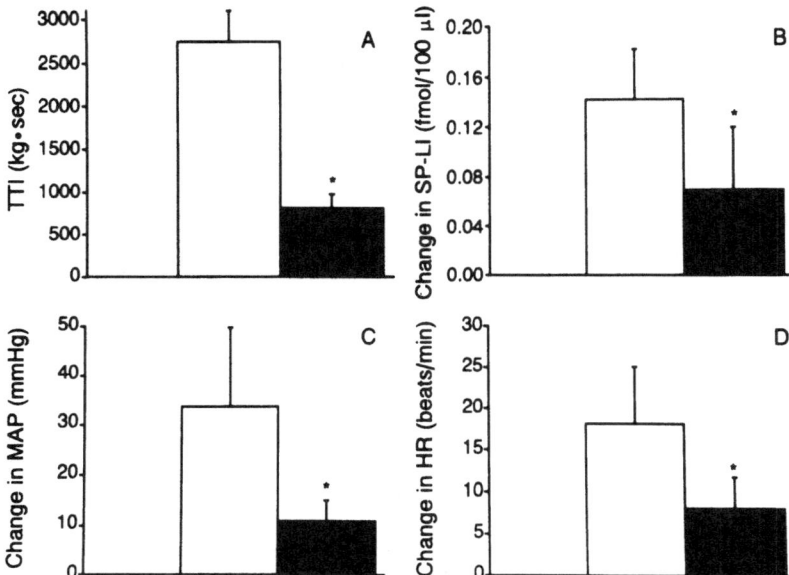

**Figure 8.** Graphs depicting the TTI (panel A), change in SP-LI (panel B), change in MAP (panel C), and the change in HR (panel D) in response to HT contractions (open bars) and LT contractions (filled bars). *-denotes significant difference compared to HT contractions. Numbers are mean±S.D. (n=8). [Reproduced with permission. *Cir Res* 73:1024-1031, 1993. Copyright 1993 American Heart Association.]

calculated for the low and the high tension contractions. With the low tension contractions, there was a small increase in the release of substance P and a small increase in mean arterial pressure and heart rate. With the high tension contractions, there was a significantly greater release of substance P, and greater increases in mean arterial pressure, and in heart rate.

## CONCLUSIONS

Studies have shown that a reflex neural mechanism (exercise pressor reflex) can be important in determining the cardiovascular response to static exercise. In addition, the present experiments suggest that skeletal muscle fiber type plays a role in this response. Static contraction of a highly glycolytic muscle elicits a greater blood pressure response than does that of a highly oxidative muscle. Further studies are needed to determine the metabolic product(s) that are responsible for these different cardiovascular responses. Also, the present studies provide considerable evidence that the release of substance P in the dorsal horn of the spinal cord plays a role in mediating the exercise pressor reflex.

## REFERENCES

1. Alam M, Smirk FH: Observations in man upon a blood pressure raising reflex arising from the voluntary muscles. J Physiol (Lond) 89:372-383, 1937.
2. Astrand P, Rodahl K: Textbook of work physiology. Third Ed. McGraw-Hill; 12-53, 1986.
3. De Biasi S, Rustioni A: Glutamate and substance P coexist in primary afferent terminals in the superficial laminae of spinal cord. Proc Natl Acad Sci USA 85:7820-7824, 1988.

4. Fallentin N, Sidenius B, Jorgensen K: Blood pressure, heart rate and EMG in low level static contractions. Acta Physiol Scand 125:265-275, 1985.
5. Frisk-Holmberg M, Essen B, Fredrikson M, Ström G, Wibell L: Muscle fibre composition relation to blood pressure response to isometric exercise in normotensive and hypertensive subjects. Acta Med Scand 213:21-26, 1983.
6. Goodwin GM, McCloskey DI, Mitchell JH: Cardiovascular and respiratory responses to changes in central command during isometric exercise at constant muscle tension. J Physiol (Lond) 226:173-190, 1972.
7. Hökfelt T, Kellerth JO, Nilsson G, Pernow B. Experimental immunohistochemical studies on the localization and distribution of substance P in cat primary sensory neurons. Brain Res 100:235-252, 1975.
8. Iwamoto GA, Botterman BR. Peripheral factors influencing expression of pressor reflex evoked by muscular contraction. J Appl Physiol 58:1676-1682, 1985.
9. Kaufman MP, Kozlowski GP, Rybicki KJ. Attenuation of the reflex pressor response to muscular contraction by a substance P antagonist. Brain Res 333:182-184, 1985.
10. Kaufman MP, Rybicki KJ, Kozlowski GP, Iwamoto GA. Immunoneutralization of substance P attenuates the reflex pressor response to muscular contraction. Brain Res 377:199-203, 1986.
11. Krogh A, Lindhard J: The regulation of respiration and circulation during the initial stages of muscular work. J Physiol (Lond) 47:112-136, 1913-1914.
12. McCloskey DI, Mitchell JH: Reflex cardiovascular and respiratory responses originating in exercising muscle. J Physiol 224:173-186, 1972.
13. Mense S, Craig Jr AD: Spinal and supraspinal terminations of primary afferent fibers from the gastrocnemius-soleus muscle in the cat. Neuroscience 26:1023-1035, 1988.
14. Mitchell JH. Neural control of the circulation during exercise. Med Sci Sports Exerc 22:141-154, 1990.
15. Mitchell JH, Kaufman MP, Iwamoto GA: The exercise pressor reflex: its cardiovascular effects, afferent mechanisms, and central pathways. Annu Rev Physiol 45:229-242, 1983.
16. Peter JB, Barnard RM, Edgerton VR, Gillespie CA, Stemple KE: Metabolic profiles of three fiber types of skeletal muscle in guinea pigs and rabbits. Biochemistry 11:2627-2633, 1972.
17. Petrofsky JS, Lind AR. The blood pressure response during isometric exercise in fast and slow twitch skeletal muscle in the cat. Eur J Appl Physiol 44:223-230, 1980.
18. Petrofsky JS, Phillips CA, Sawka MN, Hanpeter D, Lind AR, Stafford D: Muscle fiber recruitment and blood pressure response to isometric exercise. J Appl Physiol 50:32-37, 1981.
19. Pette D, Ramirez BU, Müller W, Simon R, Exner GU, Hildebrand R: Influence of intermittent long-term stimulation of contractile, histochemical and metabolic properties of fibre populations in fast and slow rabbit muscles. Pflügers Archiv 361:1-7, 1975.
20. Rotto DM, Kaufman MP: Effect of metabolic products of muscular contraction on discharge of group III and IV afferents. J Appl Physiol 64:2306-2313, 1988.
21. Rybicki KJ, Kaufman MP, Kenyon JL, Mitchell JH: Arterial pressure responses to increasing interstitial potassium in hindlimb muscle of dogs. Am J Physiol 247:R717-R721, 1984.
22. Rybicki KJ, Waldrop TG, Kaufman MP: Increasing gracilis muscle interstitial potassium concentrations stimulate group III and IV afferents. J Appl Physiol 58:936-941, 1985.
23. Salmons S, Sreter FA: Significance of impulse activity in the transformation of skeletal muscle type. Nature 263:30-34, 1976.
24. Tallarida G, Baldoni F, Peruzzi G, Raimondi G, Massaro M, Sangiorgi M. Cardiovascular and respiratory reflexes from muscles during dynamic and static exercise. J Appl Physiol 50:784-791, 1981.
25. Tallarida G, Peruzzi G, Raimondi G. The role of chemosensitive muscle receptors in cardiorespiratory regulation during exercise. J Auton Ner Sys 30:s155-s162, 1990.
26. Tuchscherer MM, Seybold VS: A quantitative study of the coexistence of peptides in varicosites within the superficial laminae of the dorsal horn of the rat spinal cord. J Neurosci 9:195-205, 1989.
27. Waldrop TG, Eldridge FL, Iwamoto GA, Mitchell JH: Central neural control of respiration and circulation during exercise. In: Handbook on Exercise: Integration of Motor, Circulatory, Respiratory and Metabolic Control During Exercise, L. Rowell and J Shepherd (Eds), In press, 1995.
28. Wildenthal K, Mierzwiak DS, Skinner Jr NS, Mitchell JH: Potassium-induced cardiovascular and ventilatory reflexes from the dog hindlimb. Am J Physiol 215:542-548, 1968.
29. Willis Jr JD, Coggeshall RE. Sensory mechanisms of the spinal cord. 2nd Edition. Plenum Press, New York, NY, 1991, pp. 79-151.
30. Wilson LB, Dyke CK, Parsons D, Wall PT, Pawelczyk JA, Williams RS, Mitchell JH. Effect of skeletal muscle fiber type on the pressor response evoked by static contraction. J Appl Physiol, submitted.
31. Wilson LB, Dyke CK, Pawelczyk JA, Wall PT, Mitchell JH. Cardiovascular and renal nerve responses to static muscle contraction of decerebrate rabbits. J Appl Physiol 77:2449-2455, 1994.

32. Wilson LB, Fuchs IE, Matsukawa K, Mitchell JH, Wall PT. Substance P release in the spinal cord during the exercise pressor reflex in anaesthetized cats. J Physiol 460:79-90, 1993.

33. Wilson LB, Fuchs IE, Mitchell JH. Effects of graded muscle contractions on spinal cord substance P release, arterial blood pressure, and heart rate. Circ Res 73:1024-1031, 1993.

34. Wilson LB, Wall PT, Matsukawa K, Mitchell JH. Effect of spinal microinjections of an antagonist to substance P or somatostatin on the exercise pressor reflex. Circ Res 70:213-222, 1992.

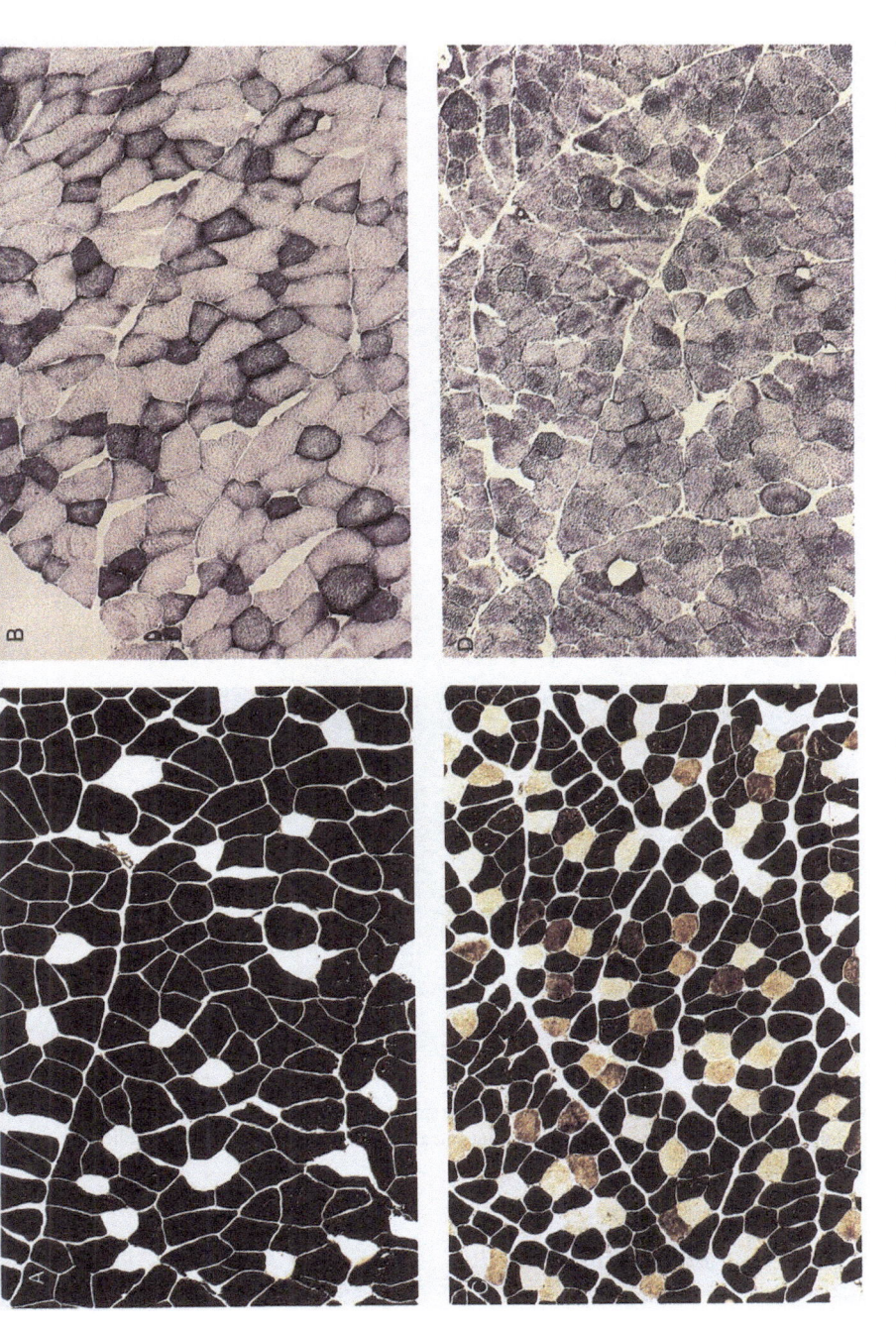

**Figure 1.** Light micrographs of the gastrocnemius muscle of a rabbit. Panels A and C show the black and white staining pattern of the myofibrillar actomyosin ATPase, which characterizes the Type I and Type II fibers. Panel A - Muscle from the control, unstimulated (glycolytic) muscle, with a predominance of fast twitch, Type II fibers. Panel C - Muscle from the contralateral, stimulated (oxidative) muscle which shows an increased number of fibers with light brown, tan, and white stain indicating reduced amounts of myofibrillar actomyosin ATPase. Panel B - Section of NADH stained muscle tissue from the glycolytic hindlimb with the variegated staining pattern typical of untrained muscle. Most cells are lightly stained indicating a highly glycolytic capacity. Panel D - Muscle from the contralateral stimulated muscle stained for NADH, showing a uniform, dense stain for the majority of fibers indicating a greater oxidative potential for these cells than those depicted in panel B. (See chapter 19 for source.)

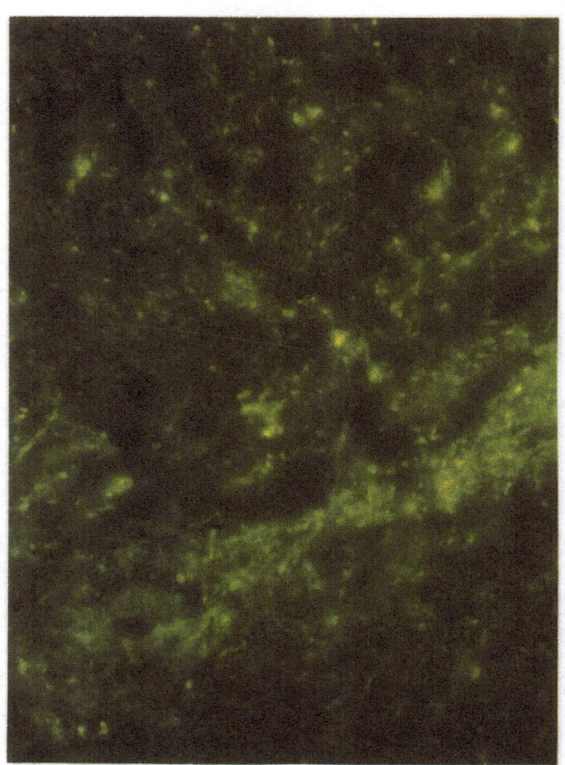

**Figure 5.** Immunofluorescence micrograph of the dorsal horn of the spinal cord of a cat incubated with antiserum to substance P. The bright green is the fluorescence of FITC conjugated to the secondary antibody, thus indicating the presence of substance P.

# 18

# A LIFETIME OF RESEARCH IN INTEGRATIVE PHYSIOLOGY

Hazel M. Coleridge and John C. G. Coleridge

Emeritus Professors of Physiology
Cardiovascular Research Institute
University of California, San Francisco
San Francisco, California 94143

We are deeply conscious of the honor afforded us by the organizers of this symposium. It is one of the happiest events of our lives and one to which our thoughts will return again and again. First we must offer our heartfelt thanks to those who made this meeting possible: to Marc Kaufman and Tissa Kappagoda, who were responsible for the nuts and bolts organization and who shouldered the heaviest burden: to John Longhurst, who gave it strong support and provided the roof over our heads: to the pharmaceutical companies listed in the front of this volume, whose financial backing made the symposium possible: and to the many friends who have joined us here at UC Davis, and whose scientific contributions and lively discussions, have ensured its success.

We should like to begin by acknowledging our debt to one who had the greatest influence on our choice of career. We regret that we were unable to attend the 1993 International Congress of Physiology at which Bjorn Folkow paid tribute to Eric Neil, one of the greatest teachers of integrative physiology. We were both fortunate, as medical students at the University of Leeds, to have Eric as our teacher. Eric was a colorful, irascible curmudgeon, roundly disliked by some students, but loved by others, always accessible and generous of his time. A great deal of his effectiveness as a teacher stemmed from the fact that he was an incredible romantic, able to clothe dull facts in colorful garments, and to provide vivid accounts of the advances in physiology that provide the basis for our understanding of the various control systems. Eric's emphasis on the blessed gift of curiosity, and on the delight of discovery, was never absent from his presentations, and it was this that hooked us firmly on Physiology from an early stage. The last time we saw him was after the second Leeds Symposium on cardiovascular receptors. We traveled down by train together, from Leeds to London. As we parted on the concourse at Kings Cross we both had a premonition that we might not see him again, and afterwards regretted that we hadn't thanked him for influencing us in the way he did. One of the things one learns with advancing years is to say thank you before it's too late.

We have spent most of our professional lives in the study of mammalian physiology. Aspects of neutral control of the heart, lungs and circulation have been the topics of research endeavor for the most of the people in this room at some time or other. For us it all started

*Control of the Cardiovascular and Respiratory Systems in Health and Disease*
Edited by C. T. Kappagoda and M. P. Kaufman, Plenum Press, New York, 1995

**199**

when John and Ron Linden began their studies in cardiovascular physiology in Albert Hemingway's laboratory in Leeds, and set out to measure intrathoracic and cardiac pressures and to examine the mechanism for the Bainbridge reflex. It soon became obvious that progress in the study of any reflex mechanism required information about the afferent nerve endings initiating the reflex. The techniques for recording from single sensory fibers were already well established, and the activity of pulmonary stretch receptors, baroreceptors and carotid chemoreceptors had been examined, but apart from Autar Paintal's pioneering studies carried out in David Whitteridge's laboratory in Edinburgh, accounts of activity in sensory receptors in the cardiac chambers, recorded from the vagus nerves, were relatively few. Several months had been spent in Hemingway's laboratory in unsuccessful attempts to record action potentials from atrial receptors when the two of us set out for Edinburgh to consult David Whitteridge. It took less than an hour for Whitteridge to show John the basic techniques of afferent fiber recording — he had learnt them from Adrian who drew a rough diagram of an envelope and jotted down a few explanatory pointers. Once this technique was in place in Leeds, the research was able to broaden its scope.

Recording sensory activity from afferent fibers of the vagus nerve can become addictive. The endings are so various, and the rhythmic discharges of cardiovascular and pulmonary mechanoreceptors, monitored by the loud speakers have a soothing effect on the investigator. Ainslie Iggo, whose interests were mainly in isolating and analyzing somatic sensory mechanisms, once asked us how we could possibly cope with the sensory receptors in the chest, because the structures they innervate don't keep still, but move about so much. In any event the sound background in cardiovascular and respiratory physiology that was the strength of the Leeds department under Albert Hemingway (he had been one of Starling's last pupils) stood us in good stead. Given some understanding of the forces that produce them, the cyclic movements of the heart and lungs are a godsend for an investigator engaged in recording from afferent nerve pathways. No need to employ sophisticated techniques to determine the threshold and mechanical transduction properties of the sensory receptors when these are revealed as pressure and volume in the cardiac chambers and lungs vary spontaneously. No need to search blindly for a given nerve ending when its pattern of discharge reveals its general location in the heart, great vessels or lungs. Taking the cue from neurophysiologists working in the somatic sensory system, the Leeds group soon found that the location of a receptor could be determined quite precisely by exploring the thoracic viscera, first with a finger and then with a fine probe, preferably post mortem, when all movement had ceased. In this way the preferential distribution of atrial receptors at the veno-atrial junctions, described by anatomists, could be confirmed, and the distribution of mechanoreceptors at the other sites. In general these intrathoracic mechanoreceptors supply information relating to the forces involved in maintaining cardiorespiratory function, and set in train reflex changes that maintain the status quo. When Cecil Kidd joined the team the presence of baroreceptors in the pulmonary artery (which hitherto had only been suspected from histological studies) was clearly demonstrated, although the functional significance of this group of baroreceptors is debated even now. For each of these categories of mechanoreceptor efforts were made to correlate the afferent properties with the reflex response in terms of threshold, sensitivity and blocking temperature.

To confirm that a given effect is reflex in nature it is necessary to show that it can be abolished when the pathway for the reflex is interrupted. Cooling the nerve that transmits the reflex has the advantage over nerve section of providing a reversible block of conduction. For this purpose the nerve is placed on a cooling platform whose temperature can be reduced progressively and then re-warmed. Studies by both Iggo and Paintal established the validity of the method, and it has been examined further by Andy Roberts, Anders Jonzon and Tom Pisarri in our laboratory in its application to pulmonary reflexes. Axonal conduction in A fibers blocks at a higher temperature than that in C fibers, partly because of the greater diameter of the A fibers, but mainly because the presence of a myelin sheath allows conduction to be saltatory, and saltatory conduction is more susceptible to cooling than conduction that takes place along the whole length of the axon. The effect of cooling on axonal conduction in both types of fiber also depends on the frequency with which the action potentials are transmitted. High frequency volleys from mechanoreceptors are blocked at temperatures at which irregular low frequency discharges can traverse the cooled region. The method allows the blocking temperature of the various afferent pathways identified in action potential studies to be compared with the blocking temperature of the reflex under investigation. In evaluating the effects of nerve cooling on a reflex, however, it is important to distinguish the temperature required to block the increase in frequency invoked by the experimental stimulus from the temperature required to interrupt all activity in the afferent fibers. Thus the temperature required to block the Hering-Breuer reflex is several degrees higher than that required to interrupt all activity in slowly adapting pulmonary stretch receptors. A vagal reflex that persists when the platform is cooled to 6-7° C is likely to be mediated by afferent C fibers. Because of the limitation on transmission frequency, however, the effect that survives cooling to 6-7 ° C is inevitably less than that obtained when conduction is unimpaired, and thus does not represent the full C fiber contribution to the reflex. We have returned to this method again and again to block conduction in the myelinated fibers that supply mechanoreceptors while allowing at least some of the action potentials from C fibers to traverse the cooled region of the nerve.

It soon became obvious that the myelinated fibers whose rhythmic activity could be recorded from the vagus represented only a small proportion of those carrying information from the thoracic viscera to the central nervous system. The bulk of vagal afferent fibers are small and nonmyelinated. The presence of these C fibers in the nerve filament placed on the recording electrode was revealed by occasional small, inconspicuous action potentials on the oscilloscope at irregular intervals between the rhythmic bursts of mechanoreceptor discharge. The origin of these action potentials was established by Paintal, who became expert in the use of chemicals to stimulate those visceral sensory endings that, unlike the conventional mechanoreceptors, have a sparse and irregular discharge. Phenyldiguanide (or, more properly, as we now know, phenylbiguanide) was the chemical he used most frequently. To guide him to the location of the C fiber endings Paintal used a technique he borrowed from reflex studies aimed at identifying a reflexogenic area: he measured the latency of the C fiber burst in response to injection of phenyldiguanide at different points in the circulation. This technique has proved invaluable in action potential studies, and we have used it repeatedly. Paintal described, in cats, the presence of small fiber endings that responded at short latency to both injection of phenyldiguanide into the pulmonary circulation and to introduction of halothane into the airways. He concluded that the endings were in the lung alveoli and, after some hesitation in the choice of a name, called them J (juxtu-pulmonary capillary) receptors. Paintal also described the dramatic reduction in breathing, heart rate and arterial pressure that results from their stimulation. Dawes and Comroe, in their review of chemoreflexes from the heart and lungs, applied the term 'pulmonary chemoreflex' to this depressor response.

Most of our early work was in dogs, rather than cats, and the pulmonary C fiber endings of dogs do not seem to be susceptible to phenyldiguanide. Thus we were unable to confirm Paintal's results, although we were convinced that his findings must be generally applicable. Cecil Kidd, browsing through the journals on the library shelves, was responsible for our next step forward, by unearthing an account of the observation by Porszasz in Hungary, that capsaicin, the active principal of red pepper, produced short-latency cardio-vascular depressor effects in dogs when injected intravenously. Porszasz generously sent us what turned out to be several years supply of capsaicin powder — a sensory stimulant so powerful that the effects of handling it were startling. The small bottle of powder was swathed in so many layers of protective padding that it came in quite a large box. Between the layers were numerous labels with warnings (in English) that the enclosed substance was of a highly irritant nature — use great caution — do not open further, etc etc! Needless to say the customs officials had opened it down to the last layer and eventually lifted the stopper from the bottle. When we opened the package we didn't stop sneezing for a week. Capsaicin stimulated pulmonary C fibers in dogs, and enabled us to confirm Paintal's results in a different species.

Our studies of these elusive afferent vagal C fibers have continued ever since, with brief detours into other aspects of cardiovascular and pulmonary innervation. For example, before we left London we began a collaboration with Alan Howe to look for a 'pulmonary chemoreceptor'. This pulmonary chemoreceptor had been described by Krahl as a small mass of glomus tissue, the glomus pulmonale, located close to the pulmonary artery and supplied by mixed venous blood. Around this time many respiratory physiologists postulated the existence of a pulmonary chemoreceptor of this type, that would stimulate breathing and explain the close match between alveolar ventilation and the composition of pulmonary arterial blood. Working with Alan was great fun, and even though our search for a mixed venous chemoreceptor proved ultimately fruitless, we at least managed to describe the detailed distribution of the small masses of aortic chemoreceptor tissue that cluster around the origin of the great vessels. By this time we had moved to the United States, and slides made from serial sections of the blocks of tissue Alan prepared continued to follow us across the Atlantic. We had little previous exposure to histology — Alan was the expert — and were amazed to find that the acreage of glass that had to be examined could have built the conservatory in Golden Gate Park. A single explanation for the precise match between alveolar ventilation and metabolically produced $CO_2$ has never been forthcoming. The $CO_2$ sensitivity of pulmonary stretch receptors was thought to make a small contribution, and we examined this with Bob Banzett, who had previously studied $CO_2$ receptors in the bird lung here at Davis, and later with Jerry Green, again at Davis. Others in this room have made more substantial contributions in this area. Jere Mitchell and Marc Kaufman, for example, have over the years described in detail the powerful respiratory drive that originates in receptors in exercising muscles. The mechanisms that link exercise and breathing still continue to fascinate integrative physiologists.

Cecil Kidd visited us briefly in San Francisco some time in the mid-seventies. On the first day he arrived in the laboratory quite late in the afternoon, and although the area was new to him as it was to us, he proceeded in the course of an hour or so to accomplish the difficult approach to recording afferent activity from the thoracic sympathetic chain and rami, and to obtain our first single unit recording. Many of these spinal cardiovascular afferent fibers were found to have multiple receptive fields, a characteristic also found in certain touch units in the skin and elsewhere. David Baker and Tone Nerdum continued the difficult task of recording from these sympathetic afferent fibers, and showed that the receptors were sensitive to the algesic autacoid bradykinin, and were sensitized to bradykinin by prostaglandins. They found a similar interaction of the two autacoids on sympathetic afferent nerve endings was an important mechanism for the production of cardiac pain. The approach to these intrathoracic recordings was daunting indeed, and we have great respect for Malliani and his colleagues in Italy who did much of the basic work in this area. A return to vagal nerve recording was always a relief. This major neural highway is placed so conveniently in the neck, remote from moving parts, as Iggo would agree, between tissue planes that can easily be separated to make an oil-filled pool, with few transversely running blood vessels to get in the way, and with many types of afferent fiber waiting to be explored. Small wonder that most of our work has been on the vagus.

Since we first set out to record afferent activity from the vagus nerves, the contribution of vagal C fibers to the sensory innervation of the thoracic viscera has held a particular attraction for us, and although we have spent a good deal of our time in studies of the afferent and reflex properties of the intrathoracic mechanoreceptors supplied by A fibers, the C fibers have remained our major interest. These nonmyelinated fibers are not by any means uniform in their afferent and reflex functions. Some of those innervating the heart and the great vessels, for example, are mechanoreceptors analogous to the mechanoreceptors supplied by A fibers, but with a higher threshold and lower discharge frequency and sensitivity. Harold Schultz, working in our laboratory, made a detailed study of C fiber baroreceptors in the carotid sinus, and compared their afferent and reflex properties with those of their A fiber counterparts. Most of the mechanoreceptors in the cardiac ventricles are supplied by vagal C fibers, and an increase in their discharge has depressor effects on the heart and circulation. In addition, we found that many cardiovascular C fibers with similar reflex effects supply endings that are primarily chemosensitive. In a collaborative study carried out in Julien Hoffman's laboratory, Jean-Paul Clozel and Tom Pisarri found that when the endings of cardiac vagal C fibers were stimulated by capsaicin, they evoked an active, vagal reflex coronary vasodilation. The function of these endings was also investigated by Marc Kaufman and David Baker, who showed that endings in both the heart and the aorta are stimulated by bradykinin, which is known to be formed and released in these areas. The mechanosensitive C fiber endings, by contrast, showed little, if any, chemosensitivity.

The C fiber innervation of the lungs and airways has always appealed to us as a potentially fruitful area for study. It is responsible for powerful pulmonary defensive reflexes and reflex depressor effects on the cardiovascular system (including vasodilation of both coronary and bronchial circulations). By analogy with C fiber innervation of the skin, it is probably responsible for the sensation of substernal itch and burning pain arising from the airways. (Sensation from the airways is known to be transmitted by the vagus nerves.) As more information became available on the various autacoids that could be released in the airways, we became convinced that at least some of their effects were reflex in nature, and were triggered by C fiber endings in the conducting airways, which, like those in the heart, were primarily chemosensitive. Paintal's J receptors (pulmonary C fibers) were clearly in the lung parenchyma. Paintal was emphatic that they were not in the bronchi, and that they received their blood supply entirely from the pulmonary circulation. A high degree of sensitivity to autacoids was not convincingly proven. We had obtained some preliminary evidence of the existence of a parallel bronchial C fiber innervation, but we searched unsuccessfully for some time for these elusive bronchial C fibers before we could confirm it. Then one day we isolated a chemosensitive C fiber with conspicuous and uniform action potentials, whose receptive field was clearly in a conducting airway. It was summer and our laboratory door was open. Julius Comroe, whose office was just beyond, was passing by. We called him in, but Julius was not impressed by our demonstration and we were unable to get him through the door. "Ah, but what does it *do*!" he said, and this was the million dollar question. (Julius loved reflexes, but in his heart he seemed to have a low opinion of the afferent nerves responsible for them. "Neurophysiology is narrow physiology" was a favorite comment.)

Shortly afterwards Heinz Ginzel came to visit, bringing with him a plentiful supply of prostaglandins, which were not easy to obtain at the time. Andy Roberts, who had previous experience in handling these compounds, examined their effects on the chemosensitive endings in the cardiac ventricles and together with David Baker, we found very low concentrations of prostaglandins of the E series stimulated bronchial C fibers. These prostaglandins are known to be released in the airways in a variety of situations, and aerosols cause cough and irritant airway sensations in humans. We concluded that bronchial C fibers were probably responsible.

Andy, Marc and Harold continued the studies of airway C fibers, and were joined by Anders Jonzon and Jerry Yu. Together they examined the reflex airway effects, in dogs, of attempting to stimulate both pulmonary C fibers and bronchial C fibers using capsaicin and bradykinin. The dog has a convenient right bronchial artery that arises from an intercostal artery. A catheter can be introduced into the latter, so that small volumes of drug solution can be injected directly into the bronchial circulation to the right lung. Two reflex end-points were chosen, contraction of airway smooth muscle and, with the collaboration of Brian Davis, tracheal submucosal gland secretion. Capsaicin had reflex effects whether injected intravenously or into the bronchial circulation. Bradykinin, by contrast, was without effect on the airways when injected intravenously, but injection into the bronchial circulation had powerful reflex effects. Later in collaboration with Ed Schlegle and Jerry Green here at Davis, we carried out a combined afferent and reflex study on the changes in respiratory pattern induced by ozone breathing. Our results showed that bronchial C fibers were primarily responsible for the resulting tachypnea, and that pulmonary C fibers were not significantly affected. This added to our conviction that the sensory endings of the bronchial C fibers supplying the conducting airways and the pulmonary C fibers supplying the lung parenchyma have different characteristics.

Gibbe Parsons, here at Davis, has also welcomed us to his laboratory. Gibbe studies bronchial blood flow in sheep, and was responsible for introducing us to this small but enthralling and highly labile division of the systemic circulation. In return we introduced him to capsaicin, and that cooling the vagus or the administration of atropine almost invariably eliminates this effect. Tom Pisarri adapted Gibbe's method of measuring bronchial blood flow to the right bronchial artery of the dog, and this enabled us to continue our investigations back in San Francisco. By this time bronchial C fibers had received a wider press, and the airway mucosa in which their endings had been described was found to contain neurokinins. It became generally accepted that neurally-induced changes in the airway function were due to local release of neurokinins from the sensory endings i.e. due to an axon-reflex mechanism, independent of the central nervous system, and that human airway disease, for example, asthma was probably of axon-reflex origin. This hypothesis, which obviously deserved to be tested, had major clinical significance, but virtually all the evidence in favor came from studies in rats, and axon reflexes were difficult to demonstrate in other species. In sheep the capsaicin-induced vasodilation seemed to be almost entirely due to an autonomic reflex arc, and the same was generally true of a similar phenomenon in dogs. However, about half the dogs in which C fibers were stimulated by capsaicin displayed a small bronchial vasodilator response that could not be abolished by interrupting the central pathway, and seemed to be attributable to an axon-reflex. Subsequently, using as the stimulus a small volume of water, warmed to body temperature and injected into a bronchus, Tom again found a few dogs in which an axon-reflex contributed to the resulting bronchial vasodilation. Why stimulation of airway C fibers provokes an axon-reflex bronchial vaso-dilation in some animals and not in others is unknown. Airway infections may predispose to this peripheral neural phenomenon. Alternatively a genetic predisposition may be the determining factor. These possibilities remain to be explored.

We cannot end this account of our research interests without acknowledging our indebtedness to the two most important members of our research team that we have not mentioned so far. Since we first set up our laboratory here in the United States we think we have had the best technical assistance available in this country in the person of Albert Dangel. Sadly, we have become convinced that his kind are dying out. Albert has been joined in the last ten years by Ron Brown. Without these two our work would not have been possible, and their good company has contributed to the well-being of all lab members.

The time has now come for us to retire, and we look back on what sometimes seems to us to have been a charmed life in physiology. Our generation of physiologists has been fortunate to work at a time when federal funding for biological studies was generous and other sources of financial support were plentiful. We have seen many changes in our professional lifetime, and perhaps, in conclusion, we may be permitted a few general comments on them. When we started out in the fifties departments were small, and the space they occupied was correspondingly limited. Even the most eminent physiologists of the day had little laboratory space compared with the suites of rooms given over to principal investigators and their associates in present day institutions. The scientists themselves spent many of their working hours in their laboratories. Whereas four or five publications a year might be regarded by principal investigators as a respectable and appropriate contribution to the literature, a present day output of forty or fifty papers would not be thought unusual. Of course this can only be done by increasing the size of laboratories and the number of investigators who staff them. Heads of laboratories may now have little time to spend at the bench, and have been known to add their names to published work even though they never set foot in the laboratory while the work was being carried out. The addition of their names certainly adds the weight of authority to those of junior authors, but this is not always justified. In any event we all publish too much for reasons that often have more to do with personal prestige and with the continuation of grant funding than with the dissemination of information.

The most important change in our professional lifetimes has been the application of the new techniques and approaches of molecular biology. One cannot deny that in recent years the great advances in physiology and in the understanding and treatment of disease stem from this discipline. Does it follow, therefore, as some suggest, that the useful life of integrative, systems physiology is drawing to a close? We think not. Not all biological problems will yield to the reductionist approaches of the cellular and molecular biologist. To take a simple analogy, no amount of study of the minute structure of the cogs, wheels, handlebars and brakes will tell us how a bicycle works, because it works according to operational principles that can only be learned from the intact machine. This is true also of living organisms. In our day integrative, systems physiology was taught as an essential basis for medical studies, and the rudiments of different diagnosis were presented during our physiology course. It was easy for us to recognize the necessity of understanding how different parts of mammalian systems fit together to form a functional whole. Medical students and graduate students in physiology are now presented with so many courses whose content they must master that the fundamental importance of integrative physiology is no longer readily apparent. We must not be too hasty in celebrating its demise, however. Many of the problems remaining to be explored concern operational principles within its realm. Some time ago we received a telephone call seeking advice from a young scientist who was testing the effect of a new drug on airway smooth muscle contraction in dogs. To obtain a baseline level of contraction, how better than to stimulate the peripheral vagus nerve electrically? He found that when he stimulated the vagus, however, the heart stopped and he had to abandon the tests. The likelihood that the heart would stop when the vagus was stimulated seemed not to have occurred to him when he was designing his experiments. He had overlooked the fact that the lungs, airways, heart and circulation are functionally connected, sharing a common motor innervation and common neurotransmitters.

In these days integrative physiology is finding fewer and fewer practitioners, and the state of physiology in general is like that of many of our cities, expanding at the periphery but leaving a neglected and decaying center. It seems possible that little will be remembered of the body of classical physiology by the time that the last of the people gathered in this room are dead. As physiologists we must find some way of keeping our established body of knowledge alive, while continuing to push out the frontiers. One remembers the slow, painful steps by which physiologists advanced during the nineteenth and twentieth centuries struggled to establish, for example, the role of the vagus nerve in autonomic function. What is happening to this established body of knowledge, and to our understanding of other aspects of integrative function?

Biologists once classified themselves by subject; they do so now by the analytical level at which they work. Thus we have molecular biologists, cellular biologists and biologists like most of the people here, who work on some organ or system at the level of the integrated whole. In the prevailing climate of reductionism the first of these disciplines has become the most prestigious, and encourages the notion that there is no biological problem that cannot be answered using the techniques of molecular biology. Some of its more arrogant practitioners appear to believe that integrative physiology is an inferior discipline that is now outmoded, and has no further contribution to make to biological science. In reality it is becoming increasingly difficult in mammalian biology for a single person using a single technique to make major advances. Collaboration between people expert in different methodologies is often required — with the consequence that integration must now occur between people, not in the head of a single individual. But in the process something valuable often slips through the cracks. We believe that in this collaboration the generalist, the integrative physiologist, is capable of bringing a special viewpoint to biological problems that is too valuable to be lost.

As science advances the growth of organized factual knowledge represents a burden that becomes well-nigh insupportable. To avoid being crushed by this burden it seems one must take refuge in specialization. The inevitable consequence of specialization is that scientists even with the single discipline of biology become less able to communicate with each other, particularly since each branch and sub-branch must develop its own methodology and compounds the problem by continually adding to its own esoteric vocabulary. Specialist journals multiply, and the literature in general proliferates exponentially. Peter Medawar expressed the optimistic view that the factual burden of science varies inversely with its degree of maturity, and that as science advances individual facts can be incorporated into general statements of increasing explanatory power. He suggested that as the factual basis of any discipline increases, the textbooks should become smaller, not larger. Such hopeful sentiments are of little use to biologists nor, we suspect, to the generality of scientists. Even those molecular biologists who claim that their specialty can answer all the truly fundamental problems can hardly be this optimistic. Thus although molecular machinery is in general the same for all cells, fundamental functions such as ligand binding and gene expression are specific to cells of individual organs and tissues, and are adapted for the function of these organs and tissues. No general laws of molecular biology can encompass these specific functions. As far as we can see, though much can be stored in and manipulated by computers, a heavy load of factual knowledge will continue to burden future generations of scientists. Facts are what science is based on, and it's probably unwise to shelve the facts even when teaching general principles.

In conclusion, scientific enquiry is a great game. We don't always play it for the prizes, but for the love of the game itself. Let's face it, our taste for enquiry is expensive, and we should all be grateful to a society that accepts scientific pursuits as a whole as a proper way for many of its members to spend their waking hours. The reason can only be that the primitive quality of curiosity is powerful in all human beings, and we're deputed to embody that curiosity. It's primitive curiosity more often than not that drives us and keeps our efforts on the straight and narrow path, when the ego, more greedy for the world's acclaim, might deflect them. As scientists we can never be certain of the truth of how things work, because however convinced we may be of the rigor with which we have tested our favorite hypotheses, the subjective element remains, and some of our theories will crumble to dust. The only true test is the test of time. The true test of the validity of our work is that those coming after can build upon it.

# SHORT COMMUNICATIONS

# 19

# MECHANISMS UNDERLYING SPINAL NEUROMODULATION OF THE EXERCISE PRESSOR REFLEX

Charles L. Stebbins[*]

Department of Internal Medicine
Division of Cardiovascular Medicine and
Department of Human Physiology
University of California, Davis
Davis, California 95616

## INTRODUCTION

Spinal neurotransmitters or neuromodulators can affect the afferent arm of the reflex cardiovascular response to muscle contraction (Hill and Kaufman, 1990; Hill and Kaufman, 1991) [i.e. the exercise pressor reflex]. Potentially important neuromodulators may be vasopressin, oxytocin and prostaglandin $E_2$ ($PGE_2$). Both oxytocin and vasopressin have been located in the spinal dorsal horn where primary sensory afferent input is received and processed (Milan et al, 1984). The origin of this oxytocin and vasopressin is cell bodies of neurosecretory neurons located in the paraventricular nucleus of the hypothalamus (Nilaver et al, 1980). These neurons project to the substantia gelatinosa (lamina II) of the dorsal horn (Nilaver et al, 1980). This area is known for processing information from group III and IV afferent nerves associated with nociception (Jessel and Jahr, 1985). A portion of these sensory fibers may also serve as ergoreceptors that evoke the exercise pressor reflex (McCloskey and Mitchell, 1972).

Lumbar spinal vasopressin can modulate nociceptive responses that are mediated by group IV afferent nerves and this also may be true for oxytocin because these neurohormones have some properties in common (Jessel and Jahr, 1985; Richard et al, 1990; Thurston et al, 1992).

Prostaglandins are distributed in virtually all mammalian tissue and significant release of $PGD_2$ and $PGE_2$ from spinal cord astrocytes has been demonstrated (Marriott et al, 1990). Furthermore, $PGE_2$ enhances nociceptive responses (Taiwo and Levine, 1988) and

---

[*] Address for correspondence: Charles L. Stebbins, Division of Cardiovascular Medicine TB 172, University of California, Davis, Davis, CA 95616: Phone (916) 752-0717, FAX (916) 752-3264

*Control of the Cardiovascular and Respiratory Systems in Health and Disease*
Edited by C. T. Kappagoda and M. P. Kaufman, Plenum Press, New York, 1995

209

this apparent sensitization of nociceptors also may apply to ergoreceptors that evoke the exercise pressor reflex.

Based on these observations, our laboratory in a series of three separate studies, has tested the hypotheses that spinal vasopressin (Stebbins *et al*, 1991) and oxytocin (Stebbins and Ortiz-Acevedo, 1994) attenuate and $PGE_2$ augments the magnitude of the reflex pressor and heart rate responses to muscular contraction via their action in the lumbar spinal cord.

## METHODS AND RESULTS

Anesthetized cats ($\alpha$-chloralose, 80-100 mg/kg, iv) of either sex were studied. Increases in mean arterial pressure (MAP) and heart rate (HR) were monitored in response to 30-s of static hindlimb contraction before and after lumbar intrathecal injection ($L_5$; 200 $\mu$l) of: 1) vasopressin (100 pmol in 9 cats ) or the selective vasopressin $V_1$ receptor antagonist d($CH_2)_5$,Tyr(Me))AVP (8-9 nmol in 6 cats) (Stebbins *et al*, 1992); 2) 300 pmol of oxytocin (n=6) or 300 nmol of the selective oxytocin antagonist d($CH_2)_5^1$,O-Me-$Tyr^2$,$Thr^4$,$Tyr^9$,$Orn^8$-vasotocin (n=7)(Stebbins and Ortiz-Acevedo, 1994); and 3) $PGE_2$ (500 ng, n=5) or the cyclooxygenase inhibitor indomethacin (1 mg, n=3). Static contraction was evoked by electrical stimulation of the sciatic nerve using stimulation parameters (<40 Hz, .025 ms, $\leq$2 times motor threshold) that do not activate group III and IV afferents (Rybicki and Kaufman, 1985).

At the end of each experiment, intrathecal injection of Evans blue dye (2%, 200 $\mu$l) was injected intrathecally into the lumbar spinal cord in order to estimate the actual region of the spinal cord to which the peptides, prostaglandins or antagonists had migrated. If the dye migrated rostral to vertebra $L_1$ then that cat was excluded from analysis.

Lumbar intrathecal injection of either vasopressin or oxytocin attenuated the pressor response to static contraction but caused no changes in the heart rate response (Table 1). On the other hand, inhibition of either vasopressin $V_1$ or oxytocin receptors augmented the contraction-induced pressor response in spite of heart rate responses that were unchanged. (Table 2). Conversely, neither intrathecal injection of $PGE_2$ nor indomethacin had any effect on the exercise pressor reflex (Tables 1 and 2). Intrathecal injection of the vehicle for vasopressin, oxytocin or their respective antagonists had no effect on the blood pressure or heart rate responses to contraction.

## DISCUSSION

Results of these studies provide evidence that centrally released vasopressin and oxytocin are capable of modulating the pressor response to static contraction. Intrathecal injection of either peptide attenuated this response while inhibition of vasopressin $V_1$ or oxytocin receptors augmented contraction-induced increases in blood pressure. Although action of other neurotransmitters/neuromodulators [i.e., opioid peptides (Hill and Kaufman, 1990) and serotonin (Hill and Kaufman, 1991)] in the lumbar spinal cord can diminish the magnitude of the blood pressure response to contraction, inhibition of their receptors does not enhance this response. Consequently, during muscle contraction, vasopressin and oxytocin may share a unique ability to tonically inhibit muscle afferent fibers that induce the exercise pressor reflex.

It is not clear why vasopressin and oxytocin had no effect on the heart rate response to contraction. One possible explanation is that the contraction-induced heart rate responses were small, which minimized the effects of vasopressin or oxytocin on this variable.

**Table 1.** Changes in blood pressure (MAP), myocardial contractility (dP/dt), heart rate (HR) and muscle tension to static muscle contraction before and after intrathecal injection of either vasopressin (AVP), oxytocin (OXT) or PGE$_2$ into the lumbar spinal cord

|  | MAP (mmHg) | HR (beats/min) | Tension (kg) |
|---|---|---|---|
| Initial contraction† | 29±3 | 9±3 | 4.0±0.4 |
| Contraction + 100 pmol AVP† | 14±3* | 7±2 | 3.9±0.3 |
| Initial contraction‡ | 42±9 | 9±2 | 2.5±0.6 |
| Contraction + 300 pmol OXT‡ | 28±10* | 7±1 | 2.5±0.6 |
| Initial contraction | 24±4 | 9±2 | 3.4±0.7 |
| Contraction + 500 ng PGE$_2$ | 26±5 | 6±3 | 3.3±0.7 |

Values are means ± S.E.M. *$P<0.05$, initial contraction vs contraction + blockade. Baseline values were not affected by any drug intervention. (†Adapted from Stebbins et al., 1992; ‡Adapted from Stebbins and Ortiz-Acevedo, 1994)

While both vasopressin and oxytocin have similar effects on the exercise pressor reflex, it appears that vasopressin is the more powerful neuromodulator of the two peptides. Evidence for this contention is the fact that the 100 pmol dose of vasopressin attenuated the contraction-induced increase in blood pressure by 32% and a dose of 300 pmol oxytocin was required to attenuate the same response by only 28%.

As previously mentioned, vasopressin and oxytocin probably expressed their effects in the dorsal horn. Both of these peptides can affect spontaneous activity of dorsal horn cells. For example, it has been demonstrated that oxytocin can inhibit 67% and vasopressin 74% of cells responding to these peptides (Tamarova, 1985). Consequently, dorsal horn modulation of group III and IV afferent nerve transmission by vasopressin and oxytocin may be similar during muscle contraction.

Dorsal horn modulation of sensory nerve activity from contracting muscle may be due to neural input from the ventrolateral medulla into the paraventricular nucleus where the cell bodies of vasopressin- and oxytocin-containing neurons that innervate the dorsal horn

**Table 2.** Changes in blood pressure (MAP), myocardial contractility (dP/dt), heart rate (HR) and muscle tension (Tension) to static muscle contraction before and after either vasopressin (AVP) V$_1$ receptor, oxytocin (OXT) receptor or cyclooxygenase blockade in the lumbar spinal cord

|  | MAP (mmHg) | HR (beats/min) | Tension (kg) |
|---|---|---|---|
| Initial contraction† | 25±6 | 9±3 | 3.9±0.3 |
| Contraction + AVP V$_1$ receptor blockade† | 34±6* | 9±2 | 4.2±0.4 |
| Initial contraction‡ | 38±11 | 11±2 | 2.2±0.4 |
| Contraction + OXT receptor blockade‡ | 45±11* | 14±11 | 2.3±0.4 |
| Initial contraction | 55±15 | 18±6 | 3.4±1.1 |
| Contraction + cyclooxygenase blockade | 55±8 | 19±4 | 3.4±1.2 |

Values are means ± S.E.M. *$P<0.05$, initial contraction vs contraction + blockade. Baseline values were not affected by any drug intervention. (†Adapted from Stebbins et al, 1992; ‡Adapted from Stebbins and Ortiz-Acevedo, 1994)

are located (Nilaver *et al*, 1980). Thus, contraction-induced neural input into the paraventricular nucleus could alter the activity of this structure such that vasopressin and oxytocin are released into the dorsal horn.

Surprisingly, neither $PGE_2$ nor the cyclooxygenase inhibitor indomethacin had any effect on blood pressure or heart rate during static contraction. We do not attribute this lack of effect to insufficient amounts of exogenous $PGE_2$ or indomethacin injected into the dorsal horn because large doses of both chemicals were used. For example, when normalized for body weight, the dose of indomethacin used in the present study was equivalent to the highest dose used in a study of rats where effects of prostaglandins on nociceptive responses were seen (Taiwo and Levine, 1988). Thus, these data suggest that spinal prostaglandins do not modulate sensory nerve transmission from contracting muscle in the cat.

In summary, intrathecal injection of vasopressin or oxytocin into the lumbar spinal cord reduced the magnitude of the pressor response to static muscle contraction. Alternatively, injection of an vasopressin $V_1$ or oxytocin receptor antagonist into this region of the spinal cord enhanced this contraction-evoked response. Similar injection of $PGE_2$ or indomethacin, however, had no effect on the cardiovascular response to contraction. These data suggest that vasopressin and oxytocin act as neuromodulators in the lumbar spinal cord that can attenuate the afferent arm of the exercise pressor reflex.

## ACKNOWLEDGEMENTS

This work was supported, in part, by an American Heart Association, California Affiliate Grant-In-Aid, #91-123A.

## REFERENCES

Hill, J.M., and Kaufman, M.P., 1990, Attenuation of reflex pressor and ventilatory responses to muscular contraction by intrathecal opioids, *J. Appl. Physiol.*, 68:2466-2472.

Hill, J.M., and Kaufman, M.P., 1991, Intrathecal serotonin attenuates the pressor response to static contraction, *Brain Res.*, 550:157-160.

Jessel, T.M., and Jahr, C.E., 1985, Fast and slow excitatory transmitters at primary afferent synapses in the dorsal horn of the spinal cord, *Adv. Pain Res. Ther.*, 9:31-39.

Marriott, D., Wilkin, G.P., Coote, P.R., and Wood, J.N., 1990, Eicosanoid synthesis by spinal cord astrocytes is evoked by substance P; possible implications for nociception and pain *in: Advances in Prostaglandin Thromboxane and Leukotriene Research*, Vol. 21, B. Sameulsson et al, eds., Raven Press, New York, 739-741.

McCloskey, D.I., and Mitchell J.H., 1972, Reflex cardiovascular and respiratory responses originating in exercising muscle. *J. Physiol. Lond.* 224:173-186.

Millan, M.J., Millan, M.H., Czlonkowski, A., and Herz, A., 1984, Vasopressin and oxytocin in the rat spinal cord: distribution and origins in comparison to [Met] Enkephalin, Dynorphyn and related opioids and their irresponsiveness to stimulus modulating neurohypophyseal secretion. *Neuroscience*, 13:179-187.

Nilaver, G., Zimmerman, E.A., Wilkins, J., Michaels, J., Hoffman, D., and Silverman, A.J., 1980, Magnocellular hypothalamic projections to the lower brain stem and spinal cord of the rat, *Neuroendocrinology*, 30:150-158.

Richard, P., Moos, F., and Freund-Mercier, M.-J., 1990, Central effects of oxytocin, *Physiol. Rev.* 71:331-370.

Rybicki, K.J., and Kaufman, M.P., 1985, Stimulation of group III and group IV afferents reflexly decreases total pulmonary resistance in dogs, *Respir. Physiol.*, 59:185-189.

Stebbins, C.L., Ortiz-Acevedo, A., and Hill, J.M., 1992, Spinal vasopressin modulates the reflex cardiovascular response to static contraction, *J. Appl. Physiol.*, 72:731-738.

Stebbins, C.L., and Ortiz-Acevedo, A., 1994, The exercise pressor reflex is attenuated by intrathecal oxytocin, *Am. J. Physiol.* 267 (*Regulatory, Integrative Comp. Physiol.* 36), R909-R915.

Taiwo, Y.O., and Levine, J.D., 1988, Prostaglandins inhibit endogenous pain control mechanisms by blocking transmission at spinal noradrenergic synapses, *J. Neurosci.*, 8:1346-1349.

Tamarova, Z.A., 1985, Effect of vasopressin and oxytocin on spontaneous unit activity of dorsal horn cells in the isolated spinal cord of young rats, *Neurophysiology*, 17:221-225.

Thurston, C.L., Campbell, I.G., Culhane, E.S., Carstens, E., and Watkins, L.R., 1992, Characterization of intrathecal vasopressin-induced antinociception, scratching behavior, and motor suppression, *Peptides*, 13:17-25.

Yetunde, O., and Levine, J.D., 1988, Prostaglandins inhibit endogenous pain control mechanisms by blocking transmission at spinal noradrenergic synapses, *J. Neurosci.*, 8:1346-1349.

# 20

# THE ROLE OF VASOPRESSIN AND ANGIOTENSIN II IN THE HEMODYNAMIC RESPONSE TO DYNAMIC EXERCISE

J. David Symons and Charles L. Stebbins *

Department of Internal Medicine
Division of Cardiovascular Medicine and
Department of Human Physiology
University of California, Davis
Davis, California 95616

## INTRODUCTION

The cardiovascular response to dynamic exercise is characterized by increases in blood pressure, myocardial contractility, and heart rate (Mitchell, 1990). In addition, cardiac output is redistributed such that blood flow is reduced in the renal and splanchnic circulations, while it is greatly increased to the heart and contracting skeletal muscles (Armstrong et al, 1987). Although central neural mechanisms (i.e. central command) and reflexes originating in working skeletal muscles are largely responsible for these alterations (Mitchell, 1990), hormones released during physical activity also may contribute to or modulate the cardiovascular response to dynamic exercise.

Vasopressin and angiotensin II (ANG II) may modulate responses to dynamic exercise since both peptides: 1) increase in the systemic circulation in an intensity-dependent manner (Stebbins et al, 1994); 2) raise mean arterial blood pressure (Cowley and Liard, 1987); and 3) elevate vascular resistance and reduce blood flow to splanchnic, renal (Liard et al, 1982), and myocardial circulations (Symons et al, 1993). Therefore, the possibility exists that vasopressin and ANG II contribute to the general sympathetic activation that occurs during physical activity. We hypothesized that elevated plasma concentrations of vasopressin and ANG II released during exercise: 1) augment increases in arterial pressure; 2) limit elevations in cardiac output and myocardial blood flow; and 3) attenuate reductions in systemic vascular resistance.

---

* Address for correspondence: J. David Symons Ph.D., Alliance Pharmaceutical Corporation, Department of Cardiovascular Pharmacology, San Diego, California 92121; PH: (619) 558-5189, FX: (619) 558-4333.

*Control of the Cardiovascular and Respiratory Systems in Health and Disease*
Edited by C. T. Kappagoda and M. P. Kaufman, Plenum Press, New York, 1995

## METHODS

Miniswine were used to examine the contribution of vasopressin and ANG II to the hemodynamic responses to treadmill running. The methods have been described previously (Symons et al, 1993; Stebbins et al, 1994; Stebbins and Symons, 1993, 1995).

## EXPERIMENTAL PROTOCOLS

At least three weeks post-surgery the relative contributions of vasopressin and ANG II to the hemodynamic responses to dynamic exercise were determined using selective and non-selective receptor antagonists to these peptides. Exercise during unblocked conditions consisted of treadmill running at 80% heart rate reserve (HRR) for 20 min. HRR was calculated by the formula: $[(HR_{max} - HR_{rest}) \times 80\%] + HR_{rest}$. Absolute workloads performed during experimental interventions were identical to those completed during each respective unblocked condition.

### Exercise and Vasopressin Receptor Antagonism (N=17)

Each animal completed three protocols. In the first, hemodynamic variables were assessed while the pig stood quietly on the treadmill, and again at 20 min of dynamic exercise at 80% HRR. Second, the protocol was repeated in the presence of vasopressin $V_1$-receptor inhibition using $[d(CH_2)_5Tyr(Me)]AVP$ (10-14 µg/kg ia). Third, an identical protocol was performed in the presence of $V_1 + V_2$ receptor inhibition ($V_1$ antagonist, as above; $V_2$ antagonist, $[d-(CH_2)_5[D-Ile^2,Ile^4,Ala-NH_2]^9AVP$; 30-40 µg/kg).

### Exercise and Angiotensin II AT₁ Receptor Antagonism (N=11)

Each animal completed two protocols. The first was identical to the unblocked condition described earlier. Second, the same workloads were performed in the presence of ANG II $AT_1$ receptor blockade using losartan (15-20 mg/kg, ia). In each study: 1) the appropriate vehicle control was given during the unblocked condition; 2) the efficacy of receptor blockade was confirmed; and 3) adequate mixing of the radioactive microspheres was determined.

## RESULTS

For both investigations, pH, $pO_2$, osmolality, hematocrit, hemoglobin, rectal temperature, and arterial blood lactate were not different during exercise in the absence or presence of receptor blockade.

### Exercise and Vasopressin Receptor Antagonism

At rest, hemodynamic variables in the absence and presence of vasopressin $V_1$ receptor inhibition were similar. During exercise, however, increases in arterial pressure were attenuated while transmural myocardial blood flow was unchanged in the presence vs the absence of $V_1$ receptor blockade (Fig 1). In addition, exercise-induced increases in cardiac output were augmented (13.4±0.6 vs 12.5±0.8 L/min), while systemic vascular resistance was similar during blocked vs unblocked conditions, respectively. No differences

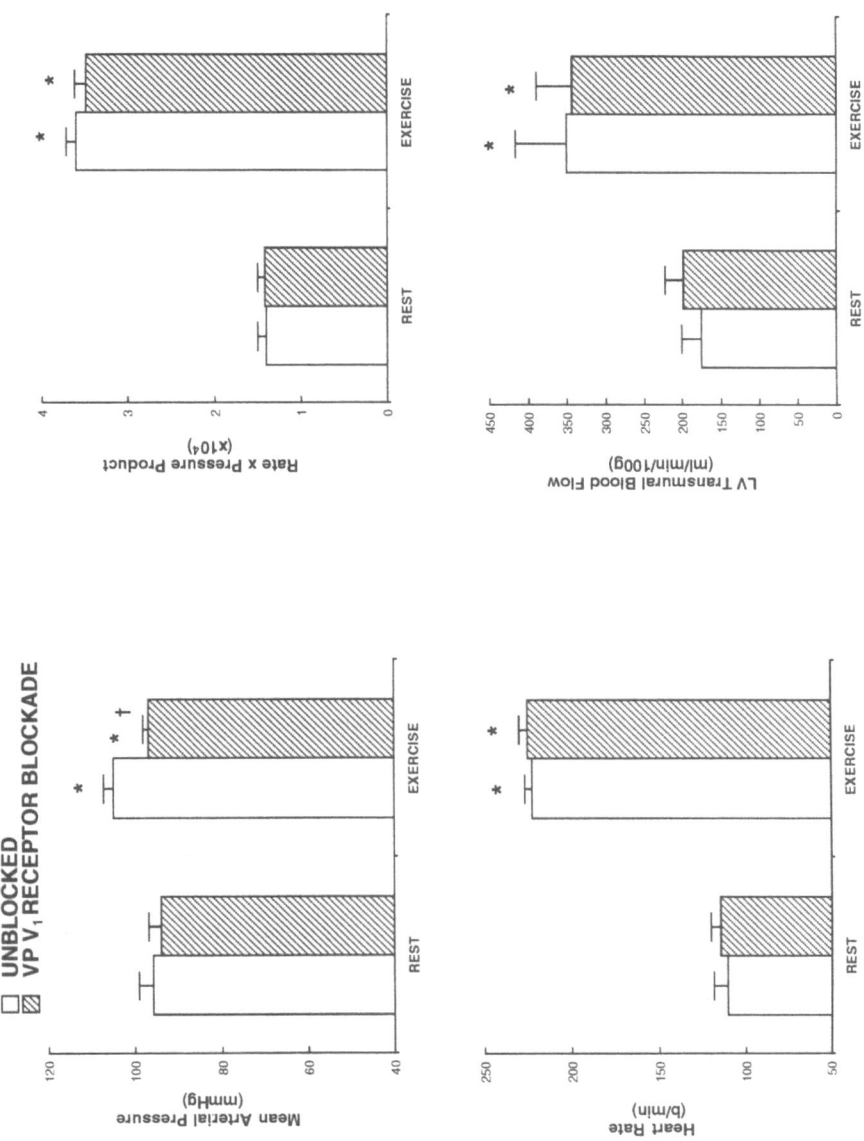

**Figure 1.** Hemodynamic responses at rest and during dynamic exercise in the absence (Unblocked) and presence of Vasopressin (VP) $V_1$ Receptor Blockade. Rate x pressure product, heart rate x systolic blood pressure product; LV, left ventricular. * $P<0.05$, Rest vs. Exercise; † $P<0.05$, Unblocked vs. $V_1$ Receptor Blockade, n=17.

occurred at rest or during exercise between selective $V_1$ or combined $V_1 + V_2$ receptor blockade.

## Exercise and Ang II AT₁ Receptor Antagonism

While ANG II AT₁ receptor antagonism did not alter resting hemodynamic variables, exercise-induced increases in mean arterial pressure and the rate x pressure product were attenuated (Fig 2). In addition, systemic vascular resistance decreased (6.7±0.3 vs 7.4±0.3 mmHg/L/min), cardiac output was similar, and transmural myocardial blood flow increased during exercise in the presence vs the absence of ANG II AT₁ receptor blockade, respectively (Fig 2).

## DISCUSSION

We confirmed our hypotheses that vasopressin and ANG II play a role in the hemodynamic response to prolonged, dynamic exercise in miniswine. Although both peptides contributed to exercise-induced increases in mean arterial pressure, the extent that ANG II and vasopressin modulated cardiac output, systemic vascular resistance, and myocardial blood flow during treadmill running was quite different. Specifically, vasopressin $V_1$ receptor blockade augmented exercise-induced increases in cardiac output, while systemic vascular resistance and myocardial blood flow were not different from unblocked exercise. In contrast, myocardial blood flow was elevated, systemic vascular resistance reduced, and cardiac output unchanged during exercise in the presence of ANG II AT₁ receptor blockade compared to unblocked conditions. Because the hemodynamic response to treadmill running during $V_1+V_2$ receptor blockade was similar to $V_1$ receptor blockade alone, we conclude that $V_2$ receptors have a minor, if any, role in the cardiovascular response to dynamic exercise in our model.

ANG II AT₁ receptors predominate in vascular tissue and are responsible for virtually all of the cardiovascular effects of this peptide (Herblin et al, 1991). Previous studies investigating the role of ANG II have used angiotensin converting enzyme (ACE) inhibitors or saralasin to prevent ANG II formation or block ANG II AT₁ + AT₂ receptors, respectively (Fagard et al, 1977, 1978, 1982). However, results from studies using ACE inhibitors may have overestimated the contribution of this peptide since they potentiate the release and/or production of vasodilators such as bradykinin and/or prostaglandins (e.g. $PGI_2$, $PGE_2$) (Swartz and Williams, 1982; Grafe et al, 1993). In contrast, the role of ANG II may have been underestimated in studies using saralasin since this peptide receptor antagonist has partial agonistic properties (Wong et al, 1990). We chose losartan because it is devoid of the confounding effects associated with both ACE inhibition and saralasin (Smith et al, 1992). Since losartan was capable of increasing myocardial blood flow and reducing systemic vascular resistance during exercise, it is likely that ANG II AT₁ receptors are primarily responsible for mediating vasoconstriction in miniswine.

ANG II receptor blockade attenuates exercise-induced increases in plasma norepinephrine (Fagard, 1977; Stebbins and Symons, 1995). This may be due to ANG II's capability to induce release of norepinephrine from postganglionic sympathetic neurons and the adrenal medulla, or inhibit its reuptake by sympathetic nerve terminals (MacLean and Ungar, 1986; Szabo et al, 1990). Consequently, reduced systemic vascular resistance observed during exercise + losartan may be due to lower concentrations of circulating norepinephrine rather than to blockade of ANG II AT₁ receptors. Thus, one mechanism responsible for the regional vasoconstrictor effects of ANG II during exercise is its facilitation of peripheral sympathetic nerve activity.

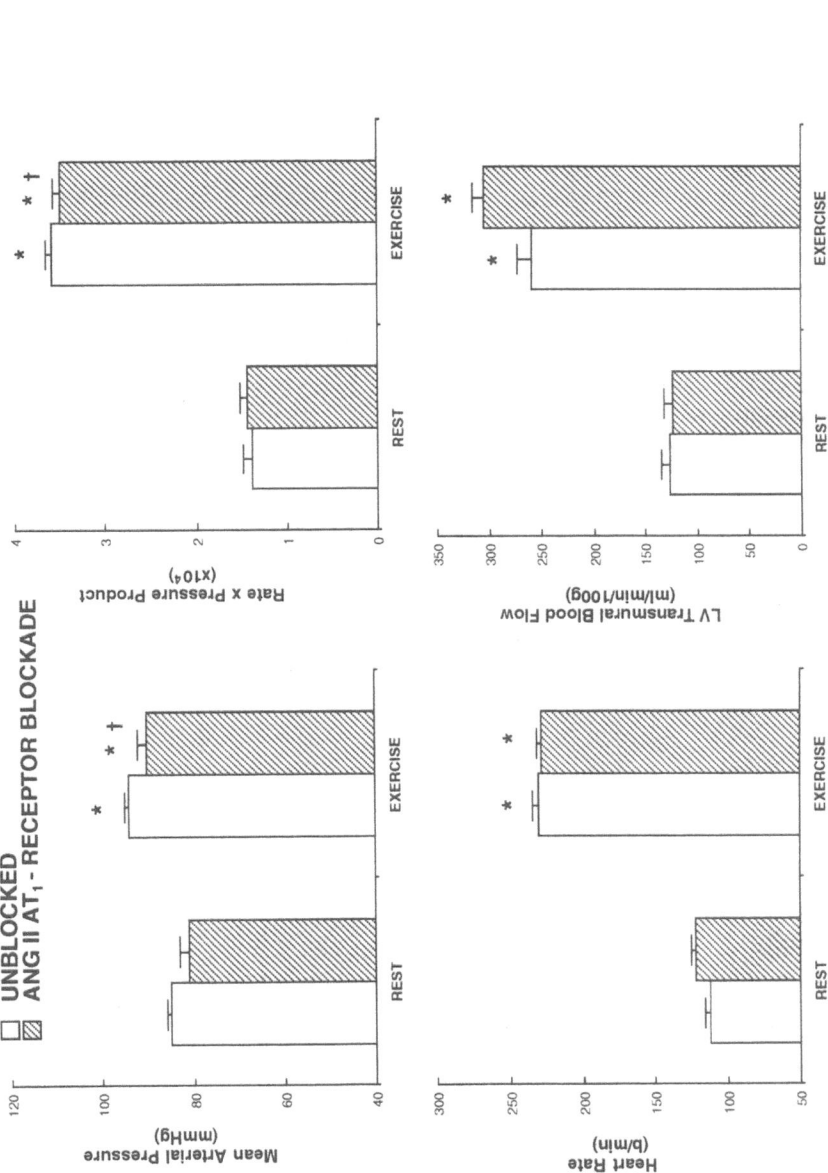

**Figure 2.** Hemodynamic responses at rest and during dynamic exercise in the absence (Unblocked) and presence of Angiotensin II (ANG II) AT₁ -Receptor Blockade. Rate x pressure product, heart rate x systolic blood pressure product; LV, left ventricular. * P<0.05, Rest vs. Exercise; † P<0.05, Unblocked vs. ANG II AT₁ -Receptor Blockade, n=11.

Our results indicate that vasopressin does not influence myocardial blood flow during exercise. In contrast, increases in myocardial blood flow and decreases in coronary resistance during exercise + ANG II $AT_1$ receptor blockade occurred even though myocardial oxygen demand was reduced compared to unblocked conditions. This effect of ANG II is similar to the limitation of coronary blood flow imposed during exercise by sympathetic $\alpha$-adrenoreceptor mediated vasoconstriction (Mohrman and Feigl, 1978). Limiting increases in coronary blood flow during periods of enhanced myocardial oxygen demand may serve an important physiological function (Feigl, 1987). For example, during exercise the endocardial to epicardial blood flow (endo/epi) ratio decreases from rest because of a disproportionate increase in blood flow to the epicardium (Ball et al, 1975). This phenomenon may facilitate blood flow to the endocardium so that reductions in the endo/epi ratio are minimized as exercise intensity increases (Huang and Feigl, 1988).

The physiological roles of vasopressin and ANG II in the cardiovascular response to dynamic exercise appear to be their contribution to elevating arterial blood pressure through their ability to redistribute cardiac output by regulating regional vascular resistances. Although the data is not reported here, both peptides contribute to exercise-provoked increases in splanchnic vascular resistance which, in turn, plays an essential role in elevating perfusion pressure during dynamic physical activity (Rowell, 1986). Since ANG II also was capable of modulating blood flow and resistance in the renal and myocardial circulations, our results suggest that ANG II makes a greater overall contribution to the cardiovascular response to dynamic exercise than vasopressin.

## ACKNOWLEDGEMENTS

These studies were supported by Tobacco-Related Disease Research Program Grant KT-57 and American Heart Association, California Affiliate Grant-In-Aid 91-123A.

## REFERENCES

Mitchell, J.H. 1990, Neural control of the circulation during exercise. *Med. Sci. Sports Exercise*. 22:141-154.
Armstrong, R.B., Delp, M.D., Goljan E.F., and Laughlin, M.H. 1987, Distribution of blood flow in muscles of miniature swine during exercise. *J. Appl. Physiol*. 62:1285-1298.
Stebbins, C.L., Symons, J.D., McKirnan, M.D., and Hwang, F.W. 1994, Factors associated with vasopressin release in exercising swine. *Am. J. Physiol*. 266 (35):R118-R124.
Cowley, A.W. Jr. and Liard J.F. 1987, *Cardiovascular actions of vasopressin*. In D.M. Gash and G.J. Boer (eds) Vasopressin: Principles and Properties, Plenum Press, New York, NY, 389-433.
Liard, J.F., Deriaz, O., Schelling P., and Thibonnier P. 1982, Cardiac output distribution during vasopressin infusion or dehydration in conscious dogs. *Am. J. Physiol*. 243 (12):H663-H669.
Symons, J.D., Longhurst, J.C., and Stebbins C.L. 1993, Response of collateral-dependent myocardium to vasopressin release during exercise. *Am. J. Physiol*. 264:H1644-H1652.
Stebbins, C.L. and Symons, J.D. 1993, Vasopressin contributes to the cardiovascular response to dynamic exercise. *Am. J. Physiol*. 264 (33):H1701-H1707.
Stebbins, C.L. and Symons, J.D. 1995, Role of angiotensin II in the hemodynamic response to dynamic exercise in the miniswine. *J. Appl. Physiol*. 78 (1):185-190.
Herblin, W.F., Chiu, A.T., McCall, D.E., Ardecky, R.J., Carini, D.J. 1991, Angiotensin II receptor heterogeneity. *Am. J. Hypertens*. 4 (4):299S-302S.
Fagard, R., Amery, A., Reybrouck, T., Lijnen, P., Moerman, E., Bogaert, M., and De Schaepdryver, A. 1977, Effects of angiotensin antagonism on hemodynamics, renin, and catecholamines during exercise. *J. Appl. Physiol*. 43:440-444.
Fagard, R., Amery, A., Reybrouck, T., Lijnen, P., Billiet, L., Bogaert, M., Moerman, E., and de Schaepdryver A. 1978, Effects of angiotensin antagonism at rest and during exercise in sodium-deplete man. *J. Appl. Physiol*. 45 (3):403-407.

Fagard, R., Lijnen, P., Vanhees, L., Amery, A. 1982, Hemodynamic response to converting enzyme inhibition at rest and exercise in humans. *J. Appl. Physiol.* 53 (3):576-581.

Swartz, S.L. and Williams, G.H. 1982, Angiotensin-converting enzyme inhibition and prostaglandins. *Am. J. Cardiol.* 49:1405-1409.

Grafe, M., Bossaller, C., Graf, K., Auch-Wchwelk, W., Baukmgarten, C.R., Hildebrandt, A., and Fleck, E. 1993, Effect of angiotensin-converting enzyme inhibition on bradykinin metabolism by vascular endothelial cells. *Am. J. Physiol.* 33:H1493-H1497.

Wong, P.C., Price, W.A., Chiu, A.T., Duncia, J.V., Carini, D.J., Wexler, R.R., Johnson, A.L., and Timmermans, P.B. 1990, Nonpeptide angiotensin II receptor antagonists. VIII. Characterization of functional antagonism displayed by DuP 753, an orally active antihypertensive agent. *J. Pharmacol. Exp. Ther.* 252:719-725.

Smith, R.D., Chiu, A.T., Wong, P.C., Herblin, W.F., and Timmermans, P.B.M.W.M. 1992, Pharmacology of nonpeptide angiotensin II receptor antagonists. *Annu. Rev. Pharmacol. Toxicol.* 32: 135-165.

MacLean, M.R. and Ungar A. 1986, Effects of the renin-angiotensin system on the reflex response of the adrenal medulla to hypotension in the dog. *J. Physiol. (London).* 373:343-352.

Szabo, B.L., Hedler, L., Schurr, C., and Starke, K. 1990, Peripheral presynaptic facilitatory effect of angiotensin II on noradrenaline release in anesthetized rabbits. *J. Cardiovasc. Pharmacol.* 15:968-975.

Mohrman, D.E. and Feigl, E.O. 1978, Competition between sympathetic vasoconstriction and metabolic vasodilation in the canine coronary circulation. *Circ. Res.* 42:79-86.

Feigl, E.O. The paradox of adrenergic coronary vasoconstriction. 1987, *Circulation.* 76:737-745.

Ball, R.M., Bache, R.J., Cobb, F.R., and Greenfield, J.C. Jr. 1975, Regional myocardial blood flow during graded treadmill exercise in the dog. *J. Clin. Invest.* 55:43-49.

Huang, A.H. and Feigl, E.O. 1988, Adrenergic coronary vasoconstriction helps maintain uniform transmural blood flow distribution during exercise. *Circ. Res.* 62:286-298.

Rowell, L.B. *Human cardiovascular control.* 1986, New York: Oxford Univ. Press.

# INDEX